# An Ecosystem Services Approach to Assessing the Impacts of the *DEEPWATER HORIZON* OIL SPILL in the Gulf of Mexico

Committee on the Effects of the *Deepwater Horizon* Mississippi Canyon-252 Oil Spill on Ecosystem Services in the Gulf of Mexico

Ocean Studies Board

Division on Earth and Life Studies

NATIONAL RESEARCH COUNCIL
*OF THE NATIONAL ACADEMIES*

THE NATIONAL ACADEMIES PRESS
Washington, D.C.
**www.nap.edu**

**THE NATIONAL ACADEMIES PRESS • 500 Fifth Street, NW • Washington, DC 20001**

NOTICE: The project that is the subject of this report was approved by the Governing Board of the National Research Council, whose members are drawn from the councils of the National Academy of Sciences, the National Academy of Engineering, and the Institute of Medicine. The members of the committee responsible for the report were chosen for their special competences and with regard to appropriate balance.

This study was supported by the National Oceanic and Atmospheric Administration under contract number 10-DELS-294-01. Any opinions, findings, conclusions, or recommendations expressed in this publication are those of the author(s) and do not necessarily reflect the views of the organizations or agencies that provided support for the project.

International Standard Book Number-13: 978-0-309-28845-3
International Standard Book Number-10: 0-309-28845-2
Library of Congress Catalog Card Number 2013952269



Additional copies of this report are available from the National Academies Press, 500 Fifth Street, NW, Keck 360, Washington, DC 20001; (800) 624-6242 or (202) 334-3313; Internet, http://www.nap.edu.



Printed in the United States of America

# THE NATIONAL ACADEMIES

*Advisers to the Nation on Science, Engineering, and Medicine*

The **National Academy of Sciences** is a private, nonprofit, self-perpetuating society of distinguished scholars engaged in scientific and engineering research, dedicated to the furtherance of science and technology and to their use for the general welfare. Upon the authority of the charter granted to it by the Congress in 1863, the Academy has a mandate that requires it to advise the federal government on scientific and technical matters. Dr. Ralph J. Cicerone is president of the National Academy of Sciences.

The **National Academy of Engineering** was established in 1964, under the charter of the National Academy of Sciences, as a parallel organization of outstanding engineers. It is autonomous in its administration and in the selection of its members, sharing with the National Academy of Sciences the responsibility for advising the federal government. The National Academy of Engineering also sponsors engineering programs aimed at meeting national needs, encourages education and research, and recognizes the superior achievements of engineers. Dr. C. D. Mote, Jr., is president of the National Academy of Engineering.

The **Institute of Medicine** was established in 1970 by the National Academy of Sciences to secure the services of eminent members of appropriate professions in the examination of policy matters pertaining to the health of the public. The Institute acts under the responsibility given to the National Academy of Sciences by its congressional charter to be an adviser to the federal government and, upon its own initiative, to identify issues of medical care, research, and education. Dr. Harvey V. Fineberg is president of the Institute of Medicine.

The **National Research Council** was organized by the National Academy of Sciences in 1916 to associate the broad community of science and technology with the Academy's purposes of furthering knowledge and advising the federal government. Functioning in accordance with general policies determined by the Academy, the Council has become the principal operating agency of both the National Academy of Sciences and the National Academy of Engineering in providing services to the government, the public, and the scientific and engineering communities. The Council is administered jointly by both Academies and the Institute of Medicine. Dr. Ralph J. Cicerone and Dr. C. D. Mote, Jr., are chair and vice chair, respectively, of the National Research Council.

**www.national-academies.org**

## COMMITTEE ON THE EFFECTS OF THE *DEEPWATER HORIZON* MISSISSIPPI CANYON-252 OIL SPILL ON ECOSYSTEM SERVICES IN THE GULF OF MEXICO

**LARRY A. MAYER** (*Chair*), University of New Hampshire, Durham
**MICHEL C. BOUFADEL**, New Jersey Institute of Technology, Newark
**JORGE BRENNER**, The Nature Conservancy, Corpus Christi, Texas
**ROBERT S. CARNEY**, Louisiana State University, Baton Rouge
**CORTIS K. COOPER**, Chevron Energy Technology Company, San Ramon, California
**JODY W. DEMING**, University of Seattle, Washington
**DAVID J. DIE**, University of Miami, Coral Gables, Florida
**JOSH EAGLE**, University of South Carolina, Columbia
**JOSEPH R. GERACI**, University of Maryland, Baltimore
**BARBARA A. KNUTH**, Cornell University, Ithaca, New York
**KENNETH LEE**, Commonwealth Scientific and Industrial Research Organisation, Perth, Western Australia
**JAMES T. MORRIS**, University of South Carolina, Columbia
**STEPHEN POLASKY**, University of Minnesota, St. Paul
**NANCY N. RABALAIS**, Louisiana Universities Marine Consortium, Chauvin
**CHRISTOPHER REDDY**,* Woods Hole Oceanographic Institution, Woods Hole, Massachusetts
**RALPH G. STAHL, JR.**, DuPont Company, Wilmington, Delaware
**DAVID W. YOSKOWITZ**, Texas A&M University, Corpus Christi

### Staff

**KIM WADDELL**, Study Director
**SHERRIE FORREST**, Associate Program Officer
**LAUREN HARDING**, Senior Program Assistant (*until October 2012*)
**HEATHER CHIARELLO**, Senior Program Assistant
**JESSICA DUTTON**, Mirzayan Fellow (*until May 2012*)
**CONSTANCE KARRAS**, Postgraduate Fellow (*from September 2012*)
**DEBRA DAVIS**, Editor

*Resigned from the committee.

# Acknowledgments

This report was greatly enhanced by the participants of the meetings held as part of this study. The committee would like to acknowledge the efforts of those who gave presentations at the committee meetings: Jackie Antalan (Operation Homecare), Natalie Bergeron (Project LEARN–LaTerre), George Crozier (Dauphin Island Sea Laboratory), Alyssa Dausman (U.S. Geological Survey), Lisa Dipinto (National Oceanic and Atmospheric Administration), Charlie Henry (National Oceanic and Atmospheric Administration), Elena Kobrinski (University of the Virgin Islands), Shirley Laska (University of New Orleans), Chris Madden (South Florida Water Management District), Maryal Mewherter (United Houma Nation), Steve Murawski (University of South Florida), Khai Nguyen (Mary Queen of Viet Nam Community Development Corporation), Denise Reed (University of New Orleans), Marc Russell (U.S. Environmental Protection Agency), and Sandra Werner (ExxonMobil)

This report has been reviewed in draft form by individuals chosen for their diverse perspectives and technical expertise, in accordance with procedures approved by the National Research Council's Report Review Committee. The purpose of this independent review is to provide candid and critical comments that will assist the institution in making its published report as sound as possible and to ensure that the report meets institutional standards for objectivity, evidence, and responsiveness to the study charge. The review comments and draft manuscript remain confidential to protect the integrity of the deliberative process. We wish to thank the following individuals for their participation in the review of this report:

**DONALD BOESCH**, University of Maryland Center for Environmental Science, Cambridge
**ROBERT DIAZ**, Virginia Institute of Marine Science (retired), Ware Neck
**MERV FINGAS**, Spill Science, Edmonton, Alberta, Canada
**JÖRG IMBERGER**, University of Western Australia, Crawley
**DAVID KARL**, University of Hawaii, Honolulu
**JUDY MCDOWELL**, Woods Hole Oceanographic Institution, Woods Hole, Massachusetts
**PAUL MONTAGNA**, Texas A&M University, Corpus Christi
**STEVE MURAWSKI**, University of South Florida, St. Petersburg
**GARRY PETERSON**, Stockholm University, Sweden
**CHRISTOPHER M. REDDY**, Woods Hole Oceanographic Institution, Woods Hole, Massachusetts
**KATHLEEN SEGERSON**, University of Connecticut, Storrs
**HEATHER TALLIS**, Stanford University, Palo Alto, California
**MICHAEL ZICCARDI**, University of California, Davis

Although the reviewers listed above have provided many constructive comments and suggestions, they were not asked to endorse the conclusions or recommendations nor did they see the final draft of the report before its release. The review of this report was overseen by **John J. Boland**, The Johns Hopkins University, appointed by the Divison on Earth and Life Studies, and **David A. Dzombak**, Carnegie Mellon University, appointed by the Report Review Committee, who were responsible for making certain that an independent examination of this report was carried out in accordance with institutional procedures and that all review comments were carefully considered. Responsibility for the final content of this report rests entirely with the authoring committee and the institution.

# Contents

# Summary

The unprecedented volume, depth, and spatial scale of the *Deepwater Horizon* (*DWH*) oil spill created many enormous challenges for policymakers. Among these was the difficult task of restoring the ocean and coastal ecosystems of the Gulf of Mexico (GoM) to the condition they would have been in but for the spill. Although multiple state and federal agencies had experience in the restoration of damaged natural resources, no agency had ever been faced with a spill affecting such a wide area and such a broad range of ecosystems.

This report aims to answer the question: in what ways might using an "ecosystem services approach" help these agencies to achieve their mission, as mandated by the Oil Pollution Act of 1990, of "making the environment and the public whole."[1] The ecosystem services approach is different from traditional approaches to damage assessment and restoration (e.g., the Natural Resources Damage Assessment) because it focuses not on the natural resources themselves, but on the valuable goods and services these resources supply to people. An ecosystem services approach can supplement traditional methods of assessing, or valuing, damage to natural resources by estimating flows of goods and services before and after an event. In addition, thinking in terms of ecosystem services can change the ways that the public and agencies conceptualize and discuss restoring natural resources to their former condition. For example, if the goal is to restore the flow of ecosystem goods and services, then how should restoration funds be spent?

The ecosystem services approach is not a panacea, although the committee believes it has great potential to improve natural resource damage assessment and restoration. This report explores both the potential benefits and the limitations of the ecosystem services approach in several ways. After explaining the conceptual basis for the approach and how it might add to current methods, the report examines the obstacles to its application. To illustrate these broad discussions, the report includes several case studies that, together with the more general material, make clear that the ecosystem services approach will add value along several dimensions.

First, by framing damage assessment and restoration as an ecosystem services issue, an ecosystem services approach can change the public's perception of natural resources and the ways agencies manage for healthy ecological systems. One of the strengths of the ecosystem services concept is that it highlights the ways in which healthy ecosystems support healthy economies. Second, as the report discusses in detail, building workable models for implementing an ecosystem services approach will require both better and novel efforts in collecting data about the interactions within ecosystems and the connections between ecosystems and benefits—both economic and nonmonetized types, such as cultural and spiritual.

These efforts, in and of themselves, will contribute to public understanding of the relation-

[1] 15 C.F.R. § 990.30 (2012).

ships between humans and the environment; in addition, they will enable the public to inform policymakers about such relationships in the unique context of their own communities. One of the key facts central to this discussion is that communities depend on ecosystem goods and services to differing extents and in very different ways. Finally, success in developing workable models for measuring natural resource damage in terms of ecosystem services will facilitate efforts to make the public whole in the wake of future disasters.

## INTRODUCTION

On April 20, 2010, an explosion on the *DWH* platform, which was drilling the Macondo well in Mississippi Canyon Block 252 in the GoM, killed 11 oil workers and injured 17 others, resulted in the largest oil spill in U.S. history, and inevitably impacted the ecosystem services of the GoM. An estimated 4.9 million barrels (±10 percent), approximately 205.8 million gallons, of oil flowed from the wellhead following the explosion.[2] For context, this amount of oil represents approximately one-third of the nation's daily consumption of oil. The *DWH* oil spill was unprecedented in both its magnitude (see Figure S.1) and the ocean depth at which oil was released, and the event captured the fears and concerns of the nation during the spring and summer of 2010. Recognizing the complexity and potential impacts of the spill, members of the U.S. Congress requested a study by the National Academy of Sciences to evaluate the impacts of the *DWH* oil spill on natural resources and ecosystem services in the GoM. The complete statement of task is given in Chapter 1.

Ecosytems contribute to our well-being and enjoyment by, for example, supplying fish for markets, wild game to hunt in coastal marshes, and powdery white sand at which to marvel. Formally, ecosystem services are defined as "the benefits provided by ecosystems to humans that contribute to making human life both possible and worth living." These services are the result of functioning ecosystems—the interactions of plants, animals, and microbes with the environment—and can be classified into four categories:

- *Provisioning services* (e.g., material goods such as food, feed, fuel, and fiber);
- *Regulating services* (e.g., climate regulation, flood control, and water purification);
- *Cultural services* (e.g., recreational, spiritual, and aesthetic services);
- *Supporting services* (e.g., nutrient cycling, primary production, and soil formation).

## THE SETTING

The GoM is a highly productive marine ecosystem that is surrounded by the United States, Mexico, and Cuba. It is the world's seventh largest peripheral sea, with a surface area of 1.51 million $km^2$ and a volume of 2.4 million $km^3$. The great habitat complexity in the GoM supports the region's high biodiversity, which consists of endemic and cosmopolitan species. The

[2] According to McNutt et al. (2012), BP's containment efforts captured approximately 800,000 barrels of oil before it reached the marine environment, making the total amount of oil to enter the water column closer to 4.2 million barrels.

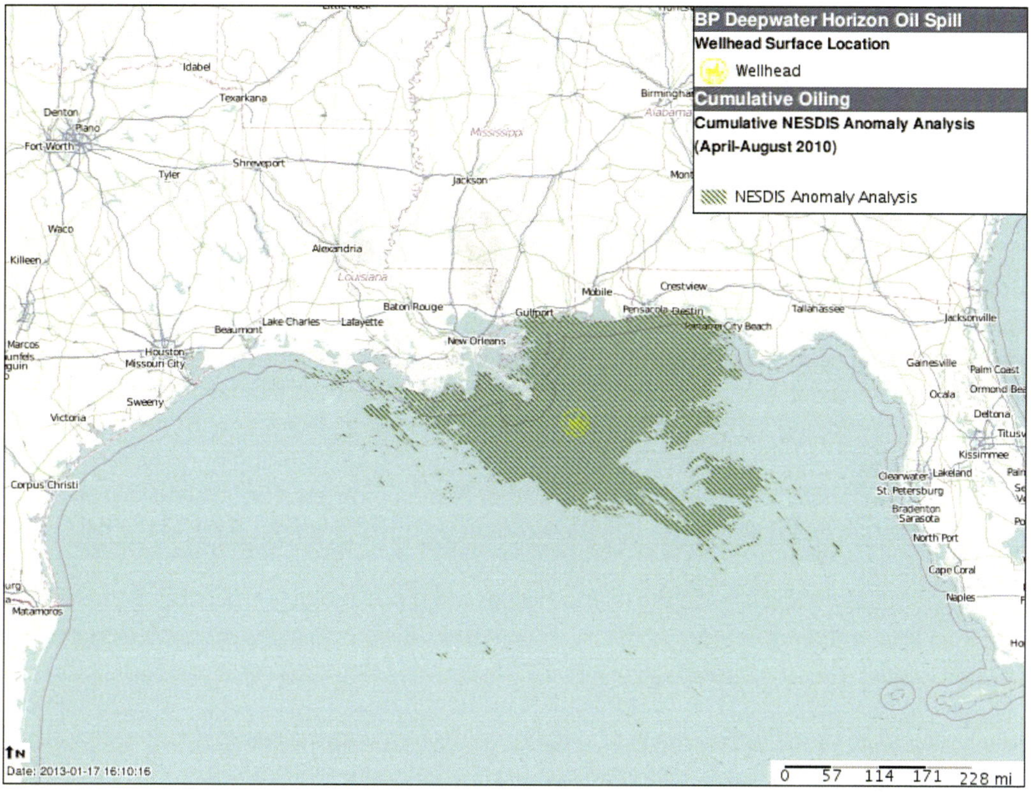


**FIGURE S.1** Location of the Macondo wellhead and spatial extent of the *DWH* oil spill in the northern Gulf of Mexico. SOURCE: National Oceanic and Atmospheric Administration.

GoM ecosystem is subject to a number of modern stressors, including habitat loss, overfishing, impacts of flood control on sediment supply, degraded water quality, and pollution resulting from activities such as oil and gas development. Harmful algal blooms and hypoxia frequently drive mobile animals from certain areas or decrease habitat suitability, and increasing coastal development and recent intense hurricanes have been destroying coastal habitats.

## ECOSYSTEM SERVICES AND IMPACTS FROM THE *DWH* OIL SPILL

Given the scale of the *DWH* oil spill, the potential for disruption of GoM ecosystem services was significant. These services provide direct and indirect benefits to the 20 million people who live in the region; they support, among many other economic activities, one of the nation's most productive commercial and recreational fisheries, a $50 billion tourism industry, and an energy sector that produces about 30 percent of the nation's oil and 20 percent of its natural gas.

This report examines the impacts of the *DWH* oil spill on GoM ecosystem services, the ways in which an ecosystem services approach could help in remediating those impacts, and the challenges and limitations of applying the ecosystem services approach. The key to feasible application of the ecosystem services approach is the development of tools capable of establishing and quantifying causal links between the event, an injury to an ecosystem, the resulting decrease in goods and services provided by that system, and the cost of that decrease to individual communities and society at large.

There are three obstacles to developing and applying these tools. First is the difficulty in establishing a baseline measurement of goods and services produced by the system just prior to the harmful event. Simply put, one cannot assess damages—that is, the difference between conditions before and after the harmful event—if one does not know what the conditions were before the event. Assuming that a baseline is established, a second major obstacle is the difficulty in developing a model that can fully predict the event's impact on the ecosystem and provide defensible estimates of the costs of reduced production resulting from the release of a given amount and kind of pollutant (or any other human-induced stressor) into that ecosystem. Although existing ecosystem models cannot do this, they might serve as the basis for models that can.

The third obstacle relates to the relative values of various ecosystem services. Some ecosystem goods and services are more easily priced than others. For example, economists are better able to price a decrease in a system's ability to produce fish for a commercial fishery than a decrease in the deep ocean's ability to regulate nutrient transport to surface waters. And, even when values can be assigned, resource limitation will force policymakers to prioritize among restoration options. The first scenario raises concerns that more-difficult-to-price services will be discounted or ignored in the decision-making process; the second highlights an important limitation to any approach to damage assessment and restoration.

## RESILIENCE AND ECOSYSTEM SERVICES

Ecosystems are subject to natural disturbances such as fires, floods, droughts, and disease outbreaks, as well as human-caused disturbances, including oil spills. Ecosystems are also subject to slowly changing long-term stresses, such as nutrient enrichment and changes in sediment supply, as observed in the GoM. These long-term stresses can affect the ability of the system to respond to a shock such as the *DWH* oil spill.

A resilience framework focuses attention on shocks (pulse disturbances), long-term stresses (press disturbances), and the response of complex systems to these shocks and stresses. Does a system recover slowly or rapidly from a shock? What factors within the system allow for more rapid and more complete recovery? Do attempts to stabilize certain components of systems lead to reduction in overall system resilience and greater potential for large changes?

Consideration of resilience is especially important in systems that can undergo persistent and fundamental shifts in structure and function following disturbances ("regime shifts").

Generally speaking, increasing resilience can reduce the risk that the system will cross critical thresholds and undergo a detrimental regime shift. However, in rare instances, decreasing resilience can increase the probability of a beneficial regime shift. The understanding of how to increase or decrease resilience places a premium on knowledge of system dynamics, including feedbacks among system components, as well as of uncertainty and variability in dynamic systems. Can factors that increase system resilience be identified and managed to increase the resilience of a system to a "desirable" state, or the inverse?

System resilience can play an important role in maintaining conditions that will sustain the provision of ecosystem services that contribute to human well-being. However, a narrow focus on managing complex systems to provide a constant flow of ecosystem services may actually reduce system resilience and increase vulnerability. An event such as the *DWH* oil spill will disrupt service provision, but a resilient ecosystem will allow for a return of services sooner rather than later or never.

Although resilience is an important concept, much like that of ecosystem services, providing practical and specific advice to managers to "increase the resiliency of ecosystem services" is a difficult task at present. As noted above, managing resilience requires understanding of the dynamics of complex and highly variable systems, which is often limited. However, without such an understanding, it can be difficult to predict how management actions will affect resilience in practical contexts. **Overall, the committee concludes that consideration of resilience provides a useful conceptual framework and important general guidance but may not always lead to specific recommendations to managers.**

## THE CASE STUDIES

Because of time and space limitations, this report could not address all of the impacts of the *DWH* oil spill on all of the ecosystem services of the GoM. Rather, the report illustrates the impacts through four case studies. The spill impacted each of the four ecosystem service categories in the GoM—that is, provisioning, regulating, cultural, and supporting services. Some impacts were ephemeral, some persist, and many are under investigation or unknown at this time. Using the ecosystems services approach, the case studies consider how key ecosystem services may have been impacted by the *DWH* oil spill, examine methods for making baseline measurements, and explore the adequacy of existing baseline data for the GoM. Additionally, for each case study, the committee offers suggestions for additional measurements that can be made to enhance an ecosystem services approach to damage assessment.

### Wetlands

The first case study focuses on coastal wetlands in the GoM, which cover a large region in the northern GoM. **Half of the nation's coastal wetlands are found along the GoM, and, of these, approximately 40 percent are in Louisiana. Unfortunately, many wetlands were among the closest land points (only about 40 miles) from the Macondo well.** Coastal wetlands, consisting of salt marshes and mangrove plant communities, provide a wealth of supporting, regulating, provisioning, and cultural services that include soil and sediment (shoreline stabilization) maintenance, nutrient regulation and water quality, food provision, recreational opportunities, and hazard moderation.

The wetland case study focuses on the regulating service of hazard moderation (specifically storm mitigation) to illustrate the opportunity that exists in using the ecosystem services approach when the underlying ecosystem science and the particular ecosystem service are well known and supported by a rich literature. The ecosystem service of storm mitigation also benefits from having been monetized; that is, the costs of storm damage and reductions in losses due to wetland buffers can be quantified.

The value of ecosystem services for GoM storm protection is directly related to the total area of wetlands and to plant community composition. **Consequently, change in total wetland area is the most direct and practical measurement of change in ecosystem services in Gulf Coast wetlands, typically measured by remote sensing by planes and satellites.** Remote sensing is used to efficiently map, monitor, and detect changes in wetlands' health and vegetation profiles, but verification of these changes at greater resolution by ground measurements is still needed.

**During the *DWH* oil spill, about 1,100 linear miles of coastal salt marsh wetland were impacted at some time during the event.** Crude oil can smother vegetation by coating leaf surfaces and can cause toxic effects, particularly the lighter fractions of the oil that are more water soluble. The *DWH*-sourced oil was already weathered and rendered heavier by the time it reached shore, so many of the impacts were associated with the weight of the oil breaking the plant stems. Oil also coated the leaves and smothered plants and their roots.

Oil spill cleanup can also have detrimental effects on marshes, including physical disturbance and compaction of vegetation and soil associated with cleanup activities. Consequently, many portions of the marshes were designated for "no further treatment" (allowing the oil to biodegrade over time) due to access, degree of oiling, and other considerations.

The expectations for long- and short-term effects of oil fouling of marshes depend on the areal extent and magnitude of the exposure. **The overall impacts in the GoM wetlands can be summarized as follows:**

- **Acute effects on marshes, where the biota are not expected to recover, appear to be confined to the edges of bays, canals, and creeks in a limited subset of the oiled wetlands.**
- **Where the vegetation has died and the root systems have been lost in heavily oiled areas, the erosion of sediment is leading to the conversion of once productive marshland to open water.**
- **Subsequent tropical storm activity resulted in additional erosion of oiled marshes.**
- **Based on numerous studies that document a rapid recovery from oiling and a relatively low sensitivity of perennial marsh vegetation to hydrocarbons, GoM marsh vegetation can be expected to suffer little or no long-term impairment**

**from the *DWH* oil spill in areas** where roots and rhizomes survived the initial impact of oil fouling. If roots and rhizomes did not survive, then an area will not recover on its own.

These impacts need to be viewed in the context of significant and continuing losses of wetlands in the GoM due to many other stressors, including subsidence, canal dredging, salt intrusion, and sediment starvation.

## Fisheries

The second case study relates to fisheries, a provisioning service with a rich literature about its valuation and assessment. **Fisheries in the GoM provide some of the most important and lucrative services through the production of seafood, industrial fish products, and recreational fishing.** In recent times, this service has been considered to be a good candidate for a holistic integration of management at an ecosystem scale that includes ecological and human components. Although the field has been developing this integration through the promotion of an ecosystem approach to fisheries management, it does not yet consider ecosystem services as a guiding principle. Fisheries, however, offer many examples of regular quantification of human impacts on the ecosystem structure and of ecological and economical productivity. This case study explores a provisioning service, the provision of seafood from the GoM, and how the ecosystem services approach may help to quantify the possible impacts of oil spills on seafood production.

The spill's potential disruption to the provision of seafood can be estimated by examining the spatial extent of the fishery closures imposed by the National Oceanic and Atmospheric Administration (NOAA) in the aftermath of the spill and entry of oil into estuarine waters (Figure S.2). The closures were intended to limit the risk of harvesting contaminated seafood. However, they underestimated the spatial extent of potential impacts, because fish can migrate through the spill area and be caught elsewhere, and larvae that survive direct impacts from the spill may end up as juvenile fish in other regions. **Still, by preventing fishermen from accessing resources, these closures alone decreased fishery landings in the GoM by as much as 20 percent and created an immediate economic hardship for fishermen affected by the closures.**

Unlike many of the ecosystem services that are provided by the GoM, fisheries have long been researched, monitored, and evaluated quantitatively. Although this research has generally focused on how to best manage individual fisheries, it has also provided valuable insight into marine ecosystem processes and fisheries baselines. **Unfortunately, impacts on fishery productivity from oil spills and other stressors are not as well understood.** Despite long-term studies and ongoing development of models, the ability to detect spatial and temporal differences in fishery productivity in the GoM is limited. Recent developments in fishery data collection, such as the introduction of vessel monitoring systems in the reef fish fishery of the GoM, could improve estimates of abundance. However, any mortality or reduction in individual

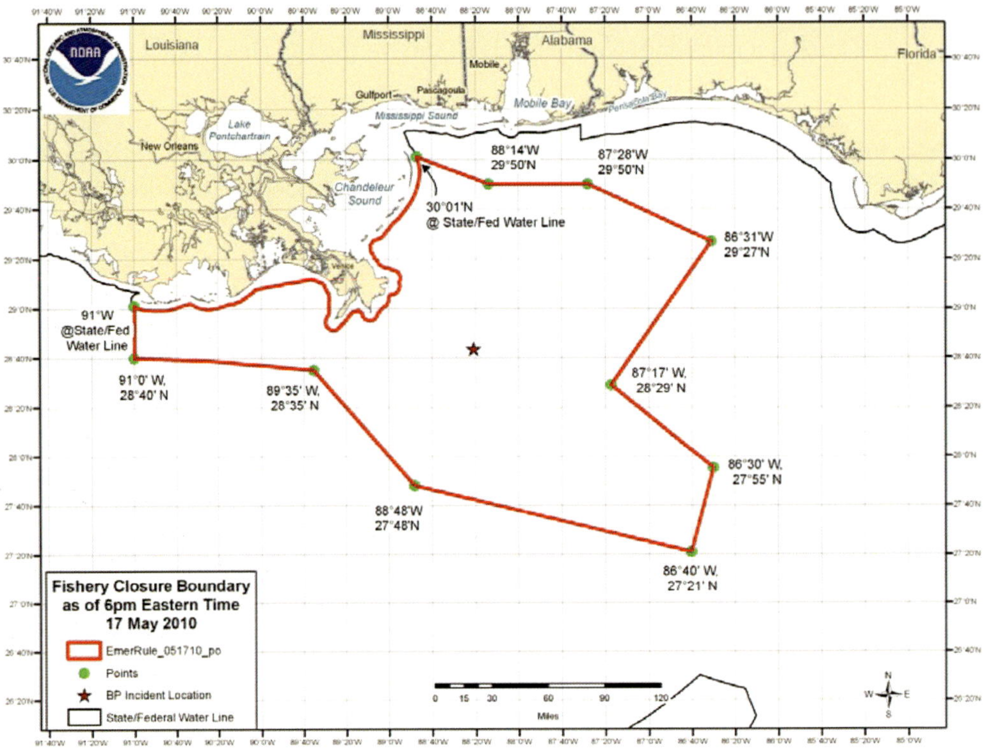


**FIGURE S.2** Fishery closure imposed by NOAA to ensure seafood safety, effective May 17, 2010. The closed areas changed daily as NOAA sampled seafood in the area and the geographical extent of the detected oil changed. SOURCE: NOAA.

fitness caused by the spill directly or indirectly may take years or, for some species, decades to transfer through the ecosystem and be observed.

Much like their health and nature, the value of fisheries in the GoM has long been considered, but available data may not provide a complete picture. Federal and state agencies charged with managing fisheries estimate the direct value of commercial fisheries to the fisherman with this straightforward formula: the dockside value of the catch (the amount that the fisherman receives) minus any expenses incurred to capture those fish. The methodology for evaluating the economic effects of an oil spill, like any other type of pollution, on commercial fisheries is also relatively simple in concept. The economic costs of pollution are derived from either reduced production (due to dieoffs or fishery closures) or by reduced consumer demand (due to the perception of reduced fish quality or safety). As stated above, the immediate economic impact was a 20 percent decrease in landings for 2010.

## Marine Mammals

Bottlenose dolphins were chosen as the subject for the third case study for numerous reasons, including their role in three of the four ecosystem services—regulating, supporting, and particularly cultural. This case study allowed for exploration of approaches to estimating the value of passive use and existence, a key, but difficult-to-establish, metric for cultural ecosystem services. Although research is limited, more is known about bottlenose dolphins than virtually any other cetacean. The world's longest-running study of a wild dolphin population, spanning five generations, focuses on Sarasota Bay in the eastern GoM. Bottlenose dolphins are capable of self-recognition, which ranks them highly on a cognitive scale. They are also apex predators, a role they share with humans, and therefore play a role in regulating the GoM food web. **As apex predators, the dolphins' health and well-being serve as important indicators of the health of the GoM and oceans in general.** Their position as the most studied and arguably the most popular and charismatic marine mammal makes them a centerpiece for conservation science, education, and ecotourism. The stranding of hundreds of dolphins in the GoM before, during, and especially after the *DWH* oil spill have stimulated considerable public concern, which speaks to our cultural needs and sensitivities regarding their "value" as an ecological resource and ecosystem service. **If the recent mortality event is determined to be linked to the *DWH* oil spill, then an opportunity may exist to establish a plan to protect and restore the dolphin habitat as well to reduce dolphin mortality due to human activities.**

## The Deep Gulf of Mexico

Finally, the deep GoM was selected as the subject for the last case study in part because of its location with respect to the *DWH* blowout and spill. **The deep sea was also selected because of the increasing concern and risk posed by the energy industry's activities as it works at the leading edge of engineering in the most poorly understood of the impacted habitats.** The biota of the surrounding seafloor at 1,500 meters depth and of the water column through which a plume of hydrocarbons and dispersants flowed received the most immediate impact of the uncontrolled discharge.

Knowledge of deep-sea processes and GoM-specific information is sufficient to begin the process of identifying ecosystem services, considering the impact of the *DWH* oil spill, examining methods to measure baselines, and identifying gaps in current databases. However, the area is so vast and sampling is so sparse that gaps in knowledge of the GoM deep sea inhibit the ability to apply an ecosystem services approach in a quantitative way. Delineating what is and is not known about this extensive subregion can be helpful for identifying relevant processes and uncertainties and thus directions for future investigation.

Based on current understanding, the primary ecosystem services of the deep GoM generally fall in the category of supporting services. The deep GoM resupplies nutrients depleted during photosynthetic activity in the photic (where sunlight is available) zone. Given that the aphotic zone is vastly larger than the photic zone, the deep GoM provides these nutrients at a rate totally independent of the photic zone's biological demand and is thus a very stable source of nutrients to overlying life. This stability provides the whole GoM with greater resilience.

**The deep GoM also provides the important regulatory service of pollution attenuation. For example, natural populations of deep-sea oil-degrading bacteria digested a significant amount of the *DWH* oil.** This ecosystem service is well documented for shallow marine and intertidal environments, but was unknown for the deep-sea water column prior to the *DWH* oil spill. The release of crude oil from the Macondo well thus created a unique opportunity to study this phenomenon.

## THE OIL SPILL RESPONSE EFFORTS

In addition to the impacts of the spill itself, the techniques that were employed in response to the spill had potential impacts on ecosystem services. These techniques included allowing or fostering the natural breakdown of oil (natural attenuation) and using chemical dispersants, in situ burning, and skimmers at sea, as well as booms, berms, and hydrologic modification closer to shore, to prevent spilled oil from reaching sensitive shoreline habitats.

**The techniques applied offshore—in situ burning, skimming, and dispersants in particular—were effective in significantly reducing the volume of oil before it went ashore.** Estimates of the reduction range from 17 percent to 40 percent. The public expressed a great deal of concern about the amount and novel subsurface application of dispersants. However, the dispersants effectively broke up the oil both at the surface and in the water column, thereby making the oil much more available for biodegradation by hydrocarbon-digesting marine microbes. Although a number of studies have shown dispersed oil to be toxic, especially to juvenile stages of many marine species, the dilution and mixing effects of the open ocean may have mitigated the toxic effects of the *DWH* oil spill.

The techniques applied near shore and onshore varied in their effectiveness, with the least effective being the construction of sand berms, primarily because of the timing and scale of the efforts relative to the oil distribution at the time. This report also discusses the operations to clean up the oil that made it to shore, including manual removal of oil, sand washing, and surf washing. **The final analysis and assessment of the impacts of the response techniques are dependent on the release of damage assessment data that are presently being analyzed, as well as of long-term monitoring data that may become available after this report is published.**

## LOOKING AHEAD

As the nation moves forward after the *DWH* oil spill, the substantial funding that is and will be available through the criminal and civil settlements offers an unprecedented opportunity to establish a comprehensive baseline and fundamental understanding of the GoM, a critical component of an ecosystem services approach. To fully implement an ecosystem services approach for the GoM, several key research needs must be addressed.

**First, there is a critical need for an overarching infrastructure for organizing and integrating the wealth of data that have been and will be collected in the GoM.** This infrastructure is needed to support comprehensive ecosystem models that can be used to evaluate the impacts of human activities on the ecosystem's ability to provide services. The development of this infrastructure must engage the participation of the academic community, federal agencies, industry, and the public. Furthermore, funding for this infrastructure and its maintenance will need to be stable and long-term.

**Second, although a substantial body of data exists to support a better understanding of ecosystem structure and function within the GoM, a comprehensive model that incorporates biophysical, social, and economic data for the GoM is needed for the long term, while models for GoM subcomponents and services are needed for the short term.** Data regarding socioeconomics and human dependencies lag far behind. Integration of such data into social-ecological models is essential. For an ecosystem services approach to be successful, this issue must be the focus of additional data collection going forward. Future research efforts must include both the collection and the synthesis of these data so that the appropriate models can include the full range of social and ecological impacts and can be used to better inform decision makers.

**Third, research and management focused on resilience both in principle and in specific applications would be useful.** Resilience provides a valuable conceptual framework for managing complex systems because it focuses attention on how systems are affected by short-term disturbances and long-term stresses, such as hurricanes and wetland losses. Resilience may be increased through management decisions that incorporate the contribution of biodiversity and the connectivity that sustain and maintain ecosystems. Resilience can also be increased by adaptive management and by improved governance from local to regional scales to enhance the functioning of institutions and to improve social cohesion. In addition, promoting the diversification of local economies will improve the resilience of communities. However, obtaining adequate data for measuring the resilience of a social system is particularly difficult, because the state or "measure" of the baseline social system has not been identified or may continuously change. Still, efforts should be made to maximize the understanding of resilience in these systems. As with ecosystem services, there are tradeoffs: managing for the resilience of a particular service may reduce the resilience of another service.

## CONCLUSION

There will be funding opportunities through multiple venues to achieve better preparedness before another event like the *DWH* oil spill happens, but there must be coordination so that innovation can be fostered without wasting funds, duplicating efforts, and causing harm. GoM communities and natural resource managers face many challenges as they contemplate and select research priorities and restoration plans. Among these are helping stakeholders to consider how best to manage multiple ecosystem services across a diverse and large marine ecosystem.

As many in the region have already realized, past decisions that were aimed at enhancing a particular ecosystem service to maximize a particular benefit—for example, energy development, fisheries, or tourism—have resulted in tradeoffs that diminished the capacity of other ecosystem services to deliver benefits. Although the fines, penalties, and outcomes of litigation will help to fund and further our scientific understanding of the GoM ecosystem and how it functions, **policymakers and the public should consider potential tradeoffs as they set priorities and goals for restoring and strengthening their communities and the GoM natural resources.** The ecosystem services approach is one tool that could be used during these deliberations to more fully capture the value of assorted services in the GoM.

CHAPTER ONE

# Introduction

Hydrocarbon extraction generates economic benefits and environmental risks. The environmental risks of offshore hydrocarbon extraction were made readily apparent on April 20, 2010, when the *Deepwater Horizon* (*DWH*) platform drilling the Macondo well in Mississippi Canyon Block 252 (MC-252) exploded. In the aftermath of the explosion, an estimated 4.9 million barrels[1] (>200 million gallons) of crude oil spilled into the Gulf of Mexico (GoM) over a period of 3 months, resulting in the largest marine oil spill and environmental response in U.S. history (Lubchenco et al., 2012; McNutt et al., 2012).

During the time since the spill (more than 3 years at the time this report went to press), numerous commissions, committees, and panels have focused on the causes of the explosion, reviewed the immediate response procedures, and offered suggestions for changes in practice, policy, and regulatory regimes to help minimize the likelihood that a disaster like the *DWH* oil spill could happen again (IOM, 2010; National Commission on the BP Deepwater Horizon Oil Spill and Offshore Drilling, 2011; NRC, 2011). In addition, the academic research community, private sector, and federal government have embarked on a range of studies aimed at understanding the impacts of the spill on the environment, economy, and people of the GoM region (NOAA, 2012c).

The spill also triggered a legal process—the Natural Resource Damage Assessment (NRDA)—to assess the "injury" (defined as the observable or measurable adverse change in a natural resource or impairment of a natural resource service), to develop, implement, and monitor restoration plans, and to establish compensation for the costs of assessment and restoration from those deemed responsible for the injury (NOAA, 2012b). The ultimate goal of the compensation and restoration process is to "make the environment and public whole."[2] The application of NRDA to oil spills is authorized under the Oil Pollution Act of 1990, legislation that was created partly in response to an earlier environmental disaster, the 1989 *Exxon Valdez* oil spill in Prince William Sound, Alaska.

The procedures for implementation of the NRDA process were developed based on experience from spills that, at the time, had been of limited volume and had been constrained to relatively shallow waters. Under typical NRDA practice, losses are generally measured in simple ecological terms (e.g., number of acres damaged or number of fish killed), and restoration generally follows relatively simple equivalency approaches (e.g., acres of habitat restored or fish stocks replaced). For some situations, sufficient pre-spill baseline data exist and the oil spill is relatively minor, and thus assessment of impact can be straightforward (e.g., the income lost from the closure of a particular fishery) and compensation easily determined. However, for

[1] According to McNutt et al. (2012), BP's containment efforts captured approximately 800,000 barrels of oil before it reached the marine environment, making the total amount of oil to enter the water column closer to 4.2 million barrels.

[2] 15 C.F.R. § 990.30 (2012).

many other situations, the complex interconnections of the ecosystem are not well understood, and thus quantification of the full impact of the spill on all components of the ecosystem may be difficult, if not impossible. Challenges to the assessment of damages under NRDA increase with the spatial and temporal scale of the injury and the complexity of the ecosystems involved, which makes it more difficult to understand and account for the full range of ecological and societal impacts. Quantification of the level of injury is further complicated when an event occurs in an ecosystem that is subject to natural and anthropogenic environmental changes that are unrelated to the specific event. In these cases, the injury caused by a specific spill must be assessed relative to dynamic or shifting baseline data.

The *DWH* oil spill was a large-scale event with impacts in the entire water column, from deep to shallow, and in open waters to coastal marshes and beaches. The ecology of the GoM system is complex and incompletely understood. It is a highly dynamic system, with a number of natural and human processes causing changes that result in a constantly shifting baseline. All of these conditions make attribution of injury and recovery planning challenging.

The interconnected nature of the coastal and marine ecosystems in the GoM makes it difficult to isolate injury to a single resource or single ecosystem process. Impacts tend to spread through the system because of physical processes, such as currents that spread oil from place to place, and biological processes that carry pollutants through the food chain and, as with migrating waterfowl, potentially cause effects far from the site of the oil spill. In addition, an understanding of social and economic processes is needed to determine how biophysical changes in ecosystems translate to injury to various segments of the public. Consequently, understanding of highly complex interconnected systems such as the GoM requires a more holistic view of ecosystems and the role of the people within them.

Two concepts articulate this more holistic view: ecosystem-based management and ecosystem services. Ecosystem-based management accounts for the complexity of interactions within an ecosystem (including those involving humans) rather than focusing on single resources or species in isolation (Christensen et al., 1996). It is a central element in the recently released National Ocean Policy Implementation Plan[3] as well as an evolving theme in current fisheries management efforts in the United States (Sissenwine and Murawski, 2004). Ecosystem services are the benefits that the public receives from natural resources and the ecological processes provided by ecosystems (Daily, 1997; MEA, 2005).

Ecosystem services are produced as a result of the normal functioning of the ecosystem—the interactions of plants, animals, and microbes with the environment. Ecosystem services include *provisioning services* (the material goods provided by ecosystems, including food, feed, fuel, and fiber), *regulating services* (climate regulation, flood control, and water purification), *cultural services* (recreational, spiritual, and aesthetic), and *supporting services* (nutrient cycling, primary production, and soil formation). These services have immense value to society and are essential to the well-being of all people.

The GoM provides a broad array of provisioning, regulating, supporting, and cultural ecosystem services. Coastal tourism (a cultural service), for example, has an estimated worth

[3] http://www.boem.gov/NOP/.

of \$19.7 billion per year (National Commission on the BP Deepwater Horizon Oil Spill and Offshore Drilling, 2011). The wetlands and marshes of the Gulf region play a critical role in storm surge protection (a regulating service) as well as nutrient cycling and water purification (supporting services) and commercial fisheries landings (a provisioning service), which account for approximately 25 percent of the seafood provided by the contiguous United States (NMFS, 2010b). Additionally, the people of the GoM region and the nation as a whole benefit from the vast hydrocarbon resources of the GoM, which, in 2009, accounted for 29 percent of the total crude oil and 13 percent of the natural gas production in the United States (DOE, 2010).

The evolving understanding of human–ecosystem interactions, which is shaped by the concepts of ecosystem-based management and particularly ecosystem services, offers an opportunity to address some of the challenges inherent in assessing the impacts of an event such as the *DWH* oil spill. By taking a more holistic view of ecosystem interactions and then following these interactions through all relevant trophic levels and spatial connections to their ultimate impact on human well-being, an ecosystem services approach to damage assessment enables formation of a more complete picture of potential impacts and a broader range of restoration options. This is particularly relevant to a spill the size, duration, depth, and complexity of the *DWH* oil spill, during which oil and dispersants were released at 1,500-m depth into a relatively poorly understood deep-sea ecosystem that includes deep-sea corals and chemosynthetic communities (organisms that derive their energy from oxidizing inorganic molecules). Oil, dispersants, and dispersed oil then traversed through the water column, interacting with fish, marine mammals, and other organisms throughout the trophic web, making their way through the photic zone, and sometimes reaching beaches and salt marshes.

All of this occurred in a region that has been subject to numerous other natural disasters (e.g., hurricanes) and human actions (e.g., levee construction for flood control, fertilizer application) that have created dynamic baselines from which to estimate the impact of the spill. Never before has the United States faced a spill of this magnitude or with the potential to impact all trophic levels of such a complex and dynamic ecosystem. To quantify the spill's impact we must first understand the interactions and linkages between and among the various components and processes (including human) of this ecosystem. Recognizing these unique and unprecedented aspects of the *DWH* oil spill and the associated complexity of the task of assessing damages, Congress requested that the National Academy of Sciences evaluate the effects of the *DWH* oil spill on the ecosystem services of the GoM. A committee was established in January 2011 and charged with addressing the questions posed in the following Statement of Task:

## Statement of Task

1. What methods are available for identifying and quantifying various ecosystem services? What are the spatial and temporal scales conducive to research that provide meaningful information for the public and decision makers?
2. What methods and types of information can be used to approximate baselines (but-for-the-spill) for distinguishing effects on ecosystem services specific to the spill?

3. What kinds of valuation methods are appropriate for measuring ecosystem services over time with regard to recovery under the following approaches: natural processes, mitigation, and restoration efforts? What baseline measures are available that would provide benchmarks for recovery and restoration efforts?
4. What ecosystem services (provisioning, supporting, regulating, and cultural services) were provided in the Gulf of Mexico Large Marine Ecosystem prior to the oil spill? How do these differ among the subregions of the Gulf of Mexico?
5. In general terms, how did the spill affect each of these services, and what is known about potential long-term impacts given the other stresses, such as coastal wetland loss, on the Gulf ecosystem?
6. How do spill response technologies (e.g., dispersant use, coastal berm construction, absorbent booms, in situ burning) affect ecosystem services, taking into account the relative effectiveness of these techniques in removing or reducing the impacts of spilled oil?
7. In light of the multiple stresses on the Gulf of Mexico ecosystem, what practical approaches can managers take to restore and increase the resiliency of ecosystem services to future events such as the *Deepwater Horizon* Mississippi Canyon-252 spill? How can the increase in ecosystem resiliency be measured?
8. What long-term research activities and observational systems are needed to understand, monitor, and value trends and variations in ecosystem services and to allow the calculation of indices to compare with benchmark levels as recovery goals for ecosystem services in the Gulf of Mexico?

Sixteen committee members were selected, representing a broad range of backgrounds and expertise (ecology, geology, geophysics, microbiology, fisheries, veterinary medicine, economics, environmental law, environmental engineering, biological oceanography, marine chemistry, biochemistry, human dimensions of natural resource management, and benthic and coastal habitats). Beginning with its initial meeting in late January 2011, the committee held six meetings and public information-gathering sessions. The committee was charged with producing an Interim Report (NRC, 2011) and this Final Report.

The Interim Report focused primarily on the first three tasks and was designed to offer early guidance to federal agencies involved in the NRDA and restoration efforts. To address these tasks, the Interim Report provided an overview of the unique physiographic, oceanographic, and ecological components of the GoM and the range of habitats that make up its ecosystem. It then introduced the concept of ecosystem services and an "ecosystem services approach" for estimating the impact of an event such as the *DWH* oil spill, contrasting this approach with the current NRDA process and its use of habitat or resource equivalency as a means to "make the environment and the public whole." Designed to overcome the challenges involved in estimating harm to natural resources, the equivalency approaches have become, in effect, surrogates for estimating how to make the environment and the public whole. By and large, they focus on estimating the implicit value of an injured habitat or organism rather than on its ultimate value to people. Consequently, equivalency approaches may not sufficiently address the human dimension (at least in a quantitative manner) and may underestimate longer-term impacts. For these reasons, equivalency approaches can fall short of making the public whole. In contrast, an ecosystem services approach accommodates the human dimension. In addition, on the ecological side, an ecosystem services approach may expand the array of possible restoration projects by designing novel ways (independent of lost habitats or resources) to restore the loss of ecosystem services. The critical question of what it really means to make the environment and the public whole (and the tradeoffs associated with these actions) is further addressed in Chapters 2 and 3 of this Final Report.

The Interim Report also introduced approaches for establishing the baseline measurements needed to understand the damages caused by the spill and highlighted key parameters that have been or should be measured in the GoM to understand the state of the ecosystems, and the services they provided, before the *DWH* oil spill. As explained in this Final Report, depending on the ecosystem service being addressed, there are vast differences in the amount and quality of data available. This fact, combined with the dynamic (and already degraded) nature of many of the ecosystems in the GoM, renders the establishment of pre-spill baseline levels one of the major challenges to any approach to damage assessment.

The final chapter of the Interim Report explored the process of estimating the impact of an event such as the *DWH* oil spill on several ecosystem components (wetlands and fisheries) through the use of ecological production functions (models that capture the mechanics of how changes in ecosystem parameters impact ecosystem services), and then it discussed economic approaches (market and nonmarket) to quantifying the monetary value of the ecosystem service. The Interim Report offered specific illustrative examples of several types of measurements that would augment and complement standard NRDA measurements and would facilitate application of an ecosystem services approach to estimating the impact of the spill. In closing, the Interim Report acknowledged that, although it offers great potential for more complete and realistic estimates of both short- and long-term impacts of an event such as the *DWH* oil spill, the ecosystem services approach is still very early in its development and faces many challenges to its implementation, the most serious of which is the lack of comprehensive ecosystem models.

Since the Interim Report was released, the analyses of thousands of additional samples have been completed, enhancing understanding of the spill's impact on various components of the GoM ecosystem. The results of many of these analyses are still not public because the NRDA process is ongoing and the studies have not been closed, but peer-reviewed publications are beginning to appear, which has allowed the committee to arrive at a clearer and better-documented picture of some of the short-term impacts of the spill. In addition, a greater understanding of the nature of ecosystem services in the GoM has evolved. Therefore, to address the first two committee tasks, Chapter 2 of this report revisits the concept of ecosystem services and explores in more detail the challenges to implementing an ecosystem services approach to damage assessment, specifically: (1) shifting or dynamic baselines in the GoM, (2) the lack of complete or validated ecosystem models that capture the full complexity of ecosystem interactions in the GoM, and (3) understanding the tradeoffs between restoration options to make the public whole. Each of these challenges is significant, yet the committee reaffirms its conclusion that an ecosystem services approach has much to offer.

Chapter 3 addresses the concept of resilience (the capacity of a system to rebound from disturbances) in the context of ecosystem services and outlines the challenges faced by managers in attempting to restore or increase the resilience of the ecosystem services in the GoM (Task 7 of the Statement of Task). Consideration of resilience can provide a useful framework for the management of complex systems such as the GoM, an area prone to major disturbances such as hurricanes, which is especially relevant when the goal is to maintain the provision of valuable ecosystem services. However, resilience can be difficult to measure, and implementation of resilience-based management strategies may be impeded by potential legislative barriers.

Chapter 4 offers a comprehensive overview of the response technologies used during and after the *DWH* oil spill. The magnitude, scale, and depth of the spill necessitated the use of many response technologies at scales never before attempted, including the application of 1.8 million gallons of chemical dispersants (Federal Interagency Solutions Group, 2010) at the surface and at 1,500-m depth. The chapter examines the technologies' effectiveness, limitations, and potential impacts on GoM ecosystem services (Task 6). In this context, Chapter 4 explores the concept of using Net Environmental Benefit Analysis and "influence diagrams" to evaluate the risks versus the benefits of using different technologies.

To bring the discussion of GoM ecosystem services into focus, Chapter 5 expands on the information provided in the Interim Report regarding the specific ecosystem services provided by the GoM (and thus addresses Tasks 1 through 5). The chapter characterizes ecosystem services within a geospatial context and describes how ecosystem services vary as a function of scale and in response to changes in physical and environmental setting. Four case studies (wetlands, fisheries, marine mammals, and the deep sea) outline some of the documented impacts of the *DWH* oil spill on GoM ecosystem services, provide examples of how an ecosystems approach may be applied to damage assessment, and discuss the opportunities and challenges inherent in such an approach. The case studies reflect a range of conditions with respect to the amount and utility of available data, our fundamental understanding of the functioning of the ecosystem subcomponents, the values of the services in market and nonmarket terms, and the range of the spill's impacts on the services. For each case study, the committee identifies key ecosystem services, considers how they may have been impacted by the *DWH* oil spill, examines methods for making baseline measurements, and explores the adequacy of existing baseline data for the GoM. Additionally, the committee offers suggestions for additional measurements that can enhance an ecosystem services approach to damage assessment. As such, they serve as exemplars of how an ecosystem services approach can add to our ability to capture the full impact of an event such as the *DWH* oil spill and, at the same time, illustrate the challenges faced while attempting this approach.

The Interim Report and this Final Report identify many areas for which data collection or a fundamental understanding of system processes are lacking, highlighting the need for additional research to take full advantage of the potential that an ecosystem services approach offers for damage assessment in the GoM. Chapter 6 reviews the post-spill actions that led to a series of legislative directives regarding response activities and a remarkable level of increased funding aimed at GoM-related research and restoration to address gaps in knowledge and/or data.

The *DWH* oil spill focused national and international attention on the environmental implications of a deep-water drilling accident. The magnitude, depth, and breadth of the spill challenged all those responsible for responding and providing for restoration. New approaches were developed "on the fly" (Lubchenco et al., 2012). Additional new approaches may be needed to fully understand the impacts and to make the environment and the public whole. Drilling in the deep GoM started in 1975, yet our understanding of the GoM ecosystem is still not at a level where the impact of an accident on the people of the region, and perhaps the nation, can be predicted. The last catastrophic spill, the *Exxon Valdez* spill of 1989, prompted the Oil Pollution Act, which was intended to address many of the important issues raised by that spill. The *DWH* oil spill differs in many important aspects from the *Exxon Valdez* spill (e.g., the *Exxon Valdez* spill originated from a tanker in shallow water and impacted coastal and rocky beaches in a cold environment). Many of the lessons learned from a spill in that environment are not necessarily applicable to a spill in the deep GoM environment. Lessons learned from the *DWH* oil spill may eventually work their way into new legislation. Perhaps more forward-thinking systematic actions could put society in a better position to react the next time a disaster strikes. The committee hopes that this broader message has been heard and that this report can, in some way, contribute to new efforts focused on developing the appropriate levels of understanding and suites of tools needed before the next disaster occurs.

CHAPTER TWO

# The Ecosystem Services Approach

An ecosystem services approach to damage assessment, which fully accounts for an event's impact on all aspects of human well-being, can improve efforts to remediate the damage to natural resources caused by events such as the *Deepwater Horizon* (*DWH*) oil spill. This approach shows promise for both phases of remediation: assessment and restoration. With respect to assessment, conceiving of a resource as a part of a system that supplies valuable goods and services to people provides a different way to measure losses—that is, in terms of changes in the flows of those goods and services. With respect to restoration, an ecosystem services approach can expand the menu of projects used to restore the public owners of the resource to their pre-event position.

This chapter first explains the concept of ecosystem services and summarizes the various definitions of the term that have been developed in an extensive literature. Next, the chapter discusses the role of "ecological production functions" in producing ecosystem services and the challenges of assigning an economic value to ecosystem services, such as uncertain and dynamic baselines for ecological and human components of the ecosystem, multiple stressors, inadequacies of data and modeling tools, and "making the public whole" in response to complex system-level disturbances. These ideas are further explored in Chapter 5, which takes a more detailed look at ecosystem services specific to the Gulf of Mexico (GoM) and uses four case studies to explore how current and expanded impact studies can be used to enable an ecosystem services approach to damage assessment.

This chapter also provides an overview of the steps required to translate the concept of ecosystem services into practical tools that can aid in remediation efforts, and it explores three major obstacles to building and applying these tools. First, baseline measures of goods and services produced by the system just prior to the harmful event are needed to assess damages; but differences in the pre- and post-spill state of the GoM ecosystem are difficult to establish because the pre-spill state is not fully known. Associated with this challenge is the added complication of dynamic or shifting baselines.

Assuming that a baseline is established, a second major obstacle is the difficulty of developing a model that can provide defensible estimates of the full impact and costs of reduced goods and services production resulting from the release of a given amount and kind of pollutant (or any other human-induced stressor or stressors) into the ecosystem. Existing ecosystem models are capable of measuring some, but not all, of the complex and intertwined social-ecological impacts of an event such as the *DWH* oil spill; however, these models might serve as the basis for models that can.

The third obstacle relates to the relative values of various ecosystem services. Some ecosystem goods and services are more easily priced than others. For example, economists have a much better idea of how to price a decrease in a system's ability to produce fish for a commercial fishery than a decrease in a system's ability to provide socioeconomic stability to small, rural communities. Furthermore, even when services can be measured, policymakers will have to prioritize among restoration options because of resource limitations. The first scenario raises concerns that more-difficult-to-price services will be discounted or ignored in decision making; the second highlights an important limitation to any approach to remediation.

The committee concludes that the key to feasible application of an ecosystem services approach is the development of tools capable of establishing and quantifying causal links among the event, an injury to an ecosystem, the resulting decrease in goods and services provided by that system, and the cost of that decreased production of goods and services to individual communities and society at large.

## WHAT ARE ECOSYSTEM SERVICES?

Society benefits from a wide variety of resources and processes that are provided by ecosystems. These benefits are known as ecosystem services, and they result from the functioning of an ecosystem—the interactions of plants, animals, and microbes with the environment (NRC, 2011).

A rich and evolving literature on ecosystem services offers a variety of definitions of ecosystem services (e.g., Barbier, 1994; Costanza et al., 1997; Daily, 1997; de Groot, 1987; de Groot et al., 2002; Ehrlich and Mooney, 1983; EPA, 2009; MEA, 2005; NRC, 2005b; TEEB, 2010; Westman, 1977; Wilson and Carpenter, 1999). The common thread through all of these definitions is the concept of a relationship between ecosystems and the value humans derive from them. In 2000, the United Nations commissioned the Millennium Ecosystem Assessment (MEA) to summarize the current status and future conditions of biodiversity and ecosystems, and to describe the consequences of ecosystem change for human well-being, including secure livelihoods, social cohesion, security, and freedom of choice and action (MEA, 2005).

The MEA defines ecosystem services as "the benefits provided by ecosystems to humans, which contribute to making human life both possible and worth living" (MEA, 2005, p. 23). Moreover, the MEA defines explicit categories of ecosystem services, including:

- *Provisioning services* (e.g., material goods such as food, feed, fuel, and fiber);
- *Regulating services* (e.g., climate regulation, flood control, water purification);
- *Cultural services* (e.g., recreational, spiritual, aesthetic); and
- *Supporting services* (e.g., nutrient cycling, primary production, soil formation).

These service categories are now widely accepted and form the basis for the discussion of ecosystem services throughout this report. When applied to the GoM, discussions of ecosystem services must take into account the geographic, oceanographic, ecological, and social context of the GoM. Details of the GoM ecosystem and the regional context for the GoM were presented in the committee's Interim Report (NRC, 2011).

## APPLYING THE ECOSYSTEM SERVICES APPROACH

Measurement of the impact of human actions on the environment—either intentional actions brought about by a policy or management change, or unintentional actions, such as an oil spill—requires an understanding of three important linkages: environmental impacts, ecological production functions (i.e., models for ecosystem interactions), and valuation (Figure 2.1).

Understanding of the linkages will be informed by answers to three overarching questions. First, what are the impacts of decisions and disturbances (whether or not caused by humans) on the structure (composition, physical, biological, and social organization) or on the basic functioning (interaction of humans, plants, animals, microorganisms and their environment) on the social-ecological systems?

Second, how do these changes in the structure and function of ecosystems lead to changes in the potential provisioning of ecosystem services? An important step in this sequence of questions is underpinned by the development of ecological production functions. Harm to ecological production functions may lead to reductions in the ecosystem's ability to generate ecosystem services. And, because most ecosystems have the ability to "right" themselves from harm, an attribute termed resilience, it is important to determine the strength of that ability when response and other management actions are taken. Chapter 3 discusses the importance of resilience in natural systems such as the GoM ecosystem and, in particular, its relevance to the *DWH* oil spill.

Third, how do changes in the provisioning of ecosystem services affect human well-being, and if the value of those changes should be estimated, then how should it be estimated (see Figure 2.1)? Economic valuation is sometimes used to estimate the value of ecosystem services in monetary terms. Summation of the estimated values across all individuals affected by the change in services can identify the overall societal value of the change in services. However, the values of some ecosystem services are difficult to measure in monetary terms. The decision-making process does not require that all ecosystem services be measured in monetary terms to be of use (Polasky and Segerson, 2009). Aspects of these three linkages are illustrated below.

### Environmental Impacts, NRDA, and Ecosystem Services

In the United States, the legal context for measuring environmental impacts of certain oil or chemical spills is defined through the Natural Resource Damage Assessment (NRDA) process and implemented under the Oil Pollution Act of 1990, a topic discussed in the Interim Report (NRC, 2011). The National Oceanic and Atmospheric Administration (NOAA) is one of several federal and state groups involved in the NRDA process; additional details on the status of its efforts can be found at NOAA (2012a).[1]

NOAA has a three-phase approach to NRDA:

[1] http://www.darrp.noaa.gov/about/laws.html#OilPollution.

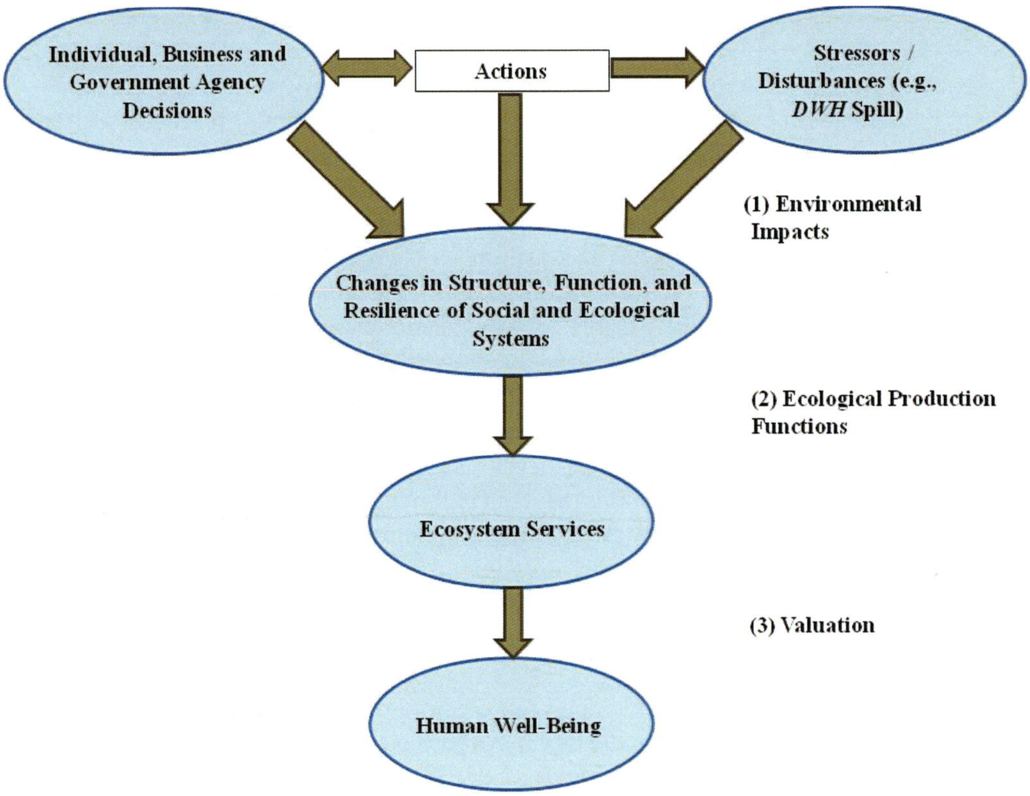


**FIGURE 2.1** The three important links from human actions to human well-being through ecosystems: (1) environmental impacts, (2) ecological production functions, and (3) valuations. SOURCE: Adapted from NRC (2005b, 2011).

1. pre-assessment,
2. restoration planning, and
3. restoration implementation.

The phases and methods are generally described in the applicable regulations[2] and in NOAA's Damage Assessment, Remediation, and Restoration Program (DARRP) guidance documents (Huguenin et al., 1996; Reinharz and Michel, 1996) (Figure 2.2).

Individuals charged with representing the public in an assessment of the damages from an oil spill (the Trustees) first need to determine if a pathway exists between the releases of contaminants or response actions (e.g., oil, dispersants, or stranded boom in tidal marshes) and potential impacts. Once a pathway is established, the Trustees will proceed to the restoration

[2] 15 C.F.R. § 990 (2012).

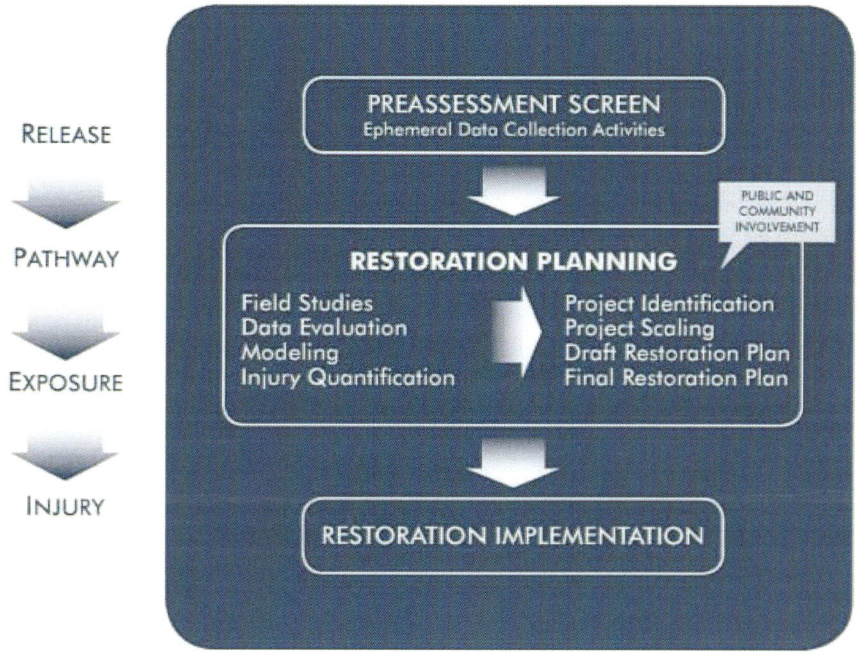


**FIGURE 2.2** Illustration of the damage assessment process. SOURCE: NOAA, 2012b.

planning phase, which includes assessing injury, soliciting public input, and scoping for appropriate restoration projects. After the injury is quantified and the appropriate amount and types of restoration activities are identified, a final restoration plan (in some cases, multiple plans) is implemented and monitored for success. Emergency and early restoration actions can occur prior to injury quantification and implementation of a final restoration plan(s) (NOAA, 2012b).

Under current NRDA practice, injuries to natural resources and losses are generally measured in ecological terms (e.g., number of acres damaged or number of fish killed), and restoration generally follows relatively straightforward equivalency approaches (e.g., acres of habitat restored or fish stocks replaced). These equivalency approaches can be applied to the *DWH* oil spill (NOAA, 2012c; NRC, 2011).

Since the *DWH* oil spill, numerous studies conducted under the NRDA process have focused on better understanding the impact of the spill on the GoM ecosystem (NOAA, 2012c). Many thousands of samples and observations have been, and continue to be, taken and analyses are under way; some of the results have already been published. Certain impacts may not become apparent until well into the future, if at all. Yet, against this backdrop, the government is legally obligated to conduct a timely assessment of the damages associated with the event.[3]

[3] 15 C.F.R. § 990 (2012).

In the Interim Report (NRC, 2011), the committee discussed the potential advantages of using an ecosystem services approach to damage assessment for the *DWH* oil spill and suggested possible mechanisms for its use within the existing NRDA process. For example, for the ecosystem service of hazard moderation (through reduction of storm surges) provided by wetlands, the committee recommended that damage assessment sampling should include documentation of plant type, height, and density, as well as estimates of the vegetation likely to experience salt burn and dieoff, measurements of the areal extent of wetlands harmed, and estimates of the ability of wetlands to recover with and without human intervention. Such a suite of measurements would then allow for the determination of an ecological production function that could relate the plant height, density, and areal extent to wave energy reduction. This approach focuses not only on restoring damaged resources (as per current damage assessment practice), but also on establishing and maintaining the usefulness of those resources to the public. It is this broader view that may be particularly applicable to an event of the magnitude and complexity of the *DWH* oil spill, which is the focus of this report.

## Ecological Production Functions

An important underpinning of an ecosystem services approach is the ecological production function (Box 2.1). In simple terms, an ecological production function represents the chemical, physical, and biological processes or elements that work collectively to provide the service that is valued by people.

**BOX 2.1**
**Ecological Production Functions**

Production functions are tools used by economists to describe how inputs can be transformed into outputs. A production function gives the feasible output of goods and services that can be produced from a given set of inputs. For example, what is the maximum amount of steel (output) that can be produced from a given amount of iron ore, energy, machinery, and labor (inputs)? The notion of production functions applied to ecological systems has a long history in agricultural economics (e.g., crop yield functions) and resource economics (e.g., bioeconomic modeling of fisheries and forestry).

Production functions have been applied recently to the provision of ecosystem services (e.g., Barbier, 2007; Daily et al., 2009; NRC, 2005b; Tallis and Polasky, 2009). An ecological production function specifies the output of ecosystem services generated by an ecosystem given its current condition. Changes in ecosystem conditions, either from natural disturbances such as hurricanes or from human disturbances such as an oil spill, can alter the amount and quality of the various ecosystem services supplied. For example, degradation of coastal marshes may reduce protection from storm surges and reduce nursery habitat for fish, in addition to other service losses.

For some ecosystem services, ecological production functions are fairly well understood and data exist that can be used to quantify the amount of a service provided. A good example of a fairly well-understood and well-studied ecosystem service is carbon sequestration in the biomass of terrestrial ecosystems, particularly forests. The U.S. Forest Service collects data on biomass in forests by stand age and tree species for different areas of the country (Gan and Smith, 2006). These data, along with knowledge of the carbon ratio in biomass, can be used to calculate carbon sequestered in forests.

In marine systems, production function approaches have been used to study the productivity of fisheries as a function of ecosystem conditions (Barbier, 2000, 2003; Barbier and Strand, 1998; Barbier et al., 2002; Ellis and Fisher, 1987; Kahn and Kemp, 1985; Lynne et al., 1981; McConnell and Strand, 1989; Parks and Bonifaz, 1994; Sathirathai and Barbier, 2001; Swallow, 1994) although there is far greater uncertainty in the functional relationship between habitat conditions and fishery productivity. The Interim Report (NRC, 2011) provided some examples of the data and analyses required to establish ecological production functions for different resource categories, expanding upon the basic NRDA approach to include information about key ecosystem services. Additional examples of production functions and the types of data required for their characterization are provided in the case studies detailed in Chapter 5.

## Valuation: Current Understanding of the Value of Ecosystem Services

As illustrated in Box 2.1, ecological production functions can be used to estimate reductions in ecosystem services. A challenge arises in attempting to quantitatively determine the impact of reductions in ecosystem services on human well-being. Economists often do this through a process that attempts to determine the monetary value of ecosystem services.

The most fully developed approach for valuing ecosystem services comes from economics. Economic analysis of ecosystem services can generate estimates of the value of services in terms of a common (monetary) metric. An economic analysis provides decision makers with clear comparisons of the benefits and costs of alternative choices. Given preferences, the value of an ecosystem service can be measured in terms of what the individual would be willing to give up to get more of the ecosystem service. By measuring what an individual is willing to give up in terms of a monetary metric, or "willingness to pay," the economic approach to valuation generates measures of the relative value of goods and services.

Valuation of ecosystem services does not need to be in monetary terms. This is simply a useful and well-recognized approach to quantify the willingness to trade one thing for another. For goods and services that are traded in markets, an individual's willingness-to-pay, or price, for a marginal change in the good or service is reflected in an individual's demand curve. Aggregating across all individuals who purchase the good or service generates an estimate of the societal value of the good or service.

A fundamental assumption of this approach is that individuals have well-defined and stable preferences. Most ecosystem services, however, are not directly traded in markets, so the direct approach to estimating value via observing market transactions is not possible.

Environmental economists have developed a set of approaches for nonmarket valuation that are applicable for estimating the value of many ecosystem services. These methods have been widely applied to value such things as carbon sequestration (Tol, 2009), water quality improvements (Johnston et al., 2005; Smith and Desvouges, 1986), wetlands (Boyer and Polasky, 2004; Woodward and Wui, 2001), and endangered species (Richardson and Loomis, 2009). The application of nonmarket methods to valuing ecosystem services has been discussed extensively in Champ et al. (2003), EPA (2009), Freeman (2003), NRC (2005b), TEEB (2010), and Chapter 4 of the Interim Report (NRC, 2011). In the case of the *DWH* oil spill, of primary relevance in applying the valuation approaches is estimating the value of the change of ecosystem services attributable to the spill.

It should be noted that there are critics of these economic approaches to valuation. Some psychologists, for example, question the assumption that people have well-defined, stable, and consistent preferences that they bring to decision making. A body of work in both psychology and behavioral economics has documented systematic departures from classic assumptions of rational behavior (e.g., Ariely, 2009; Kahneman and Tversky, 1979). A body of experimental evidence suggests that people often construct their preferences when called upon to make decisions and are therefore sensitive to the context and framing of decisions (e.g., Lichtenstein and Slovic, 2006). Sociologists question the central focus on individuals and individual decisions, which they posit does not give proper consideration to how values are shaped by larger groups, norms, and culture. Despite these criticisms and the fact that other approaches to valuation do exist, virtually all valuations of ecosystem services to date have used the economic approach.[4]

## Obstacles to Application of the Ecosystem Services Approach

Thus far, this report has focused on discussing the elements of an ecosystem services approach, illustrated in Figure 2.1, and the potential benefits of using this approach to assess damages resulting from an event such as the *DWH* oil spill. The approach has the potential to capture a more holistic picture of impacts on the ecosystem and on human well-being and thereby provide a more realistic view of the overall range of damages, as well as the potential to increase the number of restoration options. An examination of the challenges to application of this approach to the GoM and *DWH* oil spill follows.

The Interim Report (NRC, 2011) briefly identified some of the major challenges to application of the ecosystem services approach. The following section addresses three major challenges in detail, with particular focus on the GoM and the *DWH* oil spill: (1) shifting baselines of physical, chemical, and biological processes and socioeconomic conditions; (2) the lack of complete or validated models that capture the full complexity of ecosystem interactions in the GoM; and (3) understanding the tradeoffs between restoration options.

[4] See EPA (2009) for a review of both economic and noneconomic approaches.

## Establishing the Baseline

Quantification of the effects of, and recovery from, an event such as an oil spill is difficult, particularly when the effects must be measured against a changing marine environmental background (NRC, 2003; Peterson, 2001; Spies et al., 1996; Wiens, 1995). The baseline may reflect the fact that recovery from a prior disturbance is under way, which must somehow be taken into account when determining the pre-spill status.

Assessing initial damage from the acute effects of the spill, especially those of socioeconomic importance, is difficult, but assessing recovery is perhaps even more challenging. Recovery is removed in time from the acute phase of the damage and therefore may be occurring in a different environmental and socioeconomic framework (and location) than that existing at the time of the accident.

Figure 2.3 presents a simplified illustration of how stressors on an ecosystem, in this case a salt marsh, may result in short- and long-term changes in the system and ultimately in the services it provides. Under Scenario A, baseline ecological services are produced at a variable rate over time, but that variability falls within a range that is likely dependent on factors that influ-

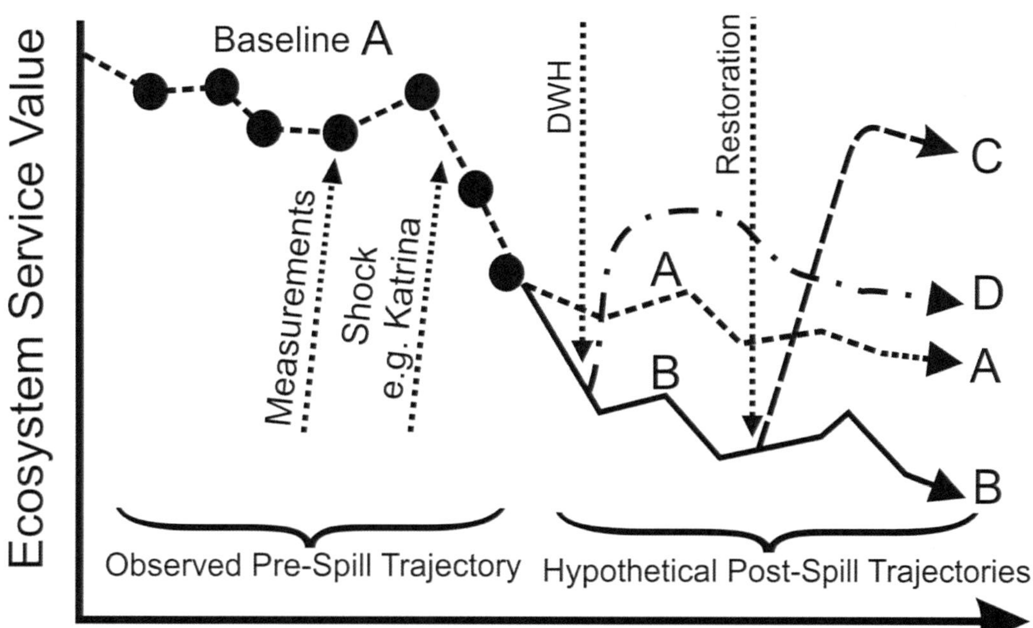


**FIGURE 2.3** Hypothetical baseline of an ecosystem (salt marsh) service pre- and post-*DWH* oil spill and possible recovery responses. Baseline A is the hypothetical trajectory in the absence of the spill. Trajectory B is a hypothetical negative response to *DWH*, while D is a hypothetical positive response. Trajectory C is the possible response following restoration. SOURCE: Committee.

ence the ecological production functions and the resilience innate to the system. After a major stress to the system—Hurricane Katrina, for example, in the case of the GoM—two outcomes are likely: the level of services continues, but is reduced (Scenario A), or another stressor like the *DWH* oil spill provides additional stress that the system is unable to withstand (Scenario B). Under both scenarios, it is possible that ecosystem services will decline and, depending on conditions that influence the underpinning ecological production functions, may or may not recover to their original level over time.

Other potential outcomes of the type of stress rendered by Hurricane Katrina and the *DWH* oil spill include Scenario D, where restoration and response actions enable the system to recover, but not to the baseline condition, and Scenario C, where restoration and response actions are highly beneficial to the extent that the ecosystem services recover to levels similar to those existing prior to the two major stressors. Figure 2.3 illustrates the importance of pre-stressor monitoring data. Without such data, it would be difficult to isolate the impact of the spill as well as to determine the success of the restoration actions.

Quantitative assessment of recovery from a spill event requires either a well-designed Before and After Control Impact (BACI) approach or an approach that compares measurements of the environmental variable of interest along a gradient of perturbation (Wiens, 1995). This gradient can be across space or time. When numbers of organisms are compared for assessment of recovery, attributes such as age or reproductive potential must be taken into account. Using the example of marine birds, young, inexperienced birds do not have the same value to the population as experienced, breeding adults.

The natural variability inherent in estimates of populations introduces considerable uncertainty in assessing impact and recovery following major stressor events such as oil spills or loss of habitat. Confidence limits in excess of 20 percent of the mean size are typical in wildlife censuses (Geissler, 1990). Such variability in the estimated mean makes it certain that population changes will be difficult to detect without a high degree of replication spatially and temporally before and after an event. More importantly, under some circumstances estimates of recovery based on the population returning to a "window" of natural fluctuation could minimize the time to true recovery. If there are no data to implement a BACI approach, then it may be possible to compare suitable data on the measurable parameters during and immediately after the spill to any suitable long-term database that may provide evidence of change in relatively well-monitored ecosystem components (e.g., the long-term fisheries data collected by NOAA Fisheries, as well as their marine mammal surveys and sea turtle surveys).

Changing baselines also make assessment of the spill impacts on the social structure and its recovery equally difficult. Economic conditions outside the control of a community, such as fuel prices, may constrain the community's ability to harvest fishery resources unless they are sufficiently abundant to cover the cost of fuel. Therefore, a spill-related reduction in the abundance of fishery resources may make harvesting economically infeasible. Underlying trends of economic stressors may mask or compound the impact of stressors that resulted from the *DWH* oil spill.

For example, after the *DWH* oil spill, the federal government imposed a moratorium on deep-water drilling pending further study of blowout prevention and response. It was reported that, during this time, deep-water drilling vessels and parts of the service fleet and logistics industry moved to Brazil to operate.[5] Such vessels and other equipment would not likely return to the initial location immediately after expiration of the moratorium. It will take more than a few years to determine the long-term economic impacts of the *DWH* spill on this sector. During this time, a global economic effect (e.g., oil embargo or new discoveries of oil elsewhere worth exploring) could readily shift the "average" baseline.

## Models and Data Needs

As briefly discussed in the Interim Report (NRC, 2011) perhaps the most notable challenge to our ability to apply an ecosystem services approach to the GoM is the lack of comprehensive models for assessing conditions in the GoM. These computer models, or simulations, are needed to better understand the potential impacts of the *DWH* oil spill on GoM ecosystem services. Ideally, a thorough ecosystem services approach would be based on a mechanistic understanding of, and model for, the complex linkages and interdependencies of the ecosystem being studied, socioeconomic factors, and the linkages of the natural ecosystem processes to sociological processes that represent the local coastal communities and the broader U.S. and global economies. Such a model would enable prediction of the provision of ecosystem services given the state of the ecosystem (i.e., the ecological production functions). Development of such a model requires either a comprehensive empirical database (from which interactions can be derived and predicted) or a thorough theoretical understanding of the complex interactions of the system—both of which are currently lacking. An ongoing challenge is that the data collected under the NRDA process are not always the most appropriate for developing a basic understanding of variability in biological and other processes, which is essential for developing appropriate ecosystem models for the GoM.

## An Example of a Thorny Modeling Issue: Multiple Stressors

Multiple stressors often affect ecosystems, populations, or communities, and present extra challenges to individuals tasked with understanding the impacts of stressors on the ecosystem (Ferenc and Foran, 2000). For instance, a Louisiana bay or wetland may be subject to changes in salinity due to freshwater diversions, eutrophication due to excess nutrients, changes in inorganic nutrient regimes, or changes in turbidity with wind mixing. The ability to determine the effects of oil on these ecosystems is complicated by these and other stressors.

The northern GoM and especially the Louisiana coast are experiencing relatively rapid sea level rises compared to other locations in the world (Donoghue, 2011). This is due to a combination of global (eustatic) sea level rise superimposed on (a) the sediment-limited and extensively manipulated Mississippi River deltaic system and (b) an isostatically sinking continental margin. Factors such as regional fluid withdrawal and sediment compaction are also thought

[5] http://www.reuters.com/article/2010/06/11/brazil-oil-rigs-idUSN1115006620.

to contribute to the relative sea level rise (Walker et al., 1987). The net result is a dramatic transition of estuarine habitats, the natural resources they support, and the human communities that depend upon them.

Multiple or single stressors may also be cumulative over time. For instance, the sediment contaminant levels of polycyclic aromatic hydrocarbons (PAHs) from produced water discharges in inshore and offshore waters represent the discharge of these materials over many years (Rabalais et al., 1992, 1998). In addition, although the range of exposure of sediments and fauna to PAHs produced from waters around oil platforms on the continental shelf of the GoM may be limited to the periphery of the production platform, the cumulative effect of many such discharges may be more concerning. Likewise, although a pulse of high nitrate-nitrogen loading from a freshwater diversion into a salt marsh may seem to stimulate aboveground biomass, longer-term exposure endangers belowground root biomass, leading to its subsequent decomposition and eventual erosion (Darby and Turner, 2008; Deegan et al., 2012; Turner, 2011).

Stressors may include socioeconomic factors that would create variability around a baseline that is a goal for recovery. The coastal county population of the GoM was approximately 14.2 million in 2010 (NOAA, 2013b). Between 1960 and 2008, the Gulf coastline counties realized an increase in population of greater than 150 percent (Wilson and Fischetti, 2010). Increases in population are not uniform across the landscape. Human populations are also shifting as more individuals move from flood-prone areas to higher elevation. Shifts in population, socioeconomics, and the ability of human communities to provide adequate support for new populations with new needs can interact as stressors that might affect the baseline sought for "recovery" of social systems. These coastline populations receive considerable economic benefit from the GoM, as evidenced by an estimated employment of more than 6.2 million people in the region in 2010.[6] Many important ecosystem services are now threatened by multiple stressors on the functions and processes of natural ecosystems.

Oil and gas development currently has the greatest economic value in the GoM ecosystem (NOAA, 2011f) through its extraction, transportation, and transformation into petroleum-based products. Fisheries, both commercial and recreational, are the most valuable, marketable living resource produced by the GoM system. In 2011, the market value of the catch from commercial fishing was worth $818 million just off the vessel; the value of the fish increases as it moves up through the value chain and is processed and sold at retail. In 2011, approximately 3 million recreational anglers took 23 million fishing trips in the GoM, worth $9.8 billion in terms of sales (NOAA, 2013a). Seafood from the GoM also supports artisanal and subsistence fishing, although data on these activities are rare or nonexistent.

It is clear that many local communities, especially in the north-central GoM coastal region, depend on a combination of energy-related economic development and natural resource harvesting. These economic pursuits often support and supplement harvesting, but they are antagonistic because of the environmental impacts on fisheries of energy sector operations in the GoM. As the fortunes of the oil and gas development and production industry wax and

[6] http://coastalsocioeconomics.noaa.gov/.

wane in the northern GoM, the economies of many coastal communities follow suit. Similarly, as the economic viability of fishing fluctuates, communities may switch employment opportunities or suffer the same economic declines as the fishery resources. Chapter 3 further explores these issues under the topic of community resilience.

The northern GoM social-ecological system that was impacted by the *DWH* oil spill is also subject to multiple external forces over which the region has, in reality, little control. A suite of global and domestic economic forces impact the region, including fluctuating fuel prices, an influx of imported goods such as cultured shrimp, and the loss of social services through federal and state budget cuts. These forces further challenge a social structure that depends on healthy, functioning ecosystems for food, jobs, well-being, and a sense of identity. An episodic impact, such as the loss of 5 acres of salt marsh and its ecosystem services, is more easily quantifiable, and the baseline more obvious, than is a suite of interacting chronic impacts amid a shifting set of baselines further influenced by global market forces.

As noted in this section, baseline conditions within the GoM, or any ecosystem for that matter, change over time, which presents a challenge to individuals attempting to parse out the natural variability in a system from the potential impacts associated with stressors originating from the *DWH* oil spill. This challenge can be alleviated through the development of models that can approximate biotic and abiotic conditions in the system and then inform understanding of how stressors impact the system and the services it can provide. As discussed in the following section, such models are currently being developed.

## Existing Models

Over the past 15 years, marine ecosystem models have been developed for a variety of purposes, but, more recently, attempts have been made to use these models to evaluate the consequences of external events on the dynamics of ecosystems. Ecosystem models can be used to study and predict future changes in the ecosystem because direct experimentation on the ecosystem is seldom possible, especially when the subject ecosystem is large and has open boundaries such as the GoM ecosystem. Few of the existing marine ecosystem models have been developed with the objective of enhancing an understanding of the dynamics in a holistic ecosystem services approach; however, many of them are useful to evaluating a particular subset of services. This section discusses some of the models that have been developed to support management of marine ecosystems and how they might be used to support an ecosystem services approach. Models that were designed for fishery applications are considered first, followed by others aimed at broader segments of marine systems.

### Fishery Ecosystem Models (FEMs)

Fishery Ecosystem Models (FEMs) are distinct from purely ecological ecosystem models in that they consider not only the biological components of the ecosystem but also the fishery, which includes a human component. The structure of a FEM reflects the priorities of fishery managers and harvest activities. Thus, the model includes the monitoring and management process related to fisheries, but often ignores components representing other marine industries.

FEMs allow researchers to narrow the set of possible hypotheses to be tested when an ecosystem is disturbed by human actions. Although typically these disturbances are related to the possible cascading effects of increases or declines in harvests (Fulton, 2010), they can also be disturbances of the type caused by direct and indirect mortality resulting from pollution (Yanez-Arencibia and Day, 2004). By reducing the set of possible hypotheses, FEMs can assist in the design of studies that will collect the necessary data to more precisely estimate the effects of pollution on future fishery and ecosystem productivity. In that regard, they can also assist in the development of effective responses to oil spills (Freire et al., 2006). Although the models' ability to predict the precise magnitude of changes is still questioned (Butterworth and Plagányi, 2004), they have proven useful in predicting the direction and order of magnitude of the change, and they have been used recently to inform management decisions related to fisheries ecosystems (Fulton, 2010; Fulton et al., 2007).

One type of FEM, Ecopath with Ecosim (EwE), is a modeling framework that integrates a wide range of biological and fisheries dynamics for multiple species and functional groups (biomass pools) over long time periods, using a trophic mass-balance approach (Christensen and Walters, 2004). Although designed as a FEM, EwE has been used to address many aspects of marine ecosystem dynamics, including the impacts of the 1989 *Exxon Valdez* oil spill in the Prince William Sound ecosystem (Okey and Pauly, 1999). Oil spill impacts can be represented as both direct mortality events on susceptible biomass pools and as bioaccumulation of oil-related contaminants, using EwE's Ecotracer routine (Christensen and Walters, 2004). EwE models have been developed at three spatial scales within the GoM: (1) bay-estuary (Rosado-Solórzano and Guzmán Del Próo, 1998; Vega-Cendejas and Arreguin-Sanchez, 2001); (2) regional shelf (Arreguin-Sanchez and Manickchand-Heilman, 1998; Arreguin-Sanchez et al., 2004; Okey et al., 2004); and (3) basin (Walters et al., 2008). Within the context of the *DWH* event, EwE could at least serve to identify the components of the model ecosystem most likely to be affected by the oil spill.

## Whole-Ecosystem Models

Atlantis is an ecosystem modeling framework with submodels that simulate oceanography, ecology, fish population dynamics, fishing fleet dynamics, economics, fisheries stock assessments, management decisions, and exploitation (Fulton et al., 2004a, b, 2011). Atlantis provides a detailed model of higher trophic-level dynamics, fisheries, and socioeconomics, which may enable tracing of the impacts of spills like the *DWH* all the way to the human impacts. The model is driven by strong oceanographic data that could be used to project the distribution of oil following a spill in the GoM. It can also interface with finer-scale and more detailed models applied to some components of the ecosystem. Plagányi (2007) found that Atlantis is the best whole-ecosystem model currently available for evaluating the management strategy for marine ecosystems, in part because its modular structure allows for great flexibility in modeling a range of ecosystems. However, one significant limitation of the current model is that modelers cannot apply a sensitivity analysis to show how the uncertainty in the output can be attributed to different sources of uncertainty in the inputs.

Several integrated models are capable of modeling the joint provision of multiple ecosystem services. These models have been developed primarily for terrestrial systems (e.g., Kareiva et al., 2011), but some models have also been applied to coastal and marine systems.

One integrated model built specifically for coastal and marine ecosystems is the Marine Integrated Valuation of Ecosystem Services and Tradeoffs (Marine InVEST). Marine InVEST is a flexible tool that is spatially explicit, can be run at different levels of complexity to account for different data availability and system knowledge, and maps the provision and value of multiple ecosystem services as a function of social and ecosystem conditions (Guerry et al., 2012). The model has been applied in a variety of sites, including Mobile Bay, Alabama, and Galveston, Texas, as well as the West Coast of Vancouver Island, British Columbia, Canada, and Belize. Marine InVEST is designed primarily for use with coastal ecosystems and does not yet model deep-sea ecosystem services. The model currently includes the ability to assess habitat susceptibility to various environmental stressors, but also includes wave energy, aquaculture, coastal protection, coastal erosion, aesthetic quality, and a simple fisheries model. Soon-to-be-released modules will cover water quality, shellfish, wind energy, recreation, and tourism. Further applications allow for valuing ecosystem services when tradeoffs are made—for example, differences in construction for rebuilding a barrier island intended for wave surge protection versus one intended for tourism.

The Multi-Scale Integrated Models of Ecosystem Services (MIMES) were developed to fully account for several relevant factors that contribute to human well-being, including built, social, and natural capital. These are conceptually complex, spatially explicit models that were created by the Gund Institute for Ecological Economics at the University of Vermont.[7] MIMES are a suite of simulation models that are designed to quantify the implications of human uses in the natural environment. These models integrate socioecological relationships, analyses, strategies, and policies to comprehend the links and interdependencies that exist between complex subsystems. The five subsystems are atmosphere, lithosphere, hydrosphere, biosphere, and "anthroposphere" (human habitats), which reflect an integrated approach to assessing the capacity of complex systems to deliver a healthy stream of benefits to people.

The Artificial Intelligence for Ecosystem Services (ARIES) tool constitutes a suite of applications that maps concrete, spatially explicit beneficiaries of ecosystem services and quantifies their demand for each service.[8] ARIES is a modeling platform, rather than a single model or collection of models, delivered to end users through an online Web tool. The ARIES approach is to map benefits, beneficiaries, and service flows to allow managers and conservationists to visualize, value, and manage the ecosystems on which the human economy and well-being depend.

[7] http://www.uvm.edu/giee/mimes/.

[8] http://www.ariesonline.org/.

By accounting for biophysical flows of ecosystem services across the landscape, ARIES can link marine and terrestrial habitats. For example, ARIES is being used to model flows of sediment, nutrients, and freshwater from land to near-shore ecosystems, allowing users to model changes in provision of marine ecosystem services based on changing land use practices in Madagascar (Wendland et al., 2010). In addition, ARIES has been implemented in several project sites located in America, Europe, and Africa.

### Application of Marine Ecosystem Models to GoM Services

Although each of the models described above offers some potential for providing critical insight into the impact of an event such as the *DWH* oil spill on GoM ecosystem services, the ability of a model to accurately predict outcomes is a function of the degree to which the model captures the physics and complex interactions of a particular ecosystem, including human communities, and the accuracy of the data used to parameterize the model. Data needed to parameterize ecosystem models to the GoM typically come from published data (e.g., Collins and Wlosinski, 1983; Leidy and Jenkins, 1977; Leidy and Plosky, 1980) from the National Marine Fisheries Service (NMFS) databases such as the NMFS recreational fishery statistics,[9] NMFS Fishery-Independent Survey System,[10] and NMFS Southeast Data Assessment and Review (a fishery stock assessment process),[11] as well as from the expertise of individuals who live and work in the GoM. Additional data on the ecosystem produced by the Gulf of Mexico Research Initiative consortia[12] and the NRDA process are likely to facilitate implementation of these models for the GoM. Input from stakeholders will also be necessary, so that the models correctly capture the components of the system that are relevant to them. Nevertheless, baseline data may be insufficient.

One of the limitations of the ecosystem services approach is the lack of socioeconomic data needed to implement a more robust and complex understanding of the human dependencies on natural systems. Although models exist to explore some of these dependencies and the capacity of systems to provide a healthy stream of benefits to citizens (e.g., InVEST models,[13] and MIMES[14]), no comprehensive model has been developed that integrates the biological, physical, and socioeconomic dimensions, and data are often lacking on social aspects of the system such as the multigenerational linkages described earlier.

Emerging ecosystem models, such as Atlantis, could be modified to support an ecosystem services approach, because they include shallow- and deep-water components of the ecosystem, chemical and physical processes, and certain components of human dimensions. The University of South Florida, the University of Miami, and NOAA are leading an effort to develop

[9] http://www.st.nmfs.noaa.gov/st1/recreational/queries/index.html.
[10] https://www.st.nmfs.noaa.gov/finss/si/siMain.jsp.
[11] http://www.sefsc.noaa.gov/sedar/.
[12] http://www.gulfresearchinitiative.org/2011/rfp-i-consortia-grant-awards-gri-years-2-4/.
[13] http://naturalcapitalproject.org/InVEST.html.
[14] http://www.uvm.edu/giee/mimes/.

such a model for the GoM with the purpose of evaluating fishery management strategies.[15] The model can be thought of as a snapshot of the best currently available information about the fishery ecosystem of the GoM. Current model implementations, including the one being developed for the GoM, however, fall short of considering all ecosystem services.

The Marine InVEST group of the Natural Capital Project consortium[16] (a consortium of institutions developing tools for facilitating the inclusion of natural capital into policy and business decisions), including The Nature Conservancy, began implementing the InVEST models of ecosystem services in the Galveston Bay area in early 2012. This project is developing methods for assessing nature-based and engineered adaptation solutions to climate change. The project is funded by the Climate and Societal Interactions Program of NOAA.

Although there are no structural impediments, neither the MIMES nor the ARIES models have been applied to GoM ecosystems. Thus, while several ecosystem models under development are relevant to the GoM, none of them incorporates all of the relevant processes describing the dynamics of the provision of ecosystem services in the GoM. Lack of a comprehensive model for the GoM ecosystem remains a major challenge to the ecosystem services approach as well as to fuller application of the NRDA process to the *DWH* oil spill. The models presented above, however, are appropriate for the evaluation of the impacts of the spill on various subsets of ecosystem services. Until a more comprehensive model for the GoM is developed, it will be necessary to consider an approach that uses multiple models to assess changes in GoM ecosystem services related to the *DWH* oil spill.

## Outside-the-Model Obstacles to the Ecosystem Services Approach

The ecosystem services approach can be thought of as a lens that allows us to view natural systems in a different way: as sources of services and goods for people (Figure 2.4). Use of the lens conceptually deconstructs each natural system into a set of ecosystem services and goods, which provide insight into its value. Identification of the goods and services provided by a natural system can also change the way we think about approaches to making the public whole when the natural system has been damaged.

Figure 2.4 helps to illustrate some high-level problems inherent in using an ecosystem services approach to remediation. These problems are not intractable; rather, they are denoted as high level because they will arise even if effective methods to identify baselines and to model the linkages between impacts and costs are developed. This section highlights three issues. First, as noted in Figure 2.4 in the valuation question related to the ecosystem service of "community stability," some services are difficult, if not impossible, to measure. The committee described this broad category of services as "indirect benefits" and provided examples of these benefits as they were brought to its attention by community representatives from around the GoM. The important point is that, even though a service may be difficult to measure, it may still

[15] http://cimage.rc.usf.edu/presentations?destination=presentations.

[16] http://www.naturalcapitalproject.org/InVEST.html.

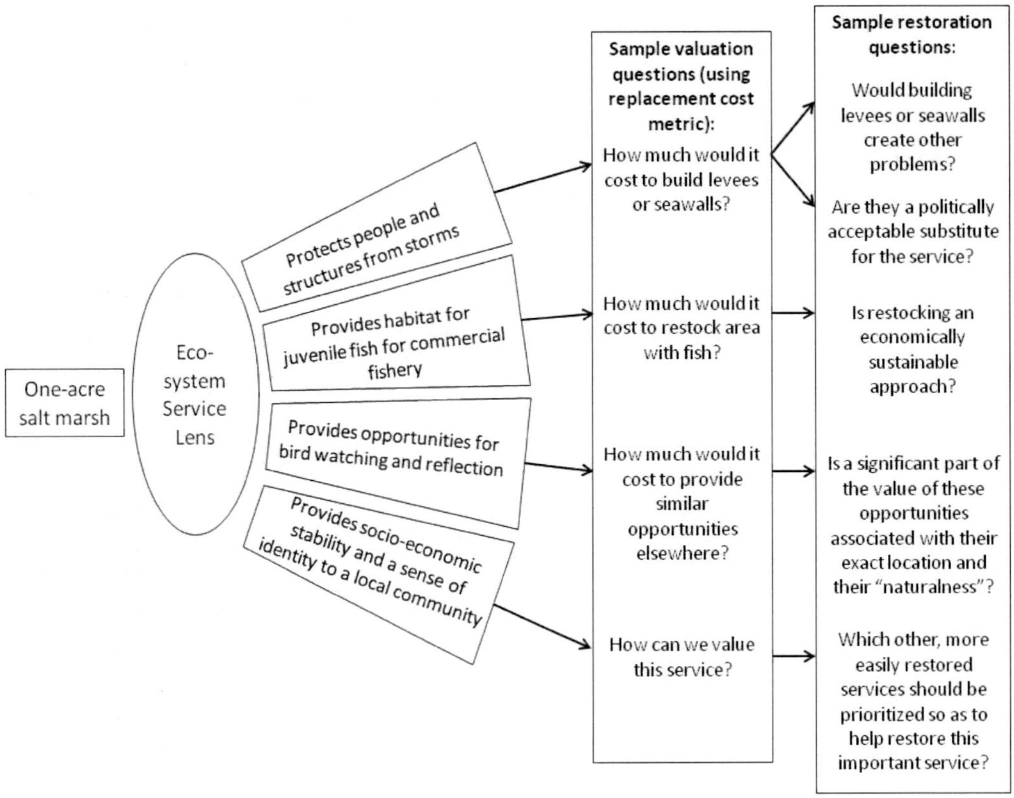


**FIGURE 2.4** An example (a salt marsh) of a natural system viewed as a resource of goods and services via an ecosystem services approach. SOURCE: Committee.

have enormous value. Policymakers should take steps to ensure that difficult-to-measure but important services are given adequate consideration in decision-making processes.

Second, even when the value of all services can be measured, policymakers will still be faced with evaluating difficult tradeoffs among restoration options. Resources are limited, and choices may have diverse impacts on different sectors of the public in different regions. Finally, using an ecosystem services approach to restoration can raise concerns about replacing nature with human-engineered substitutes. Policymakers should be clear that using such substitutes is not an objective of the ecosystem services approach, but rather a last resort—that is, an option to be employed only when there is no other way to make the public whole.

## Indirect Benefits

Economic metrics can be used to evaluate many types of ecosystem services, but they are not effective at measuring the value of important services such as the contribution of a healthy ecosystem to the social stability of rural communities. The relationship between ecosystems and communities is particularly relevant throughout the GoM region. Many coastal communities depend on access to the GoM ecosystems and the diverse services they provide, including the economic benefits that result from oil and gas exploration and the extraction and production of hydrocarbons.

It is also important to note that populations and communities are themselves diverse, and the perceptions of these ecosystem services can vary greatly and consequently influence their value and utility at quite local scales. The result of these interactions between communities and their environment over generations is a social-ecological fabric underpinned by the GoM and its multitude of ecosystem services. Hence it is not unexpected that impacts to these services would also translate, directly or indirectly, into impacts on the social-ecological fabric of this region. These impacts are explored further in Box 2.2.

By definition, ecosystems are linked to human well-being through the regulating, provisioning, supporting, and cultural services they provide. Figure 2.5 (MEA, 2005) "depicts the strength of linkages between categories of ecosystem services and components of human well-being that are commonly encountered, and includes indications of the extent to which it is possible for economic, social, and technological factors to mediate the linkage. The strength of the linkages and the potential for mediation differ in different ecosystems and regions. In addition to the influence of ecosystem services on human well-being depicted here, other factors including other environmental factors as well as economic, social, technological, and cultural factors influence human well-being." Thus, ecosystem services provide both direct and indirect benefits to humans, which in turn collectively contribute to the various human systems that support social cohesion, security, adequate livelihoods, individual and community health, and ultimately freedom of choice and action in pursuing a lifestyle of choice.

Events that disrupt or interfere with the normal functioning of ecosystems may impair ecosystem services, which has the potential to cause short- and long-term harm and loss of benefits to the individuals and communities that depend on those services (see also Adger, 2000). Understanding and quantifying the nature and level of these impacts are difficult and complex tasks, but ones that are essential for establishing the appropriate procedures for recovery, restoration, and management, and when applicable, for seeking compensation for damages caused. This latter point is particularly relevant to the ongoing NRDA process being undertaken by state and federal natural resource agencies and described in the Interim Report (NRC, 2011).

## Considering Tradeoffs in an Ecosystem Services Approach

The goal of the NRDA process is to "make the environment and the public whole for the injuries to natural resources and services" (NOAA, 1996). Disturbances to ecosystems such as the *DWH* oil spill lead to changes in ecosystems and the provision of at least some services in some locations for some period of time. This is true even if the ecosystem eventually fully recovers. Recovery that aims to "make the environment and the public whole" will therefore necessarily

**BOX 2.2**
**Additional Impacts of the *Deepwater Horizon* Oil Spill on Coastal Communities in the Gulf of Mexico**

The Gulf region comprises diverse human communities, each with its own cultural dynamics and values and each reliant on a suite of cultural, provisioning, regulating, and supporting ecosystem services. Ecosystem services, such as fisheries, not only provide commercial and recreational value, but also are integral to the lives of people within these communities. Many traditions and lifestyles of Gulf residents are based on their relationship with the waterways and land around them and have defined their way of life for generations. Because these services are often difficult to evaluate and quantify, they may be overlooked when assessing the impacts of disasters such as the *Deepwater Horizon* (*DWH*) oil spill.

During the public session of its fifth meeting, held on April 25, 2012, in Mobile, Alabama, the committee heard a series of presentations by invited community leaders and representatives on the impacts of the *DWH* oil spill. In addition to the direct economic effects, many of the presenters discussed lifestyle changes that the spill has imposed on their communities. Natalie Bergeron, Executive Director of Project LEARN–LaTerre in Chauvin, Louisiana, defined "lifestyle" as a "composite of motivations, needs, and wants ... influenced by factors such as culture, family, reference groups, and social class." Bergeron further stated that lifestyle is "expressed in both work and leisure behavior patterns, in activities, attitudes, interests, opinions, values, and allocation of income" and "reflects people's self-image or self-concept: The way they see themselves and believe they are seen by others."

She explained that jobs in the bayou communities, which in large part involve shrimping, crabbing, oyster harvesting, and charter fishing, are intertwined with the cultural history, values, and ultimately the overall lifestyles of the people in these communities. Many fishing enterprises, particularly in the shrimping industry, are family operations that were severely impacted by the fisheries closures in the GoM. Closures not only affect the people who make a living catching fish, but also impact other community members who benefit from fishing activities. As part of the social fabric of many small coastal communities, fishers routinely set aside a portion of their catch to share with community members in need—the elderly, the disabled, and the unemployed. This seafood provides much of the protein in these individuals' diets, and the closures forced them to spend a larger portion of their very limited income on other protein sources.

Bergeron also reported that the market for Gulf seafood has remained depressed since the *DWH* oil spill because of low demand due to public perception that Gulf seafood is not safe. She concluded by emphasizing that the spill's impacts compounded existing stressors on the communities such as limited access to employment, adequate housing, health care, and education.

Maryal Mewherter, representing the United Houma Nation, described the tribe's historical reliance on Louisiana's coastal environment for food, materials for handicrafts, traditional medicine, and other cultural practices, specifically for weaving marsh grasses into baskets, making jewelry from fish scales and seashells, and collecting assorted medicinal plants and mosses. The Houma Nation also relies on commercial fishing as the "economic foundation" of its community. Referring to the *DWH* oil spill, Mewherter stated that "[w]ith fishing being a fundamental component of tribal living, the cultural impacts of this disaster on our indigenous communities are difficult to assess." She emphasized that cumulative impacts—from multiple hurricanes, the recent global economic downturn, loss of wetlands due to erosion and subsidence, and the *DWH* oil spill—have collectively imposed severe challenges to the tribe's way of life. As with the other bayou communities, the oiling of the wetlands and the fishery closures greatly reduced the tribe members' ability to feed themselves and generate income from traditional crafts and medicines.

A primary challenge facing the Houma Nation is the loss of coastal wetlands; families and villages are being displaced, and the availability of land further inland and at higher elevations is limited. Although wetland loss is a historic and ongoing challenge, the *DWH* oil spill is an exacerbating factor. The Houma Nation also struggles to influence state and federal agencies—especially in the development of land use policies and restoration efforts in the GoM wetlands and barrier islands. Mewherter concluded by pointing to a tribal crest bearing the image of the red crawfish, which serves as the Houma Nation's war emblem and as an identifying symbol within its community, another example of how important local resources are to the tribe's culture.

Khai Nguyen, a business development counselor for the Mary Queen of Viet Nam Community Development Corporation, highlighted the impacts of the *DWH* oil spill on the local Vietnamese-American community. According to Nguyen, many of the first Vietnamese settlers in the region arrived in 1975 as war and political refugees after the conclusion of the Vietnam War. They were, in large part, from three fishing villages in Northern Vietnam, and they found the New Orleans area attractive because of its similar climate and proximity to rich fishing grounds. In addition to support for their families, both in the United States and in Vietnam, the GoM fisheries provide the local Vietnamese-American community with subsistence catch. As with the other coastal communities, Vietnamese-American fishermen historically have saved a portion of their catch to feed their families and neighbors, as well as to trade for vegetables and other farm products with the gardeners in their community. Nguyen stated that many of the fishermen have sought compensation for their loss of subsistence since the *DWH* oil spill, but without proof of subsistence use, they have found it nearly impossible to succeed with their claims. Still, Nguyen says that the "fishermen are willing to go through the process to stand up for their way of life."

As with the other communities, the fishery closures and subsequent loss of income forced a number of Vietnamese-Americans to apply for social services for the first time in their lives. Facing these difficulties and others, many Vietnamese-American fishermen have reiterated their commitment to see their children acquire better education and new opportunities in different fields.

In contrast to the other presentations, which focused on the loss of coastal and marine ecosystem services and subsequent community impacts, the presentation by Jackie Antalan, Director of Building Lasting Organizations in Communities, Inc., and a family services advocate on behalf of rural African-American communities, focused on the impacts of the *DWH* oil spill response and cleanup on inland communities. According to Antalan, nearly 40,000 tons of oil and dispersant-laden debris from the spill cleanup efforts were dumped in nine rural landfills across Louisiana, Mississippi, Alabama, and Florida. Public health concerns were primarily twofold: air quality (volatiles from the oiled debris were detectable in many neighborhoods) and contamination from the oil and dispersants leaching from the debris in these dumps into local water tables and supplies. Antalan asserted that a significant percentage of the debris was sent to dumps in communities with predominantly African-American, Latino, and Native-American populations. Secondary to the public health concerns was the lack of communication among the spill's responsible parties and the public-sector agencies involved in the cleanup efforts and the impacted communities. This lack of communication led to the perception within these communities that their fears and concerns—real or perceived—were neither being acknowledged nor addressed by local and regional officials.

Collectively, these presentations highlight some of the pressing, and difficult-to-measure and address, challenges that communities can face after a disaster of the magnitude of the *DWH* oil spill. The loss or degradation of ecosystem services results not only in income losses, but also in losses to community self-sufficiency, sense of identity, and independence. That several of these communities engage in traditional but not commercial transactions to provide or share resources provides an example of how typical quantitative economic impact metrics may be inadequate to capture the real value of the transaction or its loss. It is also clear that the communities that are particularly reliant on GoM ecosystem services are as vulnerable as the GoM itself. Just as estimating the impacts of the *DWH* oil spill on the diverse suite of GoM ecosystem services presents challenges, so does estimating the impacts of the spill on these diverse GoM communities and their traditions, cultures, and values.

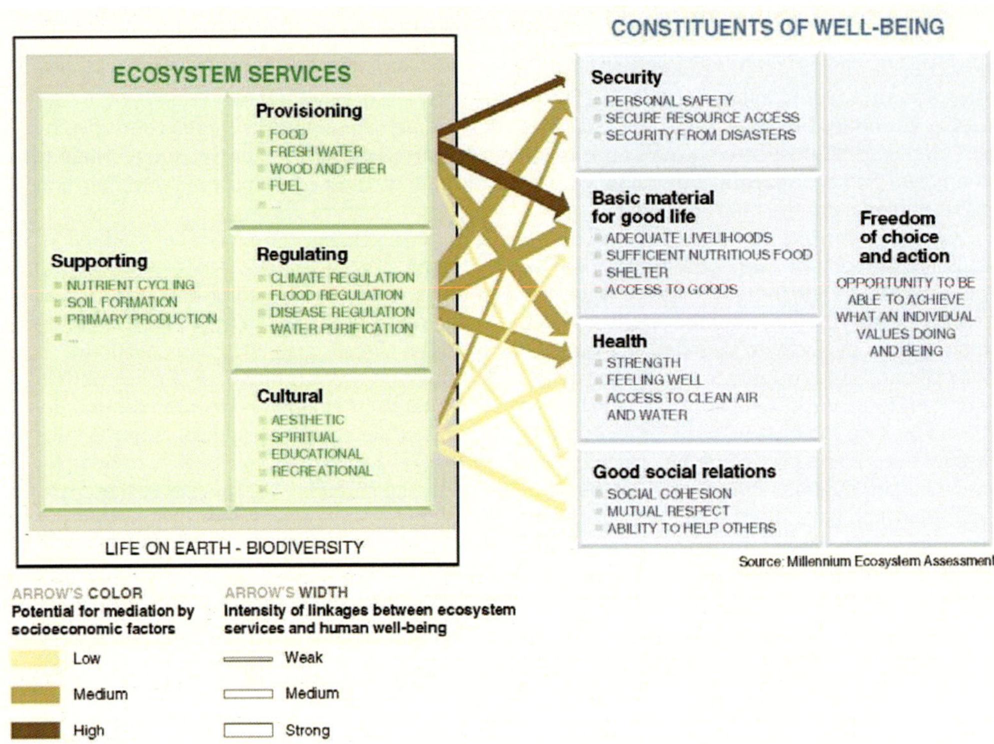


**FIGURE 2.5** Linkages between ecosystem services and human well-being. SOURCE: MEA, 2005.

involve tradeoffs. Is restoration aimed at increases in ecosystem services in the future, perhaps involving a different mix of services and locations, sufficient to offset the damages to ecosystem services? The question of "benefits for whom?" further complicates issues of making the public whole. For example, mitigation efforts that restore ecosystem processes and enhance provision of services in a location different from the one that incurred damages can generate benefits for communities that weren't negatively impacted by the disturbance.

Assessing whether recovery efforts are adequate to make the environment and the public whole is easiest when ecosystem conditions and services are restored as close as possible to original conditions. Then, the pre- and post-disturbance conditions and services can be compared to determine the success of the restoration. As discussed above, baseline conditions may not have been accurately measured or may be continually changing, which complicates the setting of restoration goals. Restoration efforts may then aim to restore proximal ecosystem conditions such as equivalent amounts of "healthy marsh habitat" under the hypothesis that restoring such conditions will lead to full restoration of pre-disturbance services.

The traditional approach under NRDA has been to restore equivalent acres of habitat, populations of species, or other resources known to have been harmed. However, it may not be possible to restore pre-disturbance conditions, or at the very least, doing so will take time. Large-scale disturbances to ecosystems can cause fundamental shifts in ecosystem processes that can make it time-consuming, resource intensive, or in some instances impossible to recover pre-disturbance conditions (see the discussion of resilience in Chapter 3). In this case, restoration efforts may by necessity aim to replace ecosystem conditions in other locations or replace damaged services via other means, raising questions about what are adequate replacements to "make the environment and the public whole."

Ecosystems provide multiple services that benefit a range of societal groups. For example, coastal ecosystems provide fish and shellfish, recreational opportunities, aesthetic beauty, habitat for species, storm protection, carbon sequestration, and other services. Ecosystem services cannot be simultaneously maximized; as one service is optimized, other services may be reduced or lost (Holling and Meffe, 1996). Tradeoffs will be inevitable when decisions are made about which ecosystem services to restore or make more resilient, even with the influx of research and resources of the scope committed to the GoM region since the *DWH* oil spill. Dynamic societal objectives combined with heterogeneous human communities make selecting or predicting the highest priority services difficult. Decision making often shows a preference for provisioning services first, then regulating, cultural, and supporting services in that order (Pereira et al., 2005), potentially disadvantaging human communities dependent on cultural services.

Even if a set of preferred ecosystem services could be agreed upon, no one service is likely to benefit all people equally; often there will be disconnects between where the benefits are experienced and where the costs are borne (Carpenter et al., 2009). Inherently, the distribution of benefits will be determined based on a wide variety of variables, including but not limited to spatial dimensions such as the geographic proximity to the ecosystems providing the services and temporal variables (Rodriguez et al., 2006) (see also Chapter 5). For example, in the case of fisheries, individuals in the fishing industry in the GoM are likely to be the most immediate beneficiaries of a recovered fishery, but fishery product consumers across the globe may also benefit. In contrast, the benefits associated with storm mitigation from wetlands are almost entirely determined by one's geographic position relative to the specific GoM wetland that has been maintained (or degraded or lost).

From a practical perspective, it is important to consider who will reap the benefits and who will endure the costs associated with particular ecosystem services. Identification of the primary stakeholders for each service assists in determining who will support and who will oppose restoration efforts. More importantly, however, it allows decision makers to consider whether or not particular groups or communities are being fairly or disproportionately harmed or enriched.

Use of an ecosystem services approach raises a potentially problematic question about what it means to "make the environment and the public whole." For example, suppose that a wetland that provided coastal protection for a community is damaged and restoration efforts are directed toward building a seawall for that community that provides equal coastal protection. In this case, the particular ecosystem service has been replaced and it could be argued that the community has been made whole. This argument, however, makes many stakeholders nervous. Policymakers should be aware that the goal of an ecosystem services approach to remediation is not to replace the lost natural system with a set of constructed replacements or a long-term period of payment to users for the lost ecosystem service value. Rather, the goal should be framed in a much more limited manner. Replacement projects for services will not "make the environment whole" but may be appropriate when there is no feasible way to restore the natural system and when the replacement is acceptable to the public. Often, the restoration of natural systems is not feasible even in a world of unlimited resources, because of our inability to clean or reconstruct ecosystem components such as salt marsh or benthic habitats. In this case, it is better to implement a publicly acceptable project that restores lost services rather than to deny any compensation to the public simply because it could not be complete.

## Summary

An ecosystem services approach to evaluating the impact of events such as the *DWH* oil spill involves measuring the impact of the event on the structure and function of the social-ecological system (typically through comparisons to baseline data), understanding how changes in the structure and function affect the provision of ecosystem services (typically through modeling), and understanding how changes in the provision of ecosystem services affect human well-being (typically through an economic valuation process). Such an approach can offer a more holistic view of impacts as well as expand the possibilities for restoration actions aimed at "making the public whole."

An ecosystem services approach to damage assessment is not incompatible with the current Oil Pollution Act's NRDA process. However, data sets more appropriate for the development of ecological production functions and for measurement of human dependencies on natural systems may be required.

Although the ecosystem services approach offers clear advantages over more traditional approaches, it presents many challenges, including:

1. The lack of adequate baseline data for some parameters, particularly for those related to the human dimension of impacts. In the absence of adequate baseline data, it will be difficult to determine the full impact of an event such as the spill, particularly in an environment such as the GoM where the baselines are dynamic and where economic conditions are often impacted by events far-removed from the local community.
2. It is difficult to parse out the impact of a single event such as the *DWH* oil spill in an environment such as the GoM that is subject to multiple natural and anthropogenic stressors. The key to addressing this challenge is the development of validated models that can be used to understand how individual stressors impact the ecosystem and the services it provides. Although numerous models for various system components are currently under development, fully comprehensive models for social-ecological linkages and interactions are not currently available.

3. Although using an economic valuation process as a proxy for impact is convenient, it does not necessarily capture all of the goods and services provided by a natural system, and replacement sources may not be a perfect substitute for the services provided by the natural system (e.g., a seawall may replace the storm protection of a destroyed marsh, but it may also have negative impacts). These tradeoffs must be carefully weighed when applying an ecosystem services approach.

Despite these challenges, the ecosystem services approach, when used prudently, can offer an enhanced opportunity to more fully achieve the goal of making the public whole in response to an event such as the *DWH* oil spill because it may provide otherwise unavailable information on the value of lost goods and services.

# CHAPTER THREE

# Resilience and Ecosystem Services

## INTRODUCTION

Chapter 2 introduced the concept of ecosystem services and outlined an "ecosystem services approach" to damage assessment for events such as the *Deepwater Horizon* (*DWH*) oil spill. This chapter addresses resilience and its relationship to ecosystem services.

Ecosystems are subject to natural disturbances such as fires, floods, droughts, and disease outbreaks, as well as human-caused disturbances, including oil spills. Ecosystems are also subject to slowly changing long-term stresses, such as nutrient enrichment and changes in sediment supply. These long-term stresses can affect the ability of the system to respond to a shock such as the *DWH* oil spill.

A resilience framework focuses attention on shocks (pulse disturbances), long-term stresses (press disturbances), and the response of complex systems to these shocks and stresses (Ives and Carpenter, 2007). Does a system recover slowly or rapidly from a shock? What factors within the system allow for more rapid and more complete recovery? Do attempts to stabilize certain components of systems lead to a reduction in overall system resilience and greater potential for large changes (Gunderson and Holling, 2002)?

Considerations of resilience are especially important in systems that undergo persistent and fundamental shifts in structure and function after disturbances ("regime shifts"). Increasing resilience can reduce the risk that the system will cross critical thresholds and undergo a detrimental regime shift. On the other hand, decreasing resilience can increase the probability of a beneficial regime shift. Understanding of how to increase (or decrease) system resilience places a premium on knowledge of system dynamics, including feedbacks among system components, as well as of uncertainty and variability in dynamic systems (Walker and Salt, 2006). Can factors that increase system resilience be identified and managed to increase (decrease) the resilience of a system to a "desirable" state?

System resilience can play an important role in maintaining conditions that will sustain the provision of ecosystem services that contribute to human well-being. However, a narrow focus on stabilizing complex systems to provide a constant flow of ecosystem services may reduce system resilience and increase vulnerability (Gunderson et al., 1995). An event such as the *DWH* oil spill may disrupt service provision, but a resilient ecosystem will allow faster recovery so that services will return sooner rather than later or never.

***Finding 3.1.*** *Resilience provides a useful conceptual framework for managing complex systems. It focuses attention on system dynamics and how systems are affected by short-term disturbances and long-term stresses.*

In the context of resilience of the Gulf of Mexico (GoM) ecosystem, the committee was charged with addressing the following questions from the Statement of Task (see Chapter 1):

> *In light of the multiple stresses on the Gulf of Mexico ecosystem, what practical approaches can managers take to restore and increase the resiliency of ecosystem services to future events such as the Deepwater Horizon Mississippi Canyon-252 spill? How can the increase in ecosystem resiliency be measured?*

Although resilience is an important concept, much like the concept of ecosystem services, providing practical and specific advice to managers to "increase the resiliency of ecosystem services" is a difficult task at present. As noted above, managing resilience requires understanding of the dynamics of complex and highly variable systems, which is often quite limited. Without such understanding, it can be difficult in practical contexts to know how management actions affect resilience. Some researchers have gone so far as to state that resilience is "too vague of a concept to be useful in planning" or ecosystem management (Lindenmayer and Hunter, 2010).

***Finding 3.2.*** *Limited data and understanding of complex system dynamics make it difficult to provide specific practical advice to managers on how to restore or increase the resilience of ecosystem services.*

The next section explores definitions of resilience, its application in ecosystems and in integrated social-ecological systems, and the relationship between resilience and provision of ecosystem services. The final sections discuss options to manage systems to enhance resilience and approaches to measuring changes in resilience.

## RESILIENCE IN SOCIOECONOMIC SYSTEMS AND ITS RELATIONSHIP TO ECOSYSTEM SERVICES

### Defining Engineering and Ecological Resilience

The study of resilience emerged primarily in ecology, with initial applications focused on the resilience of ecosystems (Holling, 1973), but resilience is now used more broadly in ecology, economics, engineering, law, natural resource management, psychology, sociology, and other disciplines. The concept of resilience has been applied to ecological systems, social systems, and more recently to integrated social-ecological systems (Berkes and Folke, 1998; Berkes et al., 2003). With this expanded use of resilience has come multiple definitions of resilience. As long as each field of study used a definition that was well defined within the contexts of that field, few problems arose. However, in an interdisciplinary context such as the resilience of ecosystem services in response to a perturbation such as the *DWH* oil spill, multiple definitions of resilience can come into play, leading to confusion (Gallopin, 2006).

For this report, resilience is defined as the ability of a system subject to disturbance to retain its essential structure, function, and feedbacks (Walker and Salt, 2006) and return to its predisturbance state. Two distinct notions of resilience fit within this general definition, depending on whether the system is characterized by a single equilibrium or multiple equilibria (Holling, 1996). With regard to a dynamic system, "equilibrium" refers to the state where the system will not change unless subjected to a disturbance, also called "steady state." More generally, one can refer to a "regime" of a dynamic system in which the system follows a given trajectory rather than being static. So, for example, a system may follow a regular cycle unless disturbed. The notion of resilience for a system with a single equilibrium focuses on the speed with which it returns toward equilibrium following a disturbance and is referred to as "engineering resilience" (Holling, 1996). The notion of resilience for a system with multiple equilibria focuses on the magnitude of disturbance the system can absorb without shifting to a new equilibrium (Walker et al., 2004). This form of resilience is referred to as "ecological resilience" (Holling, 1996).

## Engineering Resilience

Engineering resilience refers to the speed with which a system returns to equilibrium after a disturbance. For example, how fast does a material return to its original form after a shock that deforms the material? A related notion is resistance, which is the ability of a disturbed system to stay near equilibrium. A rubber band is easily stretched when force is applied (low resistance) but quickly returns to its original shape (high resilience) once the force is removed. A steel rod, on the other hand, is highly resistant to force, but it will bend once the force is sufficiently strong and remain bent even when the force is removed (low resilience). Both the rubber band and the steel rod can retain their original forms after being subject to a disturbance, but only the rubber band returns to its original form after it is deformed by the force of the disturbance. A sufficiently strong disturbance will cause the rubber band to break or the steel rod to bend, which may lead to a new type of equilibrium discussed below under the heading of ecological resilience.

Ecologists have a long history of studying disturbance and recovery. Clements (1936) held that the predictable succession of species followed trajectories toward a specific climax state after a disturbance. Considerable progress was made in the 1950s to provide a solid theoretical basis and predictive capability for this view of resilience as recovery to a climax state. This line of investigation led to seminal works on post-disturbance ecosystem recovery (e.g., Connell and Slater, 1977; Odum, 1969).

Ecologists have been especially interested in the relationship between species diversity and resilience. Intuitively, it seems that high diversity should be related to greater resilience. Similar to how an asset portfolio is diversified to reduce financial risks, highly diverse ecosystems are more likely to contain species that can respond well to particular kinds of disturbances, thereby allowing the system as a whole to better maintain functions in the face of disturbances (Tilman et al., 1996, 2005). However, May (1974) showed mathematically that greater diversity could lead to lower system stability. In principle, empirical evidence is largely supportive of the hypothesis that greater species diversity leads to greater system stability (Tilman, 1996; Tilman and Downing, 1994; Tilman et al., 2006). However, some debate remains as to whether high species diversity contributes to system stability by increasing resilience (Ives and Carpenter, 2007; McCann, 2000; Naeem, 2002; Rooney et al., 2006).

## Ecological Resilience

During the 1960s ecologists questioned the view that ecological systems tended toward a unique climax state and instead raised the possibility that systems might have multiple potential equilibria (Holling, 1965; Lewontin, 1969). Once disturbed, new conditions can foster a new set of feedbacks and prevent the system from returning to its pre-disturbance equilibrium. For example, plants absorb phosphorus and limit algal growth in shallow lakes with low levels of phosphorus. An increase in phosphorus inputs, however, can lead to algal blooms that reduce light penetration and kill plants, releasing more phosphorus for algae. Algal domination can persist even when phosphorus inputs into the lake decline back to original levels (Carpenter, 2003).

With multiple equilibria, which equilibrium a system will tend to move toward depends upon the set of system conditions. Multiple equilibria generate the potential for a system to cross critical thresholds and flip between equilibria (see Figure 3.1). In the context of multiple equilibria, resilience typically refers to the ability of a system to undergo disturbance without crossing a critical threshold and therefore return to the original equilibrium state (Carpenter et al., 2001; Gunderson and Pritchard, 2002; Holling, 1973, 2001). When a system crosses a critical threshold, it is said to have undergone a "regime shift." Once a regime shift has occurred, it may be difficult for the system to reverse back to the original equilibrium. Simply removing the

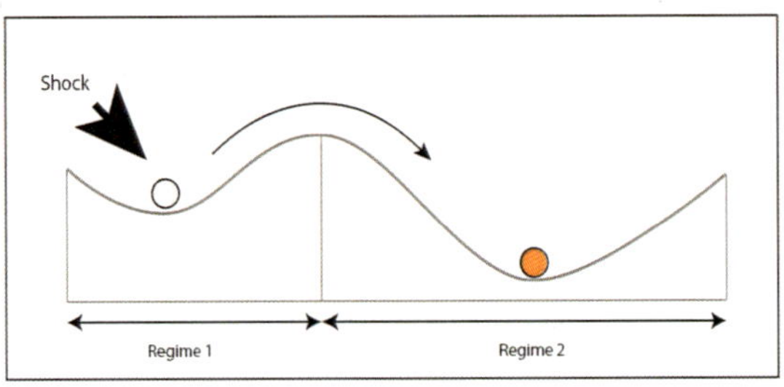


**FIGURE 3.1** Illustration of a regime shift. Initially, the system is in equilibrium in regime 1. It will remain in regime 1 unless a sufficiently strong disturbance pushes it over the tipping point and into regime 2. Once the system is in regime 2, it will need an even stronger disturbance to shift it back to regime 1. SOURCE: Biggs et al., 2012a; http://www.stockholmresilience.org/news/researchinsights/regimeshifts.106.d3c937136e935f4864514.html

stresses that lead to a regime shift may be insufficient to return the original state (Scheffer et al., 2001).

Evidence of regime shifts has been found in a several types of ecosystems (e.g., Beisner et al., 2003) including shallow lakes (Scheffer and Carpenter, 2003), grassland grazing systems (Van de Koppel et al., 1997), coral reefs (Hughes et al., 2003), and other marine systems (Nyström et al., 2012; Petraitis et al., 2009). However, some researchers have been skeptical of this evidence. Sousa and Connell (1983, 1985) employed strict criteria to judge whether an observed change represented a single system with multiple steady states and found little convincing evidence for multiple states. Schroder et al. (2005) reviewed experimental studies testing for alternative stable states and found mixed results.

## The Mechanics of Ecological Resilience

Ecological systems with strong negative feedbacks are resilient. Negative feedbacks work in opposition to any disturbance and tend to push a system back to equilibrium. A thermostat provides an example of a negative feedback. In economic systems, price responses often function as negative feedbacks. In cases of excess demand, price increases reduce demand and increase supply, thus bringing demand back into balance with supply. That said, the existence of positive feedbacks tends to reinforce disturbances, leading to instability. For example, the introduction of fire-adapted invasive grass species tends to promote fire, which in turn reinforces the competitive advantage of these grasses, leading to a new equilibrium dominated by fire-adapted invasive species (Mack and D'Antonio, 1998).

The concept of negative feedback can be traced to early 20th century work by Russian physiologist P. K. Anokhin (1935). Indeed, physiology has a long history of describing negative feedback systems that operate at the organismal level to maintain homeostasis in body temperature and chemistry. Homeostasis in physiology is the maintenance of a constant internal environment in response to a changing external environment. A great deal of research has been done at the organismal level to understand how feedback mechanisms maintain homeostasis.

Salt marshes, such as those found in the GoM, provide a good example of negative feedbacks that maintain ecosystem conditions in the face of disturbances and long-term stresses. Salt marshes accumulate sediments at rates that can keep the marsh in equilibrium with sea level even with sea level rise (Morris et al., 2002). This type of feedback has been termed ecogeomorphological (Fagherazzi et al., 2004) or ecogeomorphic feedback (Kirwan et al., 2010). The sedimentation of suspended solids carried by tides over the marsh surface increases with the duration of flooding (Friedrichs and Perry, 2001; Krone, 1985) and with the density of standing vegetation (Morris et al., 2002). Flood duration is proportional to the depth of the marsh surface below mean high water, at least until the surface falls below the lowest tide level. In addition to surface deposition, production of organic matter, primarily of roots and rhizomes, contributes to the total accumulation rate (Reed, 1995; Turner et al., 2001). Density and primary productivity of vegetation are related to flood duration and depth (Morris et al., 2002). For the dominant salt marsh species, smooth cordgrass (*Spartina alterniflora*), plant growth occurs between mean sea level and mean high water (McKee and Patrick, 1988), with maximal growth near the middle of this range (Morris et al., 2002; Figure 3.2).

Sea level rise, an issue of particular concern in the northern GoM, will decrease relative elevation of coastal marshes. If initial marsh elevation is above the elevation for optimal growth, then sea level rise will increase primary production and sedimentation (Figure 3.2). This feedback maintains the elevation of the marsh, within limits. However, if the rate of sea level rise is too great, then a decrease in elevation of the marsh will decrease primary production and sediment accretion and the marsh surface will fall further behind rising sea level, culminating in the collapse of the marsh and loss of ecosystem services. Consequently, the most stable elevation is one that is higher than optimal for maximum vegetation growth. This illustrates an important general point: aiming to maximize a particular ecosystem service can be counter-productive to other desirable properties, such as resilience, or other ecosystem services, such as storm protection.

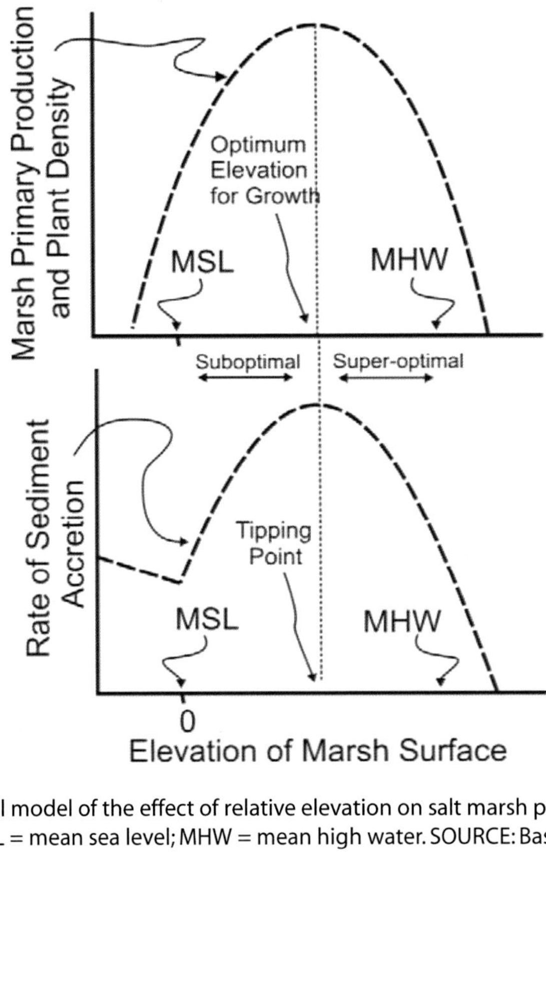


**FIGURE 3.2** Conceptual model of the effect of relative elevation on salt marsh primary production and sediment accretion. MSL = mean sea level; MHW = mean high water. SOURCE: Based on model in Morris et al. (2002).

## Resilience in Social-Ecological Systems

Humans are an integral part of virtually all ecosystems, including the GoM ecosystem. Humans have modified the flow of the Mississippi River, redirected sediment and nutrient loadings into the GoM, loaded pollutants into the water, built cities and towns along the coastline, built canals through coastal marshes, harvested various species of fish and shellfish, and extracted resources such as oil and natural gas. Inclusion of human behavior, and the feedbacks between human behavior and environmental conditions, can mean that the resilience of social-ecological systems differs in important ways from the underlying resilience of ecological systems without consideration of humans (Berkes et al., 2003; Walker et al., 2006). In thinking about the resilience of the GoM system and the ecosystem services that it provides, it is the resilience of the social-ecological system that is of greatest relevance. Knowledge of the resilience of social-ecological systems, however, is even more limited than knowledge of the resilience of ecosystems. Much of the literature on the resilience of social-ecological systems is abstract and general, making it difficult to provide specific, practical suggestions to managers.

Social scientists have begun to incorporate notions of resilience into their analyses of social systems. Perhaps the most relevant social science analysis in the context of ecosystem services is "community resilience." Community resilience refers to a community's ability to recover from losses imposed by disturbances, such as oil spills and hurricanes, without significant external assistance (Cutter et al., 2008; Mileti, 1999).

The resilience of human communities is related to the resilience of ecosystems to which communities are linked, but community resilience and ecological resilience are not perfectly correlated. Resilient ecosystems can contribute to social and economic well-being by providing a stable base for a diverse set of industries (fishing, recreation, and tourism) and stable provision of ecosystem services that contribute to quality of life (Adger et al., 2005). Resilient ecosystems can also provide protection against disturbances that would reduce income or quality of life, such as storm surge protection and flood control. However, a community may be resilient despite being linked to irresilient ecosystems. For example, a community that has diversified sources of income may be resilient to disturbances to ecosystems that disrupt specific industries such as commercial fishing. If community members have ready access to other forms of employment or to community support mechanisms that allow for retooling and retraining, then they may not suffer much long-term loss when one particular industry is disrupted (Cutter et al., 2008). In general, a highly diversified economy, including sectors that are buffered from disturbances, will tend to reduce the dependence of the social and economic systems on ecosystem resilience.

Resilience of human communities can also impact the resilience of ecosystems. Human actions resulting in changes in land use, nutrient cycling, hydrology, or pollution levels can reduce ecosystem resilience. For example, increased sediment loading and overharvesting of grazing fish can make coral reefs more susceptible to bleaching and die-offs. When human communities are overly reliant on the provision of a small set of industries, such as sole reliance on fisheries or oil and gas production, they may become locked into certain patterns of behavior that add stress to ecosystems, which can make it difficult to maintain social-ecological system resilience.

Changes in ecosystems and in social systems can be mutually reinforcing (positive feedbacks). In some cases, this can lead to a virtuous cycle where improved community well-being leads to better environmental protection that further promotes community well-being. But the cycle can also go in the other direction. For example, the decline of coral reefs may cause a decline in fisheries or tourism, with consequent decreases in community vitality and increases in social stress. Such stress may lead to reductions in coral reef protection, with further degradation of the reefs. In fishery-dependent communities, environmental shocks that affect the underlying productivity of a fishery may push fishermen to intensify their harvest to maintain their income. But intensified harvesting can further reduce stocks and harm future productivity, requiring even more harvesting pressure in the future. Such downward spirals lead to unsustainable outcomes.

The ability to change human behavior and management in light of changes in underlying conditions ("adaptability") is an essential component of overall social-ecological resilience (Folke et al., 2010). Highly functioning human systems can learn and adapt their behavior to maintain overall social-ecological system resilience. For example, good management that reduces fishing pressures when fish stocks are depleted and that allows harvests to expand when stocks are abundant can help to stabilize fish populations. In contrast, rigid adherence to constant harvest quotas set for average conditions in the face of a variable environment can lead to fishery collapse (Roughgarden and Smith, 1996). Poverty or lack of properly functioning governance can lead to maladapted behavior and rapid decline of system conditions.

Living successfully in a dynamic and changing world involves adapting to new conditions, and some authors have broadened the concept of social-ecological resilience to allow for adaptation. Folke et al. (2010) define resilience of social-ecological systems as the ability "to continually change and adapt yet remain within critical thresholds. This adaptability is part of resilience." Adaptability may require that some aspects of the system change in order for the whole system to retain essential system properties. For example, sea level rise may require shifts in the location of coastal marshes as well as human infrastructure to maintain desirable ecosystem services and to avoid future damages from flooding. Some adaptation, especially in the face of large disturbances, may require fundamental transformations (Folke et al., 2004; Walker et al., 2004). Transformations can be traumatic, as when fisheries collapse and fisherman must find a new way to make a living. Planning ahead for the possibilities of transformations can reduce the burden, such as by providing retraining support during periods of adjustment. The institutional and governance capacity of social-ecological systems is a crucial element for having capacity to adapt, transform, and innovate (Folke et al., 2010; Gunderson, 2000; Gunderson and Folke, 2011; Olsson et al., 2004).

> ***Finding 3.3.*** *Resilience in social-ecological systems depends on the feedbacks between ecological and human communities. Because of the feedbacks between human behavior, environmental conditions, ecosystem services, and human well-being, understanding of resilience of social-ecological systems requires integrated analysis that includes both nature and people.*

## The Relationship Between Resilience and the Provision of Ecosystem Services

Productive ecosystems are necessary for the supply of ecosystem services. Ecosystems that lack resilience are vulnerable to disturbances that can lead to reductions in the supply of ecosystem services. Maintaining a stable supply of ecosystem services, therefore, depends on having resilient ecosystems. The GoM ecosystem is subject to periodic disturbances from hurricanes or oil spills, which can cause temporary declines in service provision. But a resilient ecosystem will allow speedy recovery of ecosystem services provision. After the *DWH* oil spill, many fisheries in the GoM were closed because of health concerns, and tourism revenue for the GoM coastal region was substantially lower in 2010 than in 2009. However, fisheries were reopened and tourism rebounded quickly. In several locations, the numbers of visitors and tourism revenues were higher in 2011 than in 2009 (AGCCVB, 2012; Mississippi Development Authority/Division of Tourism, 2012).

Some ecosystem services are valuable precisely because they increase the resilience of social-ecological systems. Ecosystem services such as flood mitigation and coastal protection from storm surge reduce the size of disturbances and the destruction that floods and storm surge cause to human communities. Several studies have analyzed the value of coastal marshes, mangroves, and other habitats for protection of coastal communities from storm surge (e.g., Barbier et al., 2008; Costanza et al., 2008; Das and Vincent, 2009). It is often difficult to be precise about how much protection ecosystems are likely to provide given the variability of storms, including wind speed and direction, duration, and arrival of the storm relative to high tides (Koch et al., 2009), but there can be little doubt that their contributions can be significant (see Chapter 5 for additional details).

Some ecosystem services can be replaced by human-engineered services. For example, storm protection can be provided by green infrastructure (coastal marshes) or grey infrastructure (concrete barriers), and water purification for clean drinking water can be provided by forests and wetlands that filter nutrients and pollutants or by water treatment facilities. Loss of ecosystem resilience can compromise ecosystem services and may require replacement of the ecosystem service with a human-engineered service. In some instances, provision of a human-engineered service may undermine the provision of other services—for example, concrete barriers could protect one section of coastline but intensify erosion in nearby locales, resulting in increased vulnerabilities elsewhere along the coast.

Coastal communities are often highly dependent on coastal resources for their livelihood. The services provided by the ecosystem to the community may be quite diverse, as they are in the GoM, and may include commercial and recreational fishing, beach tourism, wildlife viewing, marine transportation, and other extractive uses. Any event that impacts the ecosystem, such as an oil spill, can have an impact on multiple ecosystem services, with consequent impacts on the human community (Adger, 2000). Highly dependent communities can be severely impacted by disruptions to ecosystems that cause a loss of services.

***Finding 3.4.*** *Communities with highly diversified economies and social structures are better placed to withstand disturbances to ecosystems that affect the provision of ecosystem services.*

## MANAGING FOR RESILIENCE

The health and well-being of coastal communities in the GoM are tied to the continued flow of services provided by coastal and marine ecosystems. A key question for such communities is how to maintain the flow of valuable ecosystem services in the face of long-term social and ecological changes and short-term disturbances such as oil spills and hurricanes. Resilience of the system in the face of ongoing shocks and stresses should be an important objective of resource management.

This section begins with a discussion of ways in which system resilience could potentially be influenced by management actions, both when the specific sources of disturbance can be identified (specific resilience) and when they cannot (general resilience). This section then discusses current management approaches used by federal agencies to manage the resilience of the GoM social-ecological system and then explores metrics that can be used to measure changes in resilience. Finally, the section discusses ecosystem restoration, which is a management strategy of particular interest after a human-caused disturbance such as an oil spill.

### Managing for General and Specific Resilience

When managers know the specific types of disturbances they are likely to face, they can often take specific measures to increase the system's resilience to these disturbances (Adger et al., 2005). Management for specific resilience involves using knowledge of specific types of disturbances to design plans to minimize damages and to aid recovery. For example, a coastal system subject to periodic hurricanes could increase social-ecological resilience by

- increasing the areal extent and health of coastal marshes and mangroves, which can serve as a buffer to absorb wave energy between the open water of the ocean and coastal communities (Barbier et al., 2008; Costanza et al., 2008; Das and Vincent, 2009), although there remain questions about the effectiveness of vegetation to provide protection (Feagin et al., 2010);
- improving communications and early warning systems to provide information about impending danger to people;
- investing in disaster preparedness and planning; and
- changing infrastructure and building design and location to minimize the risk of loss from storm surge or wind damage.

Building system resilience to more swiftly recover from oil spills can be aided by many of these same measures as well as by investments in safety procedures and engineering to reduce the risk of catastrophic accidents, and in emergency response capabilities should an accident occur.

Managers face a more challenging task in creating "general resilience" that allows for greater capacity to respond to many different types of disturbances, some of which will undoubtedly be a surprise (Adger et al., 2005; Anderies et al., 2006; Walker and Salt, 2006, 2012). Certain system properties are thought to increase general system resilience in a wide range of circumstances. Several authors have described properties to increase general resilience, which are summarized in five strategy categories (see Table 3-1) and described in more depth below.

**TABLE 3.1** Principles of Management for General Resilience of Social-Ecological Systems

| **Resilience Strategy** | **Levin (1999):** Eight Commandments for Ecosystem Management | **Walker and Salt (2006):** Nine Attributes of Resilience | **Biggs et al. (2012b):** Seven Generic Principles for Enhancing Resilience |
|---|---|---|---|
| Maintain diversity, variability, and redundancy | Maintain heterogeneity | Promote and sustain diversity in all forms (biological, landscape, social, and economic) | Maintain diversity and redundancy |
| | Preserve redundancy | Embrace and work with ecological variability rather than attempt to control and reduce it | |
| Manage feedbacks and slowly changing variables | Tighten feedback loops | Tighten feedbacks<br>Focus policy on slowly changing variables associated with thresholds | Manage slow variables and feedback |
| Manage modularity and connectivity | Sustain modularity | Modularity | Manage connectivity |
| Apply adaptive management, learning; anticipate potential future outcomes | Reduce uncertainty<br>Expect surprise | Emphasize learning, experimentation, and locally developed rules, and embrace change | Encourage learning and experimentation<br>Foster understanding of social-ecological systems as complex adaptive systems |
| Improve governance and increase social capital | Build trust | Promote trust, social networks, and leadership | Broaden participation |
| | Do unto others as you would have them do unto you | Include redundancy in governance structures and a mix of institutional types | Promote polycentric governance systems |
| | | Include all ecosystem services in development proposals and assessments | |

## Maintaining Diversity, Variability, and Redundancy

Sophisticated financial investors know that diversification reduces risk, or to use a more biologically based metaphor, that it isn't smart to "put all your eggs in one basket." In social-ecological systems, diversification can be accomplished through a variety of ways, including maintaining biodiversity or increasing economic diversification.

### Maintaining Biodiversity

Although May (1974) showed that increased biodiversity could reduce system stability, recent theory and empirical work have led to the conclusion that increased biodiversity generally enhances ecosystem resilience and stability (Folke et al., 2004; Tilman and Downing, 1996). Biodiversity represents a form of insurance, providing redundancy that can buffer ecosystems against losses of particular species and can reduce the chance that important ecosystem functions will be compromised (McCann, 2000; Naeem, 2002). Greater diversity also increases the probability that there is a species present that is well suited to particular conditions, thereby allowing a high level of function under varying conditions (Chapin et al., 1995; Isbell et al., 2011). Redundancy within functional groups such as primary producers is important (Levin et al., 1998), as is response diversity, which means there is a broad range of species-specific responses to perturbations (Elmqvist et al., 2003; Folke et al., 2002). Greater biodiversity should therefore increase the probability that a system will provide a consistent level of performance over a given time (Naeem and Li, 1997).

The relationship between biodiversity and ecosystem resilience may take different forms, depending on the underlying mechanism (Figure 3.3). There are three possible models for the relationship between biodiversity and ecosystem resilience: (a) the linear model; (b) the keystone model, in which the presence of one or a few keystone species is crucial for ecosystem resilience; and (c) the redundancy model, in which the functions that maintain ecosystem stability are duplicated by numerous species. The available evidence favors the redundancy hypothesis (Cardinale et al., 2011). However, species redundancy decreases over time as species sort into niches (Reich et al., 2012), and certain species may be critical for ecosystem functioning under particular environmental conditions (Isbell et al., 2011).

Although greater diversity tends to lead to greater system resilience, some ecosystems are highly resilient even though they are not particularly diverse, such as coastal wetlands. Plant biodiversity and biomass decrease with increasing salinity moving from tidal freshwater wetlands to salt marshes (Wieski et al., 2010). Salt marshes on the GoM coast are dominated by the smooth cordgrass *Spartina alterniflora*. Despite low plant biodiversity, salt marshes are extremely resilient, even in the face of the *DWH* oil spill (Silliman et al., 2012). Salt marshes exemplify extreme environments characterized by hypoxic soils, high salinity, and high sulfide concentrations (soluble sulfide concentrations of 1–5 millimolar in soil are lethal to most plant life). *S. alterniflora* also tolerates organic toxins (DeLaune et al., 1984; Li et al., 1990; Webb et al., 1985), which partially explains their rapid recovery following oil spills.

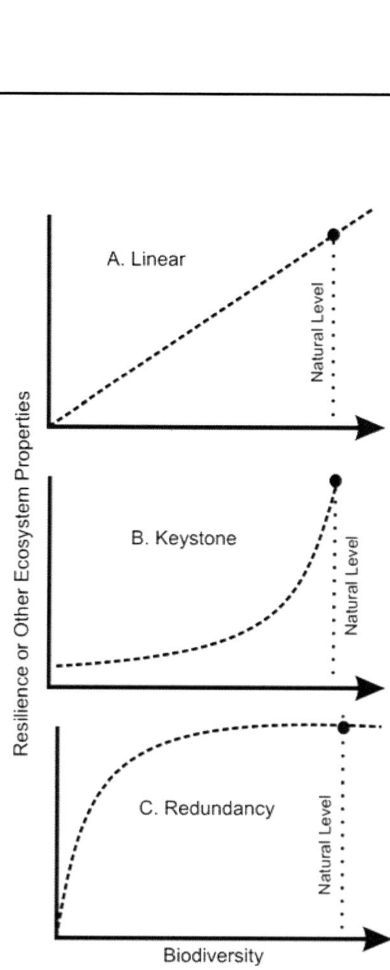


**FIGURE 3.3** Conceptual models of possible alternative responses of ecosystem resilience to biodiversity. SOURCE: Modified from Naeem (2002).

## Economic Diversification

Enhancing economic diversity is another route to increasing resilience of social-ecological systems. Communities whose economies are heavily dependent on specific natural resources are vulnerable to shocks to those resources. Highly diversified economies, such as biodiversity-rich ecosystems, can weather shocks to particular resources or particular industries, much as ecosystems can lose some species and still maintain overall system performance (Briguglio et al., 2009; Rose, 2007). Sectoral diversification is expected to have a positive influence on economic resilience (Simmie and Martin, 2010).

The reliance of its economy on specific natural resources makes the GoM region vulnerable to shocks such as the *DWH* oil spill. Fishing grounds were closed, and recreational and tourism activities suffered large losses in 2010 after the spill. Some communities along the Gulf Coast are almost totally reliant on the oil and gas industry, fishing, recreation, and tourism. When these industries experience downturns, the economies of these communities suffer. In the case of the *DWH* oil spill, however, much of the economic harm—though serious—was temporary.

The immediate aftermath of the spill in 2010 was very difficult for many GoM communities (see Box 2.2 in Chapter 2). However, oil and gas production, fishing, and tourism largely rebounded in 2011. Drilling was allowed to resume, fisheries were reopened, and tourists returned.

### Managing Feedbacks and Slowly Changing Variables

Strong negative feedbacks increase system resilience by offsetting the effect of a disturbance that pushes a system away from equilibrium. Negative feedbacks exist in ecological systems in which biological organisms perpetuate conditions favorable to their continued existence. In social-ecological systems, actions aimed at maintaining existing conditions can provide negative feedbacks. For example, fisheries management that adjusts harvest quotas based on current population size can help to stabilize fish populations by reducing harvests when stocks are low and increasing harvests when stocks are high (Reed, 1978, 1979). However, maintaining stable stocks can mean highly variable harvests from year to year. But maintaining constant harvests, especially if set to maximize yields in an average year, can lead to fishery collapse (Roughgarden and Smith, 1996). Managing systems to maintain stability for certain outcomes, such as harvests, can reduce stability in other dimensions and can affect overall system resilience (Gunderson and Holling, 2002; Gunderson and Pritchard, 2002).

Long-term changes or trends in environmental or social conditions ("slow variables") can also have a large influence on resilience. Much as an individual whose general health is declining is more susceptible to disease, ecosystems compromised by ongoing stress may be more vulnerable to rapid change from disturbances (Gunderson and Holling, 2002; Walker and Salt, 2006). For example, levees and channels to reduce flooding along the Mississippi River have changed the supply of sediment to coastal systems and have made parts of Louisiana more susceptible to damage from coastal storm surges (Costanza et al., 2006).

### Modularity and Connectivity

Connections among components within a system can allow shocks to propagate through the system so that tightly coupled systems may be more vulnerable to systemwide risks (May et al., 2008). Building in a degree of modularity may be important for preserving overall resilience (Levin, 1999). Placing booms to limit the spread of oil into coastal marshes is one example of modularity in practice. However, marine systems are inherently highly interconnected, and sometimes interconnections provide resilience, such as when local populations are reestablished through recolonization after local extinction events. An objective of an ecosystem services approach is to understand interconnections, but our ability to manage connectivity/modularity in practical situations is often limited.

### Adaptive Management and Learning

If we knew exactly what the future had in store, then it would be easier to plan for it. But as Yogi Berra said, "It's tough to make predictions, especially about the future." In adaptive management, managers try to gain information through experimentation and learning to reduce uncertainty and improve ongoing management (Holling, 1978; Lee, 1993; Walters, 1986). Adaptive management has proven difficult to implement in practice, in part because it involves risk taking that can put managers in a difficult position of justifying failures even when it provides valuable information, as well as because it requires resources for ongoing monitoring and evaluation (Lee, 1999).

Of particular importance to managing complex systems is having some ability to predict regime shifts prior to their occurrence. There are some signals of the imminent onset of a regime shift (Scheffer et al., 2009, 2012). Whether these signals can be received in time, and management revised to forestall a regime shift, is doubtful (Biggs et al., 2009). The potential for a harmful regime shift in the future should influence current management. In most cases, this potential shift should cause management to take precautionary actions to reduce the probability of its occurrence (Polasky et al., 2011). However, in some cases the impact on management of a potential shift could work in the opposite direction. For example, the threat of collapse of a resource stock may cause resource managers to use the resource more aggressively ("use it or lose it"; Reed, 1984, 1988).

## Improving Governance and Increasing Social Capital

The ability of a social-ecological system to recover from shocks such as a hurricane or an oil spill is improved by having highly functional institutions and a high degree of social cohesion. Disasters that inflict large-scale and widespread damage put ecosystems and human communities under great stress. Help from outside the stricken area, such as from the federal government, is often essential to provision of relief in the immediate aftermath of a disaster. But long-term recovery depends in large part on the resources and resourcefulness of the local community (Picou and Martin, 2006). Good governance also aids in assessing potential threats and long-term stresses and in bringing about changes in management or behavior to address them (Folke et al., 2010). Trust in institutions is a key variable in attaining cooperation to make changes, especially if the change requires at least some short-term sacrifice. Lack of good governance or social capital can lead to social, economic, or ecological decline and a downward spiral in environmental conditions and human well-being (Folke et al., 2010).

***Finding 3.5.***

- *Although systems are complex, it may be possible to manage them in ways that increase system resilience. Management itself can enhance system resilience if it creates responses to current conditions in ways that lessen the impacts of disturbances. The effectiveness of management can be enhanced by improving understanding of system dynamics, reducing uncertainty, and developing better early warning signals of approaches to thresholds.*
- *Management for specific resilience uses knowledge of specific types of disturbances to design plans to minimize damages and promote recovery.*
- *Management for general resilience is needed when the type of disturbance is unknown or when many types of disturbances are possible. Conserving biodiversity or increasing*

*economic diversity can make social-ecological systems more resilient to a range of different of disturbances, as can the other management approaches summarized in Table 3.1. The emphasis on properties to promote general resilience is important given the inability to predict the type, timing, scale, and interaction of future disturbances.*

## FEDERAL GOVERNANCE OPTIONS AND POLITICAL LIMITATIONS

Although resilience management can result in many benefits, translation of resilience theory into improved agency decision making for ecosystem service management will be difficult. Because resilience management represents a significant change from current management priorities or decision-making processes, congressional authorization in the form of new or amended laws (Ruhl, 2012) may be necessary to make it part of the process. A fundamental principle of federal administrative law, rooted in Article I of the U.S. Constitution, is that agencies cannot act in ways that exceed the statutory authority Congress has given to them (Doremus, 2001). Whether resilience management is consistent with current resource management laws depends upon current statutory authority and the details of resilience management. For example, an agency committed to incorporating resilience management would need to

- craft a firm definition of resilience for the system in question;
- identify a means of measuring the current state of resilience;
- build a model capable of predicting to some degree of certainty the resilience effects of specific decisions (or to design management experiments that could lead to the development of predictive capacity); and
- explain both the process and the results to the public and to the judges likely to review the agency's decisions (Allen et al., 2011).

In many ways, the challenges that agencies face in resilience management are similar to those in adaptive management. As J. B. Ruhl wrote: "For adaptive management to flourish in administrative agencies, legislatures must empower them to do it, interest groups must let them do it, and the courts must resist the temptation to second-guess when they do in fact do it" (Ruhl, 2012).

Some federal agencies, notably the U.S. Forest Service, the U.S. Fish and Wildlife Service, the Bureau of Reclamation, and the National Oceanic and Atmospheric Administration, have begun to recognize the importance of resilience management and to incorporate it into their broader policy objectives (Benson and Garmestani, 2011). President Obama's 2010 Executive Order 13,547[1] seeks to "improve the resiliency of ocean, coastal, and Great Lakes ecosystems, communities, and economies." It remains to be seen how agencies will translate these general policy statements into on-the-ground decision making.

Implementing resilience management could, in theory, be easier in the marine environ-

[1] Exec. Order No. 13,547, 75 Fed. Reg. 43023 (July 22, 2010).

ment than in terrestrial environments. Unlike terrestrial systems, the oceans are entirely public and are not interspersed with private holdings. The federal government thus has more freedom both to set and adjust regulations. In federal waters outside of state jurisdiction, federal agencies could manage all resource use holistically to attain broad systemic goals. Finally, the laws pertaining to resource use in the oceans have traditionally been designed to function under changing environmental conditions and substantial uncertainty. The Magnuson-Stevens Fishery Conservation and Management Act (MSA) requires regular input of scientific information and annual consideration of management measures, which enable decision makers to respond to environmentally driven stock fluctuations (Carlarne and Eagle, 2012).

Still, resilience management in the GoM ecosystem faces challenges. The United States lacks a legal tradition of managing complex systems as multiple, interacting systems. Congress has traditionally written laws that address resources individually; thus, the MSA governs the use of fisheries, the Outer Continental Shelf Lands Act regulates oil and gas and mineral use, and so on (Eagle, 2006). The resource-by-resource approach is less amenable to resilience management than is a place-based approach. That approach, in which a single agency has jurisdiction over most resource use within a single place (e.g., a national park), permits an agency to manage an integrated system. In addition, the GoM ecosystem is affected by actions taken on land that influence nutrient and sediment inputs and are controlled by agricultural and other policies (Rabalais et al., 2002, 2007).

Finally, although there are no private property interests in federal Gulf waters, there are private interest groups that would likely oppose significant changes to the current system of resource management. Such groups might fear that the "new" regulatory regime would result in significant reductions in their income or in their ability to shape management decisions. Because these groups have the incentives and resources to oppose change at all levels, a management shift would need to be accompanied by an explanation of why its benefits would outweigh the real or perceived costs (Ruhl, 2012).

***Finding 3.6.*** *Although the maintenance of resilient ecosystems has many benefits, including provision of a steady flow of ecosystem services, political, legislative, and institutional barriers may impede implementation of resilience management.*

## Option: Implement a Portfolio Approach to Management

How might governments manage publicly owned resources in a way that minimizes disruption to the flow of ecosystem services during and after a disaster? One approach is to move away from the current legal approach to managing marine ecosystems and toward a set of laws that aim to manage distinct, defined areas of the sea for narrower, specific objectives.

The direct management of marine life populations in federal waters of the GoM currently occurs pursuant to two laws: the MSA and the Marine Mammal Protection Act (MMPA). Although the efficacy of these laws in achieving their stated objectives is debatable, it is clear that each law uses a single, science-based concept aimed at achieving a single goal with respect to the GoM marine species under its auspices. The MSA attempts to achieve "sustainable fisheries" using the science-based concept of "optimum yield," while the MMPA attempts to maintain an "optimum sustainable population level" using the science-based concept of "potential biological removals."

Because of the substantial scientific uncertainty involved in estimating necessary numerical targets, such as optimum yield and potential biological removals, it might be possible to increase the resilience of marine ecosystems by diversifying management strategies and objectives on a geographical or stock-by-stock basis. So, Congress could, for example, modify the MSA and the MMPA so that managers would be required to apply a range of strategies over a set of geographically defined management areas or over a set of distinct fish or mammal populations. Some areas or species could be managed using more conservative versions of the traditional optimum yield or potential biological removal tools, and some areas or species could be managed using entirely different approaches and with different ends in mind. So, for example, one could imagine managing a particular area for "ecosystem integrity" rather than "sustainable fisheries."

The theory underlying this diversification approach is based on well-accepted principles that economists use to guide investors in the business world. The idea is that, in a realm characterized by substantial uncertainty, such as the stock market or marine ecosystems, a "portfolio approach" to investing can help to achieve an optimal combination of risk and return. As Gordon Munro observed, fishery (or ecosystem) management is similar to other forms of capital management:

> *Economists view capture fishery resources, as they do all natural resources, as a form of natural capital, assets that are capable of yielding a stream of economic returns (broadly defined) through time. Since fishery resources are capable of growth (like forests, but unlike minerals), these resources—natural capital—can be managed on a sustained basis, essentially by skimming off the growth through harvesting. This also means that the resources can provide economic benefits to society indefinitely. It means further that one can, within limits, engage in positive investment in the natural capital by harvesting less than the growth. (Munro, 2008, p. 14)*

Designing and implementing a portfolio-based law for managing marine ecosystems such as the GoM is a challenging, but not impossible, task. Conceptually, such laws offer one of the most promising approaches to enhancing system resilience and thereby reducing the impacts of future human-engineered or natural disasters on the people who rely on those systems.

> ***Finding 3.7.*** *One option for facilitating resilience management is employment of a portfolio approach to designing the system of laws regulating use of the Gulf of Mexico. Under a portfolio approach, different management goals and strategies would be applied to discrete geographic ocean and coastal areas. Such an approach, as compared with an aspatial management strategy that tries to ensure that all similar systems provide equal amounts of all services, provides a buffer against uncertainties, including future large-scale disturbances. Implementation of this portfolio approach would require congressional action because a set of new statutes would be needed.*

## Ecosystem Restoration and Resilience

Following an oil spill or other disturbance to an ecosystem, a key management objective is ecosystem restoration. Ecosystem restoration after a disturbance and ecosystem resilience are closely linked. Systems with low resilience may recover slowly or switch to a different regime and fail to return to original conditions. In the case of the *DWH* oil spill, of particular interest is how well restoration efforts will work to recover ecosystems to pre-spill conditions, both in terms of restoring the provision of ecosystem services and increasing resilience to further disturbances and ongoing stresses.

Much has been written about the restoration of the Mississippi River Delta (Boesch et al., 1994, 2006; CPRA, 2012; Day et al., 2007; USACE, 2009). Restoration efforts in the delta are taking place in the midst of long-term changes as well as periodic disturbances. Coastal wetlands in the delta have failed to maintain elevation relative to mean sea level because of an inadequate supply of sediment. Prior to European colonization and the engineering of the Mississippi River, sediment-laden water routinely spilled over the river's natural levees during flood events and nourished the wetlands. This process has been disrupted by armoring levees to prevent flooding, damming tributaries in the river basin, dredging canals, and other alterations that reduce sediment delivery to the delta and alter its hydrology. Since the 1930s, the delta has lost more than 4,800 $km^2$, an area nearly the size of Delaware (Barras et al., 2003; Dunbar et al., 1992) and could lose an additional 1,600 $km^2$ by the year 2050 (Louisiana Coastal Wetlands Conservation and Restoration Task Force and Wetlands Conservation and Restoration Authority, 1998). The U.S. Army Corps of Engineers (USACE) identified 2,460 $km^2$ of marsh or open water in five planning units within the lower delta requiring some 5.3 billion cubic yards (4 $km^3$) of sediment for restoration (USACE, 2009). A report by the National Research Council (NRC, 2009) noted, "If wetlands cannot be maintained, this implies that decision makers and citizens ultimately will have to make hard choices about where restoration can take place and where it cannot."

Much less is known about restoration efforts in other types of ecosystems. It may be difficult or impossible to fully restore some ecosystems after disturbances. Changes in landscape structure, loss of native species or invasion by exotics, changes in species dominance hierarchies, trophic interactions, and biogeochemical processes may prevent returns to pre-disturbance conditions. Moreover, restoration efforts can be altered by feedbacks among biotic and physical processes (Ehrenfeld and Toth, 1997; Suding et al., 2004). Restoration in the absence of knowledge of these feedbacks may launch an ecosystem down an unpredictable trajectory.

Restoration of ecosystem services does not necessarily follow from restoration of the ecosystem structure or the return of a habitat to a former state. Zedler (2000) concluded that numerous variables, including landscape setting, habitat type, hydrological regime, soil properties, topography, nutrient supplies, disturbance regimes, invasive species, seed banks, and declining biodiversity, can constrain the restoration process in wetlands. Zedler also concluded, "although many outcomes can be explained post hoc, we have little ability to predict the path that sites will follow when restored in alternative ways, and no insurance that specific targets will be met" (Zedler, 2000). However, this predictive capability is precisely what we seek in an ecosystem services approach.

There is no existing roadmap for restoring all ecosystem services and functions. Currently, the best we can hope to accomplish are a few practical goals. A target for a coastal wetland might be restoration to an elevation that is more resilient and that raises productivity and its value as wildlife habitat, knowing that it may not be practical or even possible to restore its original species composition. Conceptually, an ecosystem might cross a threshold in transitioning from one ecological state to another, such that restoration to a previous state is impeded by biotic and abiotic barriers (Figure 3.4). Examples of biotic thresholds could be the invasion of an exotic species (e.g., the common reed, *Phragmites* spp.) that excludes less competitive species, or a disease (e.g., chestnut blight) that eliminates a dominant member of the biological community. Examples of abiotic thresholds include saltwater intrusion into a freshwater wetland as a consequence of rising sea level. Whether recovery of ecosystem services or restoration of the ecosystem is the goal, the endpoints must be defined in terms of practical metrics

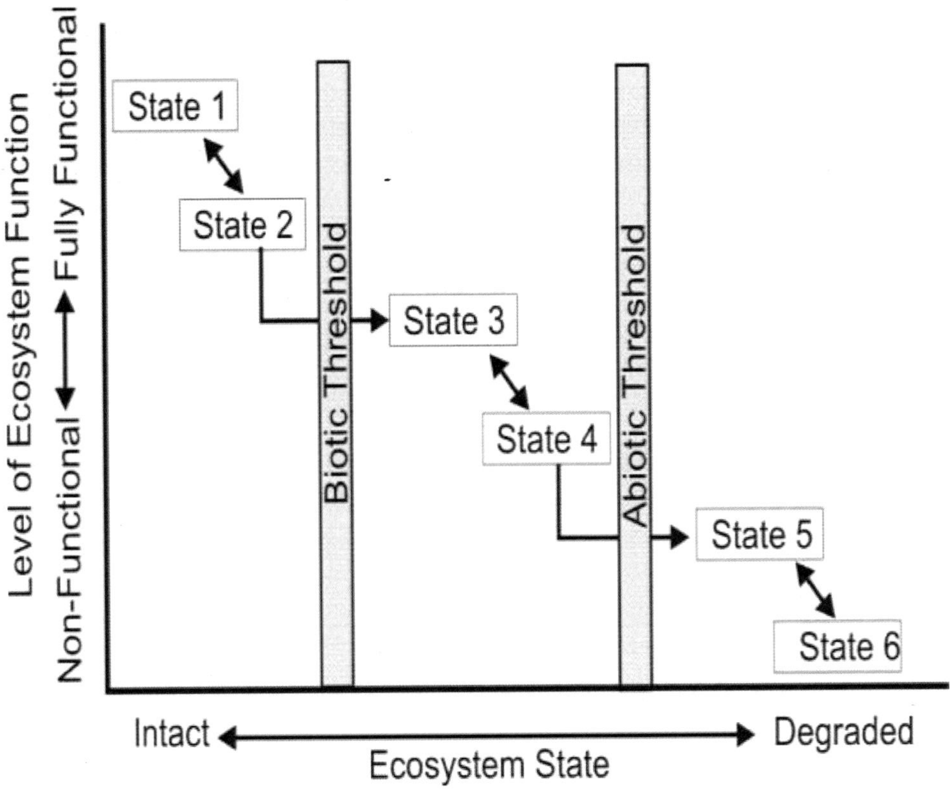


**FIGURE 3.4** Summary of the state (boxes) and transition (arrows) approach to ecosystem degradation and restoration. Hypothesized thresholds, indicated by vertical bars, may prevent recovery from a more degraded state to a less degraded state. SOURCE: Redrawn from Hobbs and Cramer (2008).

that can be monitored and a plan of action that is grounded in a thorough knowledge of the ecosystem.

> ***Finding 3.8.*** *Restoration aimed at returning a system to pre-disturbance conditions is closely related to resilience. Systems that cross critical thresholds after a disturbance can be difficult to restore, and restoration in the absence of a detailed understanding of system dynamics may launch an ecosystem down an unpredictable trajectory. Whether recovery of ecosystem services or restoration of the ecosystem is the goal, the endpoints must be defined in terms of practical metrics that can be monitored.*

## MEASURING RESILIENCE

The definitions of engineering and ecological resilience provide a basis for designing measures of resilience. Engineering resilience is the speed of recovery or the decay rate of the perturbation. Engineering resilience can be quantified post hoc as the time required to return to a pre-disturbance state after an experimental or natural disturbance. In natural systems with sufficiently rapid recovery, such as the sessile biota of the rocky intertidal, monitoring of recolonization can be done after disturbance (Conway-Cranos, 2012). One could also attempt to measure the speed of recovery of social-ecological systems after disturbances. For example, how fast or complete is recovery within a community or an industry after a shock such as a hurricane or an oil spill?

Engineering resilience can also be modeled with various degrees of mathematical sophistication (DeAngelis, 1980; Neubert and Caswell, 1997; Pimm and Lawton, 1980). With a model of system dynamics, an estimate of engineering resilience may be simulated for various severity levels of disturbances.

Although straightforward in theory, measurement of engineering resilience presents at least three practical challenges. First, there is the question of what particular measures or metrics will be used. Some system components or ecosystem services may recover more quickly than others. For example, species diversity may well return to a previous state at a different rate than does primary productivity, or not at all. Hence, quantifying resilience may be heavily influenced by the choice of what is measured. Second, measurements of the speed of return require accurate baselines of conditions prior to the disturbance. The issue of baselines was discussed in the Interim Report (NRC, 2011) and is discussed further in Chapter 5. Third, disturbances that differ in type or scale are likely to result in quantitatively or possibly qualitatively different degrees of responses relevant to resilience (Carpenter et al., 2001). Systems may be highly resilient to some kinds of shocks but not others.

Measurement of ecological resilience is a more complex matter. Social-ecological systems are often too complex to enable researchers to accurately model potential regime shifts or to have much confidence in simulation results. However, attempts to capture critical indicators of ecological resilience have been made. Borrowing the concept of "critical slowing down" from physics, an indicator of ecological resilience might be the rate of recovery from small perturbations when the recovery rate decreases as a regime shift is approached (Scheffer et al., 2012; van Nes and Scheffer, 2007). Changes in other system properties, such as variance, spatial correlation, autocorrelation, and skewness, may also prove useful in detecting approach to critical thresholds (Scheffer et al., 2012). But this field is quite new and does not yet have a body of well-established results to guide management or policy.

Other research has explored the concept of functional diversity (Allen et al., 2005), response diversity (Petchey and Gaston, 2009), and population densities in stochastic systems (Ives, 1995). These approaches require species-level ecological information about function and response. For some groups of organisms, such as birds, this information may be inferred from size and morphology (Cumming and Child, 2009).

When a system is well understood, it may be more straightforward to define appropriate metrics for ecosystem resilience. With respect to tidal wetlands, for example, it seems prudent to adopt practical metrics that are inclusive of numerous ecosystem services and overall ecosystem function. Practical metrics with these characteristics would be (1) change in total wetland area by plant community type and (2) relative elevation (relative to mean sea level). Similarly, critical components of the ecosystem service under study that can be used to measure resilience should be identified. The ability to do this links directly to the understanding of the ecological production functions associated with the service and links directly back to the understanding of ecosystem dynamics.

***Finding 3.9.*** *The measurement of resilience poses a number of conceptual and practical challenges. Measures of speed of recovery to pre-disturbance conditions (engineering resilience) depend on having good baseline data and may vary depending on what ecosystem service or system component is measured and what type and severity of disturbance is considered. Measuring how likely a system is to cross a critical threshold and undergo a regime shift (ecological resilience) is a topic that is at the frontier of science. Consensus on practical measures that can be used to predict the location of critical thresholds or the probabilities of regime shifts does not yet exist. However, for specific systems, it may be possible to define a set of metrics that measure key conditions or processes linked to system dynamics that can predict the resilience of the system and the return of provision of ecosystem services.*

Resilient ecosystems can lead to the resilience of specific ecosystem services. However, ecosystems typically generate multiple ecosystem services. Changes in ecosystem structure and function will generally not affect all ecosystem services in the same manner. Therefore, different services could have different outcomes after a disturbance. As noted by Carpenter et al. (2001), evaluation of a system's resilience requires identification of the initial state of the system (regime) and the disturbances that can impact that system. If the system is viewed through the lens of ecosystem services, then there is a need to identify the most important services and how they may be affected by potential alternate states of the system. For example, coastal marsh habitat of the GoM provides several significant services, two of which are storm surge protection and fishing opportunities, both recreational and commercial. The quality of these services is directly linked to the structural condition of the marsh. Fragmented marsh, characterized as having a high ratio of edge to area, provides an environment for high-quality recreational fishing and also habitat for commercially important species, but it is less effective for storm protection. In contrast, continuous or connected marsh provides higher quality storm surge protection, but it is not as good for recreational and commercial fishing (Minello et al., 1994; Peterson and Turner, 1994; Zimmerman et al., 2002). The quality of each of these services is strongly connected to two alternate states of marsh habitat. Focusing on the resilience of a particular state of a habitat (e.g., fragmented versus continuous marsh) can also lead to resilient ecosystem services that are connected to a specific state of the habitat. As a society we would prefer having more of all ecosystem services, but given that only one condition can occur at one time, there will be tradeoffs between services.

CHAPTER FOUR

# *Oil Spill Response Technologies*

## INTRODUCTION

In addition to exploring the viability of an ecosystem services approach to damage assessment and approaches to measuring and increasing the resiliency of the Gulf of Mexico (GoM) ecosystem, the committee was also charged with evaluating the effectiveness of the various spill response technologies used during the *Deepwater Horizon* (*DWH*) oil spill and, where possible, the effect of those technologies on GoM ecosystem services (Task 6 of the Statement of Task; see Chapter 1). Figure 4.1 illustrates the U.S. government's estimate of the fate of the spilled oil.

Given the magnitude, location, and depth of the *DWH* oil spill, multiple response technologies were applied at scales never before attempted. The goal of response efforts was to contain, remove, and remediate the oil at sea prior to its transport into sensitive wetland and intertidal regions (where cleanup is particularly difficult and costly) by employing a wide array of technologies, including mechanical operations to contain and recover the spilled oil, chemical treatments to disperse oil before it reached sensitive coastal habitats, and a number of shore-based mitigation strategies. In the case of chemical dispersants, the 1.8 million U.S. gallons used in surface and subsurface applications (Federal Interagency Solutions Group, 2010) elicited public concerns about their potential impacts on ecosystems and the services they provide.

To prevent oil from making landfall, several berms and reaches were constructed near shore in Louisiana at a cost of $220 million (National Commission on the BP Deepwater Horizon Oil Spill and Offshore Drilling, 2011). To clean oil from beaches, sand washing and surf washing were performed. To physically remove oil from marshes, manual labor, floating machinery, and manual cropping of oiled vegetation were employed.

Natural processes, such as abiotic weathering and microbial degradation, also played a major role in oil removal. The monitoring of natural recovery processes after an oil spill without active intervention (natural attenuation) must be considered a valid response option under certain circumstances. If oil has impacted a sensitive shoreline such as a wetland, which precludes the use of invasive cleanup operations, then natural attenuation would be the response of choice once accessible floating oil has been removed using absorbents. For areas in which a spill is logistically inaccessible for reasons of remoteness (e.g., the Arctic), stormy weather, or lack of equipment and manpower, natural attenuation might be the only option available.

This chapter reviews the suite of spill response technologies used during the *DWH* oil spill event, discusses how they were applied in response to the spill, presents the effectiveness and risks associated with their use, and, where possible, assesses their potential impacts on the GoM ecosystem and its services. The discussion begins with potential responses at sea that can

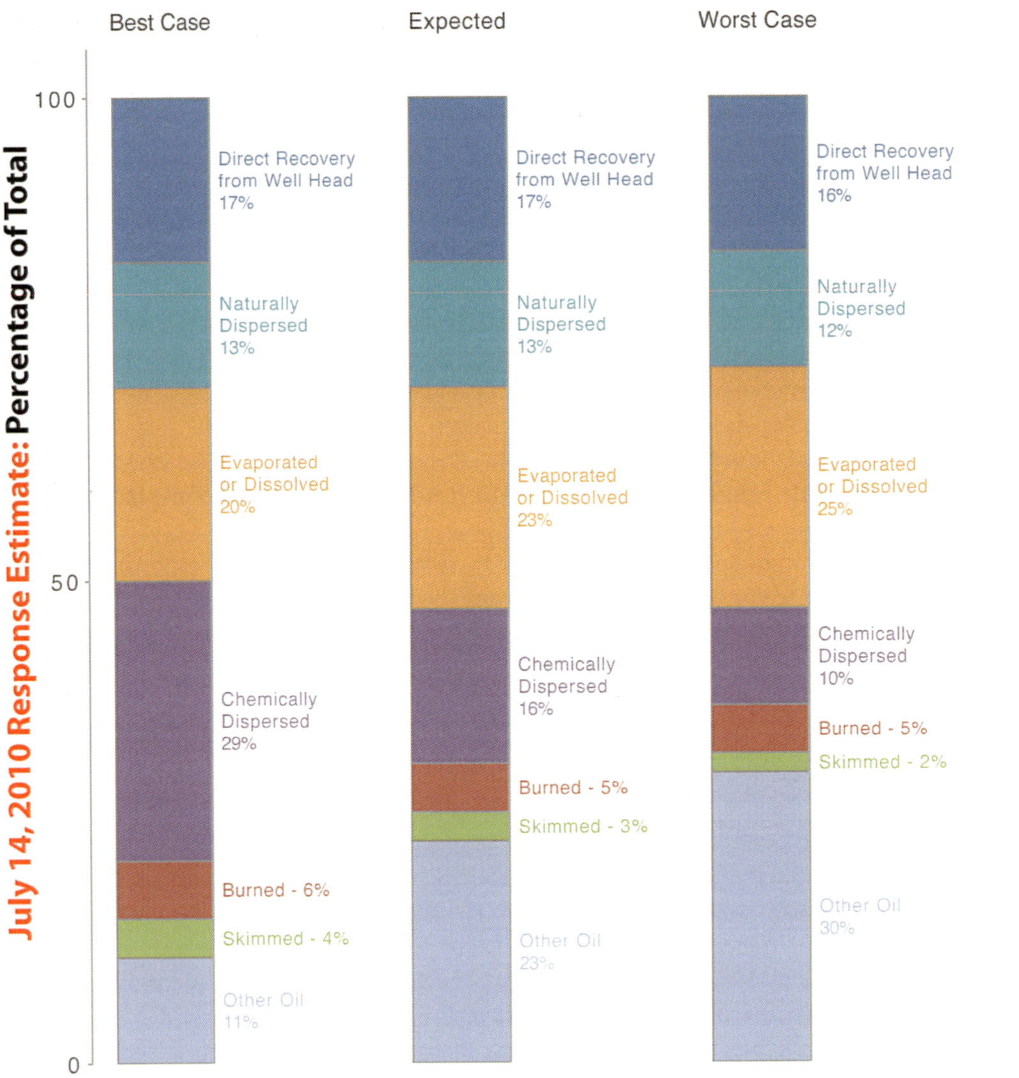


**FIGURE 4.1** *Deepwater Horizon* oil budget calculator. SOURCE: Federal Interagency Solutions Group, 2010.

prevent adverse impacts, moves to potential responses on and near the shore that can prevent adverse impacts, and then moves to methods to treat impacted shoreline habitats. The final section deals with natural attenuation—the natural fate of oil in the marine environment.

Understanding of the efficacy of response technologies used during the *DWH* oil spill is rapidly evolving, and papers and reports continue to be published. The results presented in this chapter may be verified, challenged, or negated by further findings. The extrapolation of results from lab-based studies to ecological relevance is tenuous at times, but lab-based studies do provide information about the potential cause and level of effect under certain circumstances. Studies of long-term changes in populations and community structure and function under the Natural Resource Damage Assessment (NRDA) remain undisclosed or have not been completed. Given these constraints, the committee acknowledges that it may never be possible to fully understand how some technologies have affected ecosystem processes and functions.

## CHEMICAL DISPERSANTS

Chemical oil dispersants are proprietary mixtures of surfactants and solvents that are applied directly to an oil spill in order to reduce the natural attractive forces within the oil. The use of chemical oil dispersants offshore can be effective in preventing heavy oiling of sensitive coastal environments, such as beaches and wetlands, and in mitigating risks associated with marine and terrestrial wildlife coming into direct contact with a surface oil slick. Some researchers (Lehr et al., 2010) have estimated that the use of dispersants helped to keep 500,000 barrels of the *DWH* oil away from the highly productive and sensitive coastal areas, but there is no feasible way to verify this estimate.

When an oil slick is sprayed with dispersant and exposed to mixing energy, typically from wave action, some of the oil is broken up into small droplets, which may then become entrained in the water column (Li et al., 2008, 2009a,b; Lunel, 1995). Because the chemically dispersed oil droplets may be small enough to be neutrally buoyant, diffusion and advection transport processes are expected to dilute the plume over a large volume to concentrations below toxicity threshold limits—a key objective for the use of dispersants. Furthermore, as microbial degradation of oil spilled at sea primarily occurs at the oil-water interface, oil biodegradation rates should be enhanced, provided that the dissolved oil concentration is not so large as to be toxic to the microbes. At the large scale, the overall biodegradation rate is increased when dispersants are used effectively (Venosa and Holder, 2007; Venosa and Zhu, 2003). In addition, small droplets enhance dissolution of soluble and semivolatile compounds into surrounding waters, resulting in fewer airborne volatiles, which enhances safety for cleanup workers. However, one consequence of dispersant use is that oil concentrations may increase in the water column and potentially increase exposure for pelagic species.

> ***Finding 4.1.*** *Chemical dispersants have long been applied to oil spills to break up oil into smaller droplets. From a purely physical perspective, chemical oil dispersants can reduce oil concentrations at the surface by increasing horizontal and vertical mixing of the oil. This tends to enhance biodegradation and mitigate the adverse effects of the oil at the surface including the reduction of vapors (thus enhancing the safety of response personnel). However, the overall volume of ocean impacted by the oil will increase, which may have adverse effects especially for subsurface biota.*

## Use of Dispersants During the *DWH* Oil Spill Response

### Approval and Monitoring of Dispersant Application

About 1.8 million gallons of dispersants were used during the *DWH* oil spill response. Both product formulations used in the GoM—Corexit 9527 and 9500—were approved by the U.S. Environmental Protection Agency (EPA) under the National Contingency Plan for the treatment of oil spills. A Centers for Disease Control and Prevention report (CDC, 2010) concluded that "because of the strict guidelines that must be followed to utilize dispersants, it is unlikely that the general public will be exposed (directly) to (the) product." The report further stated that "ingredients are not considered to cause chemical sensitization; the dispersants contain proven, biodegradable and low toxicity surfactants."

The use of chemical dispersants at the *DWH* oil spill site was regulated under the Clean Water Act. During the spill response, approximately 1.8 million gallons of Corexit brand chemical oil dispersant were applied (Figure 4.2): 1.07 million gallons of Corexit 9527 and 9500A at the surface and 772,000 gallons of Corexit 9500 via subsurface injection (Federal Interagency Solutions Group, 2010). Figure 4.3 shows the geographical area where approximately 90 percent of the aerial surface dispersant spraying occurred.

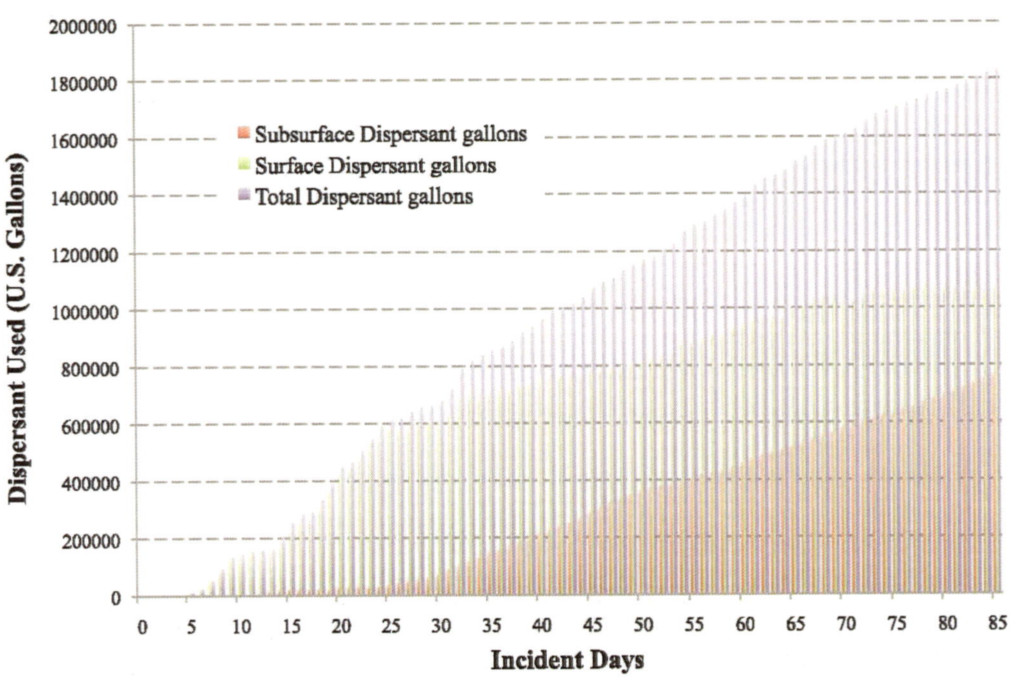


**FIGURE 4.2** Cumulative surface and subsurface dispersant use by day of spill response.
SOURCE: Federal Interagency Solutions Group, 2010.

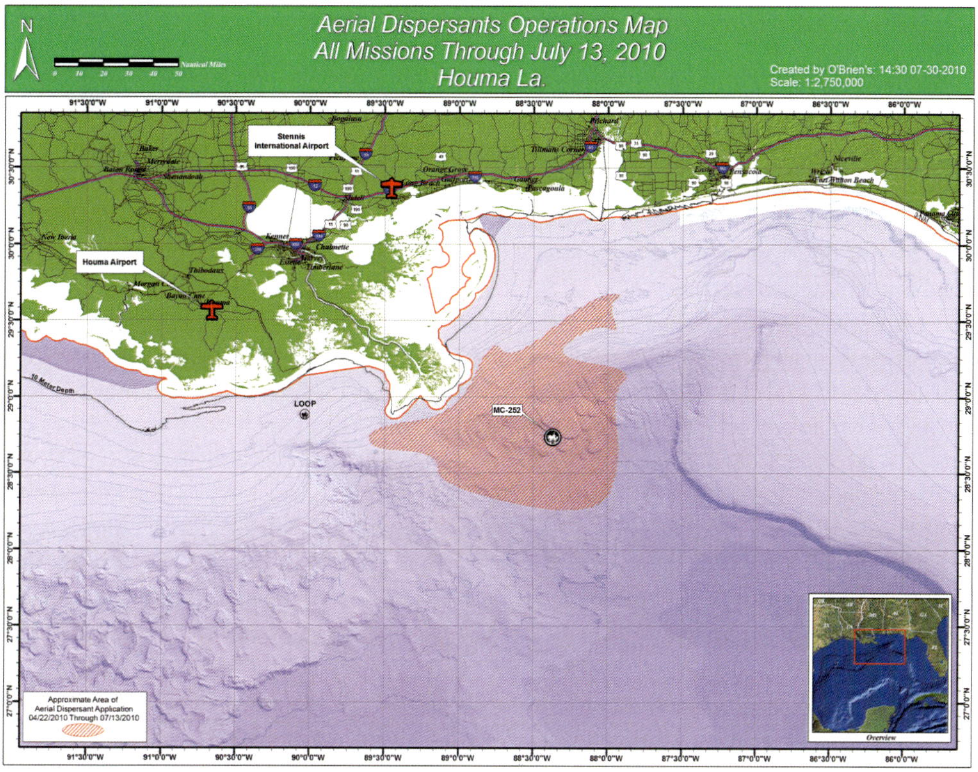


**FIGURE 4.3** Region in which 90 percent of aerial dispersant spray passes occurred. SOURCE: Created by O'Brien's Response Management.

## Dispersant Application Methods

Chemical dispersants are most often sprayed directly onto a surface oil slick from an airplane or ship. During the *DWH* oil spill response, chemical oil dispersants were also injected directly into the stream of oil and gas flowing from the damaged wellhead located 1,522 m below the surface of the ocean. The impacts of these application methods are discussed in detail below.

*Subsurface Injection of Dispersant at the Wellhead* One of the most innovative and controversial aspects of the immediate response to the *DWH* oil spill was the injection of 770,000 gallons of Corexit 9500 at the wellhead. Some experts questioned the necessity of injecting dispersants at the wellhead to keep some of the oil from reaching the surface. A plume of oil formed at depths between 1,100 and 1,200 m below the surface (Reddy et al., 2011). Some researchers (e.g., Peterson et al., 2012) have suggested that the subsurface dispersants had little impact in dispersing the spill or preventing oil from reaching the surface. They argue that the turbulence in the oil jet was sufficient to induce massive dispersion. They offer anecdotal evidence from the BP videos taken at the wellhead during the *DWH* spill that the dispersant injection wand sometimes missed the main oil plume, therefore making the dispersant ineffective.

Paris et al. (2012) argue that the application of subsurface dispersants did little to alter the vertical distribution of oil. However, their argument depends heavily on the use of an algorithm to estimate the oil droplet size that is based on oil-in-water emulsions with debatable relevance to the blowout problem. Furthermore, their calculated droplet size for the *DWH* oil spill in the absence of dispersant is so small that it could not explain the observed surface slick that formed immediately above the well.

The American Petroleum Institute (2012) provides an argument for the efficacy of subsurface dispersants. The Institute provides aerial photographs taken during the *DWH* oil spill that show that the surface slick nearly disappeared when subsurface dispersant injection was initiated, then reappeared when the injection was stopped. Other evidence for the efficacy of subsurface dispersant use is provided by Johansen et al. (2013) and Brandvik et al. (2013), who describe studies of droplet size in simulated blowout plumes conducted in a 6-m laboratory tank. Their results show roughly a threefold decrease in modal droplet size when dispersant is applied at ratios used in the *DWH* oil spill. From the laboratory data, they develop an algorithm for droplet size based on a modified Weber law that accounts for the effects of dispersant. The algorithm gives a reasonable fit to their laboratory studies as well as several other relevant studies. This algorithm suggests that dispersant application in the *DWH* oil spill dramatically reduced the droplet size. As discussed earlier in this chapter, the formation of smaller oil droplets would enhance oil biodegradation rates.

Of course, even if subsurface dispersant application does keep more oil below the surface, the concern is that dispersing the oil will simply increase the concentration of bioavailable oil in one environmental compartment, the water column, and decrease it in another, the shoreline. Despite this obvious tradeoff, the Presidential Oil Spill Commission concluded that dispersant application had reduced the overall risks associated with the spill (National Commission on the BP Deepwater Horizon Oil Spill and Offshore Drilling, 2011). The alternative to using dispersants would have been to allow additional oil from the Macondo wellhead to impact GoM coastal marshes and beaches.

*Dispersant Effectiveness* The effectiveness of a dispersant for the treatment of oil spilled at sea is largely dependent on a number of physicochemical factors such as oil properties, turbulence (e.g., waves for dispersant applied at the surface), temperature, oil weathering, salinity of the sea water, and the hydrophilic-lipophilic balance of the dispersant (Chandrasekar et al., 2005, 2006; Mukherjee and Wrenn, 2009).

*Biodegradation of Chemically Dispersed Oil* Dispersants can enhance the initial rate of petroleum hydrocarbon degradation by being the first substrate utilized by the hydrocarbon-degrading bacteria to grow and colonize dispersed oil droplets (Varadaraj et al., 1995). However, the observed effects of chemical dispersants on the rate of oil biodegradation have varied among studies (NRC, 2005a). Chemical dispersion stimulated biodegradation rates in some studies (Swannell and Daniel, 1999; Traxler and Bhattacharya, 1978), but chemical dispersion had no effect or inhibited biodegradation in others (Foght and Westlake, 1982; Lindstrom and Braddock, 2002).

Corexit concentrations of 1 to 10 milligrams per milliliter, which were likely encountered during the actual dispersant application, promoted the biodegradation of *DWH* oil in laboratory tests, but also caused significant reductions in production and viability of the hydrocarbon degraders, *Acinetobacter* and *Marinobacter* (Hamdan and Fulmer, 2011). *Marinobacter* was most sensitive to the dispersant, with nearly 100 percent reduction in viability and production. Dispersant concentrations within the immediate vicinity of the wellhead may have reduced the capacity of some microorganisms in those waters to bioremediate the dispersed oil because of the elevated concentrations of dispersant and oil above toxicity threshold limits, which became reduced as the plume moved farther away.

Attempts have been made to predict the rate of oil biodegradation in the environment from the results of kinetic models based on laboratory data (Venosa and Holder, 2007; Zahed et al., 2011). Lee et al. (2011) highlighted two important challenges to consider when designing experiments. The first challenge is to conduct tests at the low dispersed oil concentrations representative of field conditions where rapid dilution occurs. Many previous biodegradation studies were conducted at unrealistically high concentrations of dispersed oil in closed microcosms. Studying dispersed oil biodegradation at concentrations several orders of magnitude above expected at-sea concentrations in closed systems could limit biodegradation rates and total degradation by exhausting the available nutrients. Some researchers have attempted to address this by adding nutrients to the system, but that action can lead to unrepresentative modification of the microbial community.

The second challenge is to maintain a stable dispersion within the test system. Under normal sea-state conditions, oil droplets have a strong tendency to become more dispersed over time, especially for subsurface plumes where droplets can freely move in all three dimensions. Even for surface slicks, Langmuir cells (Langmuir, 1938; Leibovich, 1983) and the action of waves (Boufadel et al., 2006; Elliot et al., 1986) can temporarily overpower the oil buoyancy and drive the droplets several meters down into the water column, thus dramatically reducing the concentration and the chance of coalescence. It is difficult to simulate these kinds of spreading mechanisms in a laboratory setting. Until these issues are addressed in experimental research, actual rates for the biodegradation of chemically dispersed oil at sea will remain unknown.

*Biodegradation of Dispersants* The biodegradation of dispersants and the surfactants used in them are relatively well studied (Baumann et al., 1999; Garcia et al., 2009; Lindstrom and Braddock, 2002; Liu, 1983; Odokuma and Okpokwasili, 1992; Una and Garcia, 1983). In general, results indicate that most surfactant formulations are readily biodegraded under aerobic conditions by marine bacteria (Lee et al., 1985; Liu, 1983; Una and Garcia, 1983), with the rate of biodegradation under anaerobic conditions being much lower (Berna et al., 2007; Ying, 2006).

Different types of surfactants, and even individual surfactants of the same class, can biodegrade at different rates, depending on environmental conditions and the structural complexity of a dispersant's chemical branching. For example, Una and Garcia (1983) observed extensive biodegradation of the ethoxylated nonionic surfactants used in Corexit 9527 and 9500 in pure cultures of bacteria isolated from an estuary. These nonionic surfactants are sorbitan mono- and tri-(9*Z*)-9-octadecenoate poly(oxy-1,2-ethanediyl) derivatives, known commercially as Tween 80 and Tween 85. The authors also noted that poor biodegradation (<20 percent) of Span 80, the unethoxylated nonionic surfactant constituent in Corexit products, may have been caused by substrate inhibition due to the extremely high surfactant concentration (5 grams per liter), which likely would have impacted the integrity of bacterial membranes (Garcia et al., 2009). Campo et al. (2013) analyzed biodegradation of Corexit 9500 and Louisiana sweet crude (similar to the Macondo oil in properties and hydrocarbon constituents) at both sea surface and seafloor temperatures, and they observed complete degradation of specific surfactants such as dioctyl sodium sulfosuccinate (DOSS) within 28 days.

***Finding 4.2.*** *Existing studies of the biodegradation rate of dispersants alone show that they do degrade, although at rates that depend strongly on the concentration of dissolved oxygen and the composition of the surfactant. The biodegradation rate of chemically dispersed oil can also vary widely, but some, if not all, of this variation reflects the lack of realism in the underlying experiments. Further studies are needed that use realistic initial dispersed oil concentrations and avoid boundary effects that increase oil-dispersant concentrations.*

## Toxicity and Bioaccumulation of Dispersed Oil, Dispersants, and Surfactants

Although a great deal of work has been done under controlled laboratory conditions to evaluate the toxicity of chemical oil dispersants and dispersed oil in the marine environment, differences in results have been observed because of factors such as species-specific sensitivity, the particular dispersant formulation used, and experimental conditions (George-Ares and Clark, 2000). For example, with exposure of Atlantic herring (*Clupea harengus*) to dispersed oil, toxicity has been found to be related to concentration, duration of exposure, and life stage exposed (McIntosh et al., 2010). Some controversy remains over the use of dispersants (Chapman et al., 2007) because they may elicit toxic responses. Corexit 9500 and Corexit 9527 result in low to moderate toxicity to most aquatic species in laboratory tests (George-Ares and Clark, 2000). However, there is evidence that oil dispersed as small droplets in the water column might be more bioavailable, and therefore more toxic, than the oil alone. For example, fish studies with newly hatched mummichog (*Fundulus heteroclitus*) in 96-hour static renewal tests have shown that dispersant additions induced changes in aqueous concentrations of polycyclic aromatic hydrocarbons (PAHs) that affect larval survival, body length, or ethoxyresorufin-*O*-deethylase activity, an indicator of contaminant uptake (Couillard et al., 2005). At 0.2 gram per liter, the addition of dispersant caused twofold and fivefold increases in the concentrations of total PAHs and high-molecular-weight PAHs (three or more benzene rings), respectively. Recent studies have focused on the evaluation of chronic effects from short-term exposures to dispersants and dispersed oil (Hannam et al., 2009; McFarlin et al., 2011b). Much consideration has also been given to the environmental persistence of dispersants, dispersed oil, or dispersant by-products at low concentrations (Lee et al., 2011).

Fuller et al. (2004) conducted a study using two fish species, *Cyprinodon variegatus* and *Menidia beryllina*, and one shrimp species, *Americamysis bahia* (formerly *Mysidopsis bahia*), and the luminescent bacterium, *Vibrio fischeri*, to evaluate the relative toxicity of test media prepared with dispersant, weathered crude oil, and weathered crude oil plus dispersant. The study results indicated that the toxicity of chemically dispersed oil preparations was equal to or less than that of the oil alone. However, the real source of toxicity was a function of the soluble crude oil components, which would likely not persist over time. An EPA study by Hemmer et al. (2010) of the toxicity of Louisiana sweet crude (LSC), chemical oil dispersants, and chemically dispersed LSC on *Menidia beryllina* and *Americamysis bahia* (two species used by Fuller et al., 2004) revealed that the toxicity of the dispersant alone was lower than that of LSC or dispersed LSC, which both showed moderate to high toxicity. Similarly, Milinkovitch et al. (2011) found no significant difference between the toxicity of naturally and chemically dispersed oil when looking at a series of biomarker responses in the gills of golden grey mullet (*Liza aurata*).

Judson et al. (2010) used in vitro (cell-based) assays and concluded that, although some dispersants showed low potential for endocrine disruption, most (including Corexit) did not show any significant effect. Research into the toxicity of a series of chemical oil dispersants (the Corexit series of dispersants were not tested) on two Indo-Pacific branching coral species found that the two corals were sensitive to chemical oil dispersants and chemically dispersed oil, but that toxicity decreased sharply with dilution (Shafir et al., 2007). During this study, the water-soluble oil fraction was also tested and showed relatively low toxicity, but naturally dispersed crude oil was not evaluated. Using the frog (*Xenopus*) embryo teratogenesis assay, Smith et al. (2012) reported a range of abnormalities due to Corexit 9500A (concentration range: 1 to 1,000 microliters [μL] per liter) that included edemas of the pharynx, thorax, and eye, and incomplete and reversed gut coiling.

Fish are highly mobile and will encounter dispersed oil. Greer et al. (2012) showed that brief exposures of Atlantic herring (*Clupea harengus*) embryos to low concentrations of chemically dispersed Alaskan North Slope and Arabian light crude oils for 2.4 hours could induce blue-sac disease and reduce the percentage of normal embryos at hatching. Furthermore, toxicity increased with exposure time. However, fish are not expected to be exposed for long time periods to dispersed oil, because of natural dilution processes under dynamic open ocean conditions (Lee et al., 2011).

The media have raised concerns about the potential effects of the crude oil and dispersants on the food web dynamics in the GoM (Mascarelli, 2010). Microcosm experiments performed in situ in North Inlet Estuary near Georgetown, South Carolina, using *DWH* oil and a Texas crude oil noted a decrease in chlorophyll *a* in phytoplankton as crude oil concentration increased from 10 to 100 microliters per liter (Gilde and Pinckney, 2012). Researchers interpreted this change as a decrease in biomass rather than in the pigment itself, because of the short duration of the experiments. Diatom, cyanobacteria, euglenophyte, and chlorophyte abundances were unaffected or increased with increased oiling, whereas cryptophyte abundance decreased. The authors suggested that oiling could result in changes in phytoplankton community composition within salt marsh estuaries impacted by the *DWH* oil, thereby affecting higher trophic levels such as zooplankton, which selectively feed on phytoplankton that might be killed by the oil (Gilde and Pinckney, 2012). Graham et al. (2010) found that $\delta^{13}C$ depletion, used as a tracer of oil-derived carbon in mesozooplankton and suspended particulate samples in near-surface and bottom waters from four stations in 8- to 33-m water depth in the northern GoM, corresponded with the arrival of surface slicks from the *DWH* oil spill and demonstrated that carbon from the spill was incorporated into both trophic levels of the planktonic food web.

Ortmann et al. (2012) conducted mesocosm experiments to determine the response of the surface microbial community off Alabama to Macondo well oil and Corexit 9500A. The results suggest that the oil alone increased the abundance of ciliates (predators of the carbon-digesting bacteria), and therefore the likelihood of carbon being transferred up the food web. However, when dispersant or dispersed oil was used, an increase in heterotrophic prokaryotes with a significant inhibition of ciliates ensued, which suggests that the prokaryotes were able to use the dispersant and dispersed oil as a carbon source and that a decrease in the number of grazing microbes was occurring, with a corresponding decrease in carbon transfer to higher trophic levels (although the carbon transfer was not measured). The researchers hypothesized that the addition of dispersant and dispersed oil to the northern GoM from the *DWH* incident may have reduced the carbon transfer to higher organisms and potentially decreased zooplankton and fish production on the shelf. However, in a previous mesocosm study conducted on the Pacific coast off Canada, Lee et al. (1985) noted that the presence of crude oil dispersed with Corexit 9527 may have reduced the grazing pressure by bacterivores such as ciliates on potential oil-degrading bacteria.

Satellite measurements of algal blooms in the GoM suggest that the spill may have triggered an increase in these blooms. Initial results of changes in the Moderate Resolution Imaging Spectroradiometer (MODIS) fluorescence line height as a proxy for phytoplankton biomass and a source of carbon for trophic transfer[1] showed a statistically significant chlorophyll *a* anomaly in an area exceeding 11,000 km$^2$ in August 2010, about 3 weeks after the Macondo well was capped (Hu et al., 2011). The values of fluorescence line height were higher than in any similar August imagery from 2002 to 2009 (Hu et al., 2011). A second bloom from December 2010 to January 2011 seemed to have been from strong upwelling and mixing events during late fall, but data were insufficient to support or reject the hypothesis that subsurface oil from the *DWH* oil spill may have contributed to its occurrence (Hu et al., 2011). These studies highlight the need to fully understand the long-term effects of residual oil and dispersed oil in the marine environment.

Dispersant application during the *DWH* oil spill occurred at a time when birds in the Gulf region were nesting. Some birds and their eggs may have been exposed to oil, Corexit, or dispersed oil. To test the embryotoxicity of Macondo well oil dispersed with Corexit 9500 to a model bird species, 50:1 and 10:1 mixtures of weathered Macondo well crude oil (collected from the GoM) and Corexit were applied to the shells of mallard duck (*Anas platyrhynchos*)

[1] National Aeronautics and Space Administration (NASA), http://modis.gsfc.nasa.gov/about/.

eggs in ranges of 0.1 to 59.9 mg for the 50:1 mixture, and 0.1 to 44.9 mg for the 10:1 mixture (Finch et al., 2012). No deformities were observed, which is consistent with a previous study using weathered MC252 oil (Finch et al., 2011). It appears that the weathering of the oil reduces the potential toxic effects on bird eggs; in previous studies, fresh crude oils have been shown to be embryotoxic and teratogenic (Hoffman and Albers, 1984), because they contain lighter-molecular-weight components, including PAHs (Barron et al., 1999), that can penetrate the eggshell and membrane.

After the *DWH* oil spill, the EPA issued a number of directives (EPA, 2010) regarding the monitoring of the efficacy of subsurface oil dispersant injection and the fate and transport of dispersed oil and operational shutdown criteria. Based on the results of daily field data reports, the EPA had the authority to suspend the application of dispersants because of a significant reduction in dissolved oxygen (National Incident Command Joint Analysis Group, 2010), high mortality in rotifer toxicity tests, and other indicators related to human health. In support of these mandates, the EPA conducted a series of toxicity experiments to determine the potential threat posed to marine organisms by the large volumes of dispersant being applied during the spill response (Hemmer et al., 2010). The acute toxicity and capacity for endocrine disruption and other biological activity of eight chemical oil dispersant formulations in the presence of Louisiana Sweet Crude oil, at recommended application rates, were evaluated with GoM mysid shrimp, *Americamysis bahia*, and the inland silverside fish, *Menidia beryllina*. Acute toxicity testing was conducted using the $LC_{50}$ metric, which identifies the lethal concentration (LC) at which 50 percent of the test species die over a given period of time. The capacity of the dispersants to act as endocrine disruptors, or engage in other biological activity, was determined using a series of standardized in vitro (cell-based) assays. Overall, the dispersant-LSC oil mixtures were considered to be highly to moderately toxic, depending on the test species and dispersant. Corexit 9500A, which was used for both surface and subsurface application during the *DWH* oil spill response, was considered moderately toxic for both species. The toxicity of the dispersant alone was lower than that of LSC or dispersed LSC.

In light of the varied and somewhat circumstantial results being obtained for toxicity studies of chemical oil dispersants, George-Ares and Clark (2000) suggested that factors such as toxicity of dispersed oil, dilution and degradation in the environment, presence of species or resources requiring priority protection, potential adverse effects of all response options, and the potential for recovery of sensitive habitats and populations should weigh more heavily in the decision-making process than dispersant toxicity alone.

Some consideration has also been given to the bioaccumulation of surfactants. There is evidence that surfactants of all classes are readily taken up across the gills of some marine animals (EOSCA, 2000) and that nonionic and ionic surfactants (such as those found in most oil dispersants) may be effectively biotransformed and eliminated via the gall bladder (Tolls et al., 1994). A review of the bioaccumulation potential of surfactants conducted by the European Oilfield Speciality Chemicals Association (EOSCA) concluded that although surfactants and their metabolites can be found in aquatic organisms following exposure, there was no evidence to support biomagnification of surfactants through the food web (EOSCA, 2000).

Numerous studies of the toxicity of dispersants and dispersed oil, some with conflicting or inconclusive results, are ongoing. The body of evidence, however, implies that, in general, dispersant and dispersed oil biodegrade in the marine environment. The fate of dispersant surfactants in laboratory studies is highly dependent on the concentration and chemical characteristics of the surface-active compounds, the microbes available, and the methods used to monitor biodegradability. The degradation of surfactants in combination with the crude oil hydrocarbons remains a challenge in analytical chemistry.

> ***Finding 4.3.*** *There is some evidence that chemically dispersed oil and some dispersant compounds are toxic to some marine life, especially those in early life stages. There is contradictory evidence as to whether chemically dispersed oil is more or equally toxic to marine life compared to undispersed oil. The use of dispersants does reduce the amount of oil available to reach shorelines and shallow water environments and impact additional marine life. These facts may be weighed when considering the net environmental benefit to having used dispersants to respond to the* DWH *oil spill.*

## Dispersant Degradation During the *DWH* Oil Spill

The Operational Science Advisory Team (OSAT, 2010) chose the chemical indicators 2-butoxyethanol, dipropylene glycol *n*-butyl ether (DPnB), propylene glycol, and dioctyl sodium sulfosuccinate (DOSS) of Corexit dispersants 9500 and 9527 to monitor the expanse of Corexit and the *DWH* dispersed oil. Corexit chemical indicators were observed in 79 percent of offshore water column samples in a wide range of concentrations, but only in a small fraction (<10 percent) of all other (near-shore and inland) samples. Propylene glycol was the only dispersant indicator detected in the near-shore sediments (OSAT, 2010). A research team studying the distribution of propylene glycol, 2-butoxyethanol, and DOSS found at near-shore and inland water sampling sites of Orange Beach, Alabama, between September 2010 and January 2011 concluded that these Corexit constituents probably did not originate from dispersant use during the *DWH* oil spill, but were likely related to stormwater discharge (Hayworth and Prabakhar Clement, 2012).

DPnB was the most commonly detected dispersant indicator. It was found in 57 of the 60 water samples collected offshore, and its concentrations decreased over time. Peaks in DPnB were found in surface waters and also between 1,100 and 1,300 m. Water samples from 200-m depths had DPnB concentrations ranging from 0.0170 to 113.4 micrograms per liter, with a mean of 4.3 micrograms per liter (OSAT, 2010). Concentrations of DPnB decreased over time after dispersant application stopped in mid July; all values were less than 5 micrograms per liter by July 30 (OSAT, 2010).

Detection of the DOSS Corexit indicator was measured during dispersant injection and up to 300 km from the wellhead 64 days after injection at the wellhead had ceased (Kujawinski et al., 2011). The majority of the DOSS associated with the subsurface injection of Corexit 9500 moved into 1,000- and 1,200-m depth rather than rising to the surface. Kujawinski et al. (2011) concluded that although biodegradation might have occurred, the most significant factor causing a decrease in DOSS concentration at depth was dilution. Campo et al. (2013) observed rapid biodegradation of DOSS with and without the presence of LSC by microbial cultures isolated from the surface of the GoM, but they also observed a significant lag in the biodegradation response from the cultures isolated from deep GoM. The authors hypothesized that, because DOSS is a xenobiotic compound in that environment, they would not find cultures able to degrade DOSS rapidly at 5°C, except after a long adaptation period (Campo et al., 2013).

## The Fate of Chemically Dispersed Macondo Well Oil

Other than the study of the fate of DOSS (Kujawinski et al., 2011), no studies have specifically investigated the fate of chemically dispersed Macondo well oil (see Box 4.1 for the fate of oil from a comparable spill event). Several studies, however, have investigated the submerged plume of dispersed oil and gas droplets less than several millimeters in size that extended southwest from the wellhead between about 1,000 and 1,200 m below the surface (Camilli et al., 2010; Diercks et al., 2010; Hazen et al., 2010; Kessler et al., 2011; Valentine et al., 2010, 2012). Modeling predicted that the volume of hydrocarbons released by the Macondo well was sufficient to double the bacterial population in the deep plume over an area of 30,000 km$^2$ within the deep-water layer. The modeling results suggested that mixing processes recirculated water containing microbial oil degraders around the wellhead and spill site, causing rapid biodegradation of the suspended oil (Valentine et al., 2012) and decreased oxygen concentrations due to respiratory processes. Chakrabarty et al. (2012) reported that the concentrations of Corexit 9500 used during the *DWH* response were not toxic to indigenous microbes in the deep-water plume that included bacterial species capable of degrading the dispersant components, which include hydrocarbons, glycols, and DOSS.

**BOX 4.1**
**Fate of Chemically Dispersed Oil from the *Ixtoc I* Oil Spill**

The *Ixtoc I* well blowout in the Bay of Campeche in the southern GoM off Mexico in 1979-1980 in 50-m water depth expelled about 150 million gallons of oil (more than half of that attributed to the *DWH* oil spill), yet 5 years later there appeared to be little evidence of residual toxicological impacts (Aldhous and McKenna, 2010). It took almost 10 months to cap the *Ixtoc I* wellhead, during which time the oil was transported along the shore and as far north as Galveston, Texas. Bulk oil incorporated into the burrows of burrowing beach organisms remained largely intact (undegraded) and reappeared in successive years from wave action in the surf zone (Amos et al., 1983), and asphaltenes, an oil fraction with limited bioavailability, can still be uncovered along the Mexican coastline (Wes Tunnell, unpublished data).

The submerged plume was identified based on fluorescence (Camilli et al., 2010; Diercks et al., 2010; Hazen et al., 2010), light scattering (Diercks et al., 2010), or the concentrations of specific hydrocarbons (Camilli et al., 2010; Diercks et al., 2010; Kessler et al., 2011; Valentine et al., 2010) and was detectable up to 35 km from the MC252 wellhead (Camilli et al., 2010).

Most of these studies also observed a local dissolved oxygen (DO) decrease in the vicinity of the submerged hydrocarbon plume. Camilli and colleagues attributed this decrease, in part, to hydrocarbon interference with the in situ DO probes that were used because data from the Winkler titration tests for DO did not show oxygen depletion within the plume at the time of their study (Camilli et al., 2010). Earlier and later studies in the region showed good agreement between data from the in situ DO probe and Winkler titrations (Kessler et al., 2011; Valentine et al., 2010), suggesting that the rate of aerobic microbial metabolism within the plume was higher than in the surrounding water despite its average temperature of about 5°C (Camilli et al., 2010; Hazen et al., 2010).

Evidence of lower DO concentrations in support of biodegradation of gaseous alkanes (Kessler et al., 2011; Valentine et al., 2010) was obtained based on compositional changes that reflected preferential loss of specific compounds and changes in temperature (for ethane and propane) (Valentine et al., 2010). One study estimated that about 70 percent of the oxygen depletion observed within the plume was due to microbial metabolism of ethane and propane (Valentine et al., 2010). Microbial degradation of other hydrocarbons, including butane and longer-chain alkanes, was modeled and thought to be responsible for the additional oxygen depletion.

Although the media expressed fears that the oxygen depletion within the water column in the vicinity of the Macondo wellhead might cause hypoxic conditions or anoxic "dead zones" and threaten marine organisms, the maximum depletion observed was 40 percent, whereas 60 percent is considered detrimental to fish (Walsh, 2011). Studies are now under way to assess the potential long-term ecological impact of MC252 residual oil buried in deep-sea sediments.

Half-lives for higher-molecular-weight normal alkanes within the plume were estimated to be between about 1.2 and 6.1 days, based on in situ and microcosm data (Hazen et al., 2010). Flocculant material isolated from samples collected within the plume between May 25 and June 2, 2010, were rich in microbes (Passow et al., 2012), oil, and oil degradation products, and bacterial counts were elevated within the plume (Hazen et al., 2010). Microbial genes involved in hydrocarbon degradation were significantly increased ($p < 0.05$ or $p < 0.01$) in plume samples, and the relative abundance was correlated with the concentrations of some low-molecular-weight components of the oil, suggesting that the composition of the bacterial community changed in response to the presence of oil (Hazen et al., 2010).

The most abundant bacterial species in samples from within the plume (comprising about 90 percent of sequences) belonged to a single operational taxonomic unit that was closely related to the order *Oceanospirillales* (Hazen et al., 2010). The species dominance of this simple aliphatic hydrocarbon degrader was observed to shift to *Colwellia* and a PAH degrader, *Cycloclasticus* (Abbriano et al., 2011), in samples collected from the same area by another research group about 2 weeks later while oil was still being released from the wellhead (Valentine et al.,

2010). The putative hydrocarbon-degrading bacteria in these later samples were thought to be growing on propane, ethane, and butane (Valentine et al., 2010).

Lu et al. (2011) showed that the *DWH* oil spill dramatically altered the functional composition and structure of the microbial community in the deep sea. A variety of metabolic genes involved in aerobic and anaerobic hydrocarbon degradation were more highly enriched inside than outside the plume. Various other microbial functional genes that are associated with carbon, nitrogen, phosphorus, sulfur, and iron cycling, metal resistance, and bacteriophage replication were also enriched within the plume. The authors suggested that the indigenous marine microbial communities could have a significant role in biodegradation of oil spills in deep water.

Microbial activity is essential to the maintenance of marine ecosystem health because it controls primary processes—for example, nutrient regeneration, the production of biomass, and the biodegradation/biotransformation of contaminants. Thus, changes in the structure and function of microbial communities may have had a profound effect on the recovery and maintenance of ecosystem services within the GoM. Soon after the spill, the microbial community responded by becoming dominated by hydrocarbon degraders, resulting in reduced community diversity. DNA surveys for bacterial 16S rRNA genes from samples collected in June revealed the dominance of *Cycloclasticus* and *Colwellia*, likely degrading propane and ethane preferentially (Kessler et al., 2011; Valentine et al., 2010). Among these were *Oleispira antarctica*, *Thalassolituus oleivorans*, and *Oliphilus messinensis*, all of which are bacteria known to degrade hydrocarbons and tolerate low temperatures that occur in the deep sea. Samples collected later (September 2010) indicated a shift away from these hydrocarbon degraders to methanotrophs, including *Methylococcaceae*, *Methylophaga*, and *Methylophilaceae*. The enhanced abundance of methanotrophs and bacteria containing the particulate methane monooxygenase gene indicated that methane was consumed later in the spill sequence by a different bacterial assemblage (Kessler et al., 2011).

## Summary of Dispersant Use During the *DWH* Oil Spill Response

About 1.8 million gallons of dispersant were used during the *DWH* oil spill response, of which about 42 percent was injected at the wellhead (Federal Interagency Solutions Group, 2010). The size and composition of the GoM microbial community was altered as microbes responded to the presence of *DWH* crude oil and dispersed oil, which contained light, readily biodegradable hydrocarbons (Atlas and Hazen, 2011). Bacterial cell density was significantly higher within the deep subsurface plume at $5.51 \times 10^4$ cells per milliliter than outside of it at $2.73 \times 10^4$ cells per milliliter, when measurements were taken 5–7 weeks after the start of the spill (Hazen et al., 2010). Valentine et al. (2010) used measured oxygen and hydrocarbon component anomalies in the plume several weeks later, along with sample incubation experiments that included dissolved inorganic carbon (DIC) measurements, to deduce that 70 percent of the depleted oxygen was due to microbial respiration (not necessarily growth) of two components of the plume: propane and ethane. They presumed that respiration of other components could account for the full amount of depleted oxygen, as others demonstrated later (Kessler et al., 2011). The occurrence of natural seeps in the area of the spill may have supported the development and persistence of microbial communities capable of degrading hydrocarbons.

The short- and long-term impacts of dispersants and dispersed oil from the *DWH* oil spill on the food web and other GoM ecosystem services are still undetermined (see Chapter 5 for discussion of specific examples). Additional research is recommended to enhance understanding of the overall effectiveness and fate of dispersants and chemically dispersed oil. Of particular interest is the role that subsurface dispersant use played in generating smaller oil droplets, which ultimately controlled the volume of oil in the subsurface oil plume. This information is required for the determination of the overall effectiveness of the dispersants and the establishment of operational guidelines (e.g., establishment of optimal dispersant-oil ratios) for their use. The work of Brandvik et al. (2013) and Johansen et al. (2013), specifically focusing on this issue, offers strong evidence that subsurface dispersants can greatly decrease droplet size and the volume of surfacing oil. However, questions still remain concerning the effects of oil temperature and high pressure and the scaling of laboratory results to the real world.

There is also a need for additional research on the potential impacts from the various dispersant remediation strategies available. At this time, impacts to ecosystem services of the deep GoM, such as chemosynthetic productivity and microbial carbonate sequestration as affected by the deep dispersed oil plume (see Chapter 5), and the fate and effect of the major dispersant constituents on deep-water species, remain unresolved.

***Finding 4.4.*** *The use of dispersants helped to keep an estimated 500,000 barrels of the oil away from the highly productive and sensitive coastal areas. The use of chemical dispersants was controversial from a public point of view because of concerns about dispersant toxicity, the potential for dispersants to make oil more bioavailable (thus making dispersed oil more toxic than nondispersed oil), and the persistence of dispersants, dispersed oil, or dispersant byproducts. The long-term impacts of dispersants and dispersed oil from the* DWH *spill on the food web and other ecosystem services of the GoM are still undetermined, and additional research is necessary in this area.*

## IN SITU BURNING

Burning oil at sea (in situ burning, ISB) is a well-established practice and is considered to be a relatively low-cost, simple method to rapidly remove large quantities of oil at sea without the generation of excessive wastes that would require further disposal (NOAA, 2011a). However, there is limited scientific data and knowledge of the effects of burning oil and how best to perform in situ burns (Fingas, 2011a). ISB can be an efficient mode of removing oil from the ocean, thus reducing the risk of an oil slick impacting vulnerable habitats such as wetlands and other high-amenity shoreline sites (Valentukevičienė and Brandvall, 2008) that provide valuable ecosystem services by habitats such as mangrove stands, salt marshes, and beaches.

ISB has also been shown to be capable of restoring salt marsh ecological structure and function (Baustian et al., 2010), but it was not employed for this purpose during the *DWH* oil spill. As of August 2010, little free-floating oil was remaining, and most of the residual oil was found along the marsh fringe, where low-pressure flushing was deemed more appropriate (see Appendix D in SCAT, 2010). ISB is not appropriate for mangroves, and it was not used on sandy beaches, because it "would result in significant adverse habitat impact" (see Appendixes C and D in SCAT, 2010).

## Containment and Ignition of Oil

Burning oil at sea may appear to be an easy method to implement, but it is actually a complex process (Fingas, 2011b). A number of tools and techniques aid in the ignition process, such as a simple rag soaked with diesel or other flammable fluid dispensed by a Heli-torch flame thrower suspended from a helicopter. Volatile compounds that evaporate from the slick provide the fuel, and they must be in sufficient concentrations to maintain the burn. The ignitability of the oil on the sea surface is reduced if the oil is emulsified and has high water content; <25 percent water is typically required for ignition (Walton and Jason, 1999). As the lighter fractions of oil evaporate, leaving asphaltenes and heavier fractions that are harder to ignite (Putorti et al., 1994), the slick may become fragmented and spread over a wider area, further complicating the scenario. Containment booms can minimize the spreading and thinning of the oil on the sea surface. Offshore booms usually consist of a flotation cavity (bladder) that protrudes from the water and a skirt that hangs down in the water column. Fire-resistant booms are the obvious choice to contain the oil.

## In Situ Burning During the *DWH* Oil Spill

ISB was selected as a response technology during the *DWH* event because of the need to supplement skimming operations to capture more oil, the fact that the oil slick was initially far from populated areas, and the understanding that controlled ISB could be conducted safely and effectively to eliminate large volumes of oil with minimal environmental impact (Allen et al., 2011). Oil was contained within a variety of fire-resistant booms, then ignited with a hand-held igniter containing gelled diesel in a plastic half-gallon bottle attached to a float and road flare. The application of an open-apex deflection ("funnel") technique facilitated the burning of oil within a fire boom as newly captured oil was fed into the burn. New, improved techniques and equipment were able to match the changing oil and weather conditions. For example, controlled burning of oil immediately outside the fire boom was found to be advantageous in extending burn duration, while maintaining a reasonably sized burn area (Allen et al., 2011).

The Occupational Safety and Health Administration/National Institute for Occupational Safety and Health air monitoring specialists evaluated the air quality on vessels working near burn sites to safeguard the health and safety of responders. In addition, an EPA Aerostat team evaluated emissions in and around smoke plumes (Aurell and Gullett, 2010). No cases of harmful or prolonged smoke exposure were reported.

## Effectiveness of In Situ Burning During the DWH Oil Spill Response

The Controlled In Situ Burn Group conducted 411 oil collection and ignition attempts between April 28 and July 19, 2010, of which 376 burns were large and long enough to be included in the oil budget estimates (OSAT, 2011a). About 220,000 to 310,000 barrels of oil were burned (Allen et al., 2011) within 5 to 24 km from the wellhead, and therefore, that oil did not impact shallow waters, shorelines, and sensitive coastal resources. In comparison to other technologies, according to the oil budget calculator (Figure 4.1), the amount of oil released from the well was estimated at 4.9 million barrels, and 5–6 percent of all of the spilled oil was burned, about double the removal of oil with skimmers (see "Skimmers" section below).

## Byproducts Associated with In Situ Burning

Contaminants and byproducts resulting from the ISB were released into the atmosphere. Black particulate carbon in the atmosphere (Perring et al., 2011) was an issue for coastal populations downwind from burn sites. The emissions plume also contained light alkanes and aromatics, but no large concentrations of PAHs (Middlebrook et al., 2011), and were measured across 4 km up to 47 km downwind. The light alkanes and aromatics consumed by ISB were one-tenth of the total evaporated fraction from all the oil spilled (Middlebrook et al., 2011).

Dioxins, which can be detrimental to human health, are formed from the incomplete combustion of organic matter in the presence of chlorine; therefore, they are a possible compound of emissions from ISB of oil at sea. The EPA measured dioxin levels between July 13 and 16, 2010, using the Aerostat protocol involving a sampling device attached to a balloon released into the air above burning oil (Aurell and Gullett, 2010). Dioxin levels were above ambient background concentrations and were similar to a value slightly higher than that associated with residential woodstoves. The risk to workers exposed to dioxins near the burn sites was estimated to be below the level of concern of 1 in 1 million—an EPA risk threshold of one cancer case for every 1 million people exposed. Numerical models (AERMOD) showed air concentrations of dioxins lower onshore than at sea near the burn sites.

Dioxins can settle into sediments, where they degrade very slowly and can be ingested by organisms. Most human exposure to dioxins occurs through dietary intake of animal fats, where the chemicals accumulate. Regarding the risk of exposure from consuming contaminated seafood, Schaum et al. (2010) calculated the maximum factor by which dioxin concentrations in fish could increase due to the in situ burns to be approximately 0.024 pg TEQ/g, which is less than 5 percent of the background levels of dioxin in marine fish (0.5 pg TEQ/g). On the basis of this analysis, even the most at-risk population of subsistence fish consumers who eat on average 300 g of fish per day would not have an increased cancer risk exceeding 1 in 1 million, the top of the acceptable range of risk according to the EPA (Schaum et al., 2010).

## Ecological Impacts of In Situ Burning

At present, there are no known effects on marine organisms resulting from ISB during the *DWH* event.

> ***Finding 4.5.*** *Several technical innovations led to in situ burning being the most successful response method in terms of removing oil from the sea surface offshore (estimated to be as high as 300,000 barrels or 6 percent of the total volume of oil spilled) and thus keeping oil away from sensitive coastal habitats. Although black particulate carbon clouds were clearly visible and subsistence consumers of local fish may have had a slight increase in the risk of cancer (1:1 million), there are no other documented negative impacts at this time. Further long-term research is needed to continue to look for future potential impacts.*

# SKIMMERS

## Skimmer Technology

A skimmer is a mechanical device that removes oil from the surface of the water without the addition of chemicals. Fingas (2011b) provides a recent and detailed overview of skimmers and categorizes them into five types, depending on the physical mechanism employed: sorbent surface, weir, suction, elevating, and submersion. Of these, the skimmers using an oleophilic sorbent surface in the form of discs, drums, belts, brushes, or ropes are most suitable for rougher water that is characteristic of the open sea and coastlines. Skimmers can use many different oleophilic sorbent surfaces. Once the oil is attached to the sorbent surface, a wiper blade or pressure roller removes the oil, after which it is deposited into a barge or other type of container.

The recovery rates achieved by skimmers depend on oil properties, slick thickness, weather (especially wave conditions), and the size of the skimmer. At the large end, a skimmer like the HOSS Barge[2] can recover about 30 barrels per minute under ideal conditions. Sustained over a full 24-hour day, this would amount to 40,000 barrels per day, but conditions are rarely perfect, so the recovery rates in practice are usually considerably less. To increase their efficiency, skimmers are almost always joined to booms deployed in the shape of a "V" to funnel and concentrate the oil.

Booming and recovery of oil with skimmers has been considered the first option of choice in response operations because this approach can often recover substantial volumes of oil before it has a chance to strand onshore or spread. Furthermore, unlike some other cleanup methods, skimmers have little potential negative environmental impact. They do not introduce any chemicals into the environment, nor do they physically disturb the seabed or coastlines under normal operations. Some skimmers recover a little seawater along with oil. There are potential

[2] http://www.myfloridahouse.gov/FileStores/Web/HouseContent/Approved/Web%20Site/uploads/documents/energy_exploration/Clean%20Gulf%20Associates.pdf.

risks of entrapping smaller marine creatures and colliding with or physically damaging larger species; however, such risks are small, particularly in light of the fact that the creatures would have to be in close contact with the oil and would already be at much higher risk of acute or toxic oil impacts. Some skimmers have trained observers onboard to recover any entrained sea turtles and to report impacts.

### Skimmers in the *DWH* Oil Spill

During the *DWH* oil spill, skimmers were operated offshore, in bays, marshes, protected waters, and along beaches (USCG, 2011). Details regarding skimmer operations during the *DWH* oil spill are difficult to obtain because no accurate records were kept of the number or types of skimmers deployed or the volume of oil recovered. The only readily available information comes from the federal government's *Restore the Gulf* website,[3] which includes occasional press releases and summaries known as "Operations and Ongoing Response" reports. The official government estimate of recovery by the skimmers is 2–4 percent of the total 4.9 million barrels of oil released (Figure 4.1), which translates to an estimated 147,000 barrels. This is a calculated value because the Incident Command only tracked the total volume of recovered oily water.

Taking into account errors in the assumptions used in the data analysis, skimmers likely recovered somewhere between 75,000 to 300,000 barrels of oil during the *DWH* oil spill (Graham et al., 2011a). Several reasons likely contribute to this low oil recovery rate, including the following: the encounter rate was low due to the expanse of the spill, the oil was likely diluted and dispersed by the time it reached the sea surface, cleanup vessels had to be directed by spotter aircraft, it was difficult to distinguish reasonably thick slicks from oil sheen, and containing and concentrating a slick with booms for skimming are time-consuming tasks.

***Finding 4.6.*** *Skimmers can be an effective means of recovering oil with little impact on the environment. However, given the diluted and dispersed nature of the* DWH *oil spill, skimmers were only able to capture between 75,000 to 300,000 barrels of oil. Nonetheless, skimming prevented this oil from reaching sensitive coastal ecosystems and impacting the services they provide.*

## NEAR-SHORE AND ONSHORE PROTECTION FOR SHORELINES

The construction of obstacles to prevent oil from reaching inland of shorelines is a common practice in oil spills (Davidson et al., 2008) and generally falls into three subcategories of fixed barriers: sand barriers (berms), sediment-filled containers, and rigid structures (booms). Diversion of a watercourse to repel oil from the shoreline was also a response strategy implemented during the *DWH* oil spill.

[3] http://www.restorethegulf.gov.

## Berms in Response to the *DWH* Oil Spill

The western side of Dauphin Island, Alabama, is of low elevation (less than 2 m above sea level), and a 2.5 km portion of it (known as the Katrina Cut) was eroded by hurricane Katrina. It was argued that filling the 2.5-km stretch would stem the influx of oil from the GoM into Mississippi Sound, while decreasing the salinity in the sound, which would help to maintain the oyster population (Webb et al., 2011). Approximately 235,000 tons of sediment were used during the project between July 2010 and April 2011. The inlet to Little Lagoon was filled with 4,500 $m^3$ of sand. This manipulation of the shoreline at Dauphin Island prevented oil from washing over the island, prevented oil deposition under elevated houses, and blocked the oil from entering the Mississippi Sound (Douglass et al., 2011). Two additional sand berms were built. The first one was approximately 4 km long and was completed in several days. It was not extended over roads or driveways, but enough material was put in place so that bulldozers could fill these gaps, if necessary. The second berm was south of the first, at the highest elevation of the dry beach, with the goal of confining any stranded oil to the beach. A total of 267,000 $m^3$ of sand was placed on the island between May and July 2010. Weather conditions during the summer of 2010 were such that sand accretion occurred south of the second barrier.

A second berm-related project involved the placement of massive berms in Louisiana that consisted of six "reaches" totaling 64.5 km in length. By the time BP capped the well on July 15 (day 44 of the berm construction project), Louisiana's contractor estimated that 6 percent of the total project had been completed. It took 5 months to build roughly 16 km of berms, at a cost of about $220 million (Rudolf, 2010). The more than 15 million $m^3$ of dredged sand came from sources or locations that were not proposed for future coastal restoration, and of this, more than 9.2 million $m^3$ came from the Mississippi River.[4] Other sources were Pass a Loutre Hopper Dredge Disposal Area, Pilottown Anchorage, Cubit's Gap, South Pass, and Hewes Point (north of Chandeleur Islands). Estimates of how much oil the berms collected vary, but none collected much more than 200 tons. On November 1, the state of Louisiana announced plans to convert the berms into part of a long-term coastal restoration project.

The installation of berms changes barrier island sediment transport, causes a loss of sand, and adversely impacts benthic and pelagic flora and fauna (Martínez et al., 2012). The extent of impacts from the berm operations conducted during the *DWH* oil spill is unknown.

Dynamic barrier islands shift with sediment transport mechanisms, especially those associated with summer tropical storms. As a part of a long-term investigation of the area, the U.S. Geological Survey conducts periodic surveys of Chandeleur Islands, Louisiana (Miselis et al., 2012). The most recent survey of the area was completed just 1 month prior to Hurricane Isaac in 2012. Subsequent data from airborne optical sensors that determine volume of lost sand indicated that most of the oil spill mitigation sand berm created in 2010–2011 was swept away in the storm.[5] A seafloor survey planned for 2013 will help to pinpoint where lost sediment

---

[4] Coastal Protection and Restoration, State of Louisiana, http://coastal.louisiana.gov/index.cfm?md=pagebuilder&tmp=home&pid=131.

[5] http://coastal.er.usgs.gov/hurricanes/isaac/phtot-comparisons/ and http://www.nola.com/hurricane/index.ssf/2012/09/isaacs_surge_waves_wiped_out_b.html.

was deposited and whether or not it will be naturally available to the islands for post-storm recovery.

An evaluation of the efficacy of the Louisiana berms project (and the other oil spill response strategies used during the *DWH* oil spill) by the staff of the Oil Commission found that the berms were not an effective response technology based on a cost-benefit analysis. They concluded that the scope and scale of the berms proposed were far too extensive and could not be designed, approved, and built in the time needed to be effective. The value of the berms as a spill response technology depends on how much oil they trapped. Estimates vary, but none of the estimates is much greater than 1,000 total barrels (Graham et al., 2011b).

## Booms

Onshore booms create barriers that keep oil from affecting sensitive areas and from intruding further inland. There are three main types. A shore-sea deflection boom for high-current areas has an air tube that rides on one or two water-filled tubes. Once deployed and anchored at the selected location, water is pumped into the lower tubes to form a skirt. This water acts as ballast when the boom is floating. When the boom is grounded, these flexible tubes follow the sediment surface to form a seal. Oil is deflected toward pooling areas where it can be recovered from land (USCG, 2011). The second type is a containment boom that is used to surround pooled oil so it can be removed by skimmers. This type of boom can also be used as a barrier (protection) when laid parallel to the shore to prevent oiling of sensitive areas (USCG, 2011). The third type is a snare (sorbent) boom, which is constructed of a long fabric sock filled with oleophilic material, or a series of oleophilic polypropylene "pompoms" tied to a line. Unlike hard booms, sorbent booms do not have an attached skirt. A sorbent boom is typically anchored parallel to the shoreline to absorb oil slicks as they wash toward the shore.

### Use of Booms During the *DWH* Oil Spill

During the *DWH* oil spill, booms were extensively used to protect sensitive coastal marshes and were later removed. About 1.16 million m of hard (containment) boom and 2.95 million m of sorbent boom were deployed[6] at critical points to protect wildlife refuges, bird nesting areas, estuaries, beaches, marshes, and other environmentally sensitive and economically significant lands throughout the Gulf Coast. All of the hard booms on shorelines in Mississippi, Alabama, and the Florida Panhandle were removed by September 7, 2010, and they were inspected, cleaned, repaired if possible, and stored at sites along the Gulf Coast for further redeployment should the need arise.[7]

Response crews deployed more than 2 million feet of containment boom in Louisiana, secured by Danforth-style anchors, which were designed to embed in the sediment and collapse

[6] http://www.restorethegulf.gov/release/2011/08/19/operations-and-ongoing-response-august-17-2011.

[7] http://www.restorethegulf.gov/release/2010/09/07/all-operational-hard-boom-removed-mississippi-alabama-and-florida-panhandle.

flat when not in use. Upon boom retrieval, it was discovered that some of the anchors remained embedded in the sediment. The U.S. Coast Guard commissioned a net environmental benefit analysis (NEBA) to determine the tradeoffs, including the potential loss of ecosystem services, associated with the recovery of the anchors. The June 2011 Orphan Anchor report (Gulf Coast Incident Management Team, 2011) noted that 46 anchors were buried at an average depth of 1.9–2.1 m, where the risk of damage to personal and commercial vessels was extremely low. There had been no reports of accidents involving these anchors, and decomposition of the steel and galvanized steel components was not considered to be a risk to humans, aquatic life, or the environment. The NEBA team recommended that leaving the remaining buried anchors in place to naturally degrade was the best option (Gulf Coast Incident Management Team. 2011). Apart from the buried anchor information, no additional data on the impact of booms on local habitat or ecosystem services exist.

### Hydrology Modification via Mississippi River Diversions

The Davis Pond Freshwater Diversion was built in 2004 to reduce salinity throughout the length of Barataria Bay, enhance fisheries, and add additional nutrients and sediment to the area wetlands.[8] The diversion was opened by the state of Louisiana on April 30, 2010, during the *DWH* oil spill to prevent oil from contaminating the marshes of Barataria Bay (Bianchi et al., 2011). The initial flow of 113 $m^3/s$ was increased to 212 $m^3/s$ on May 7 and reached its full capacity of about 300 $m^3/s$ on May 10—21 days after the *DWH* explosion.[9] In July, the discharge rate was generally maintained above 200 $m^3/s$. By September, discharge was reduced to <25 m3/s. The rapid and extreme changes in salinity caused 50 percent mortality in oyster beds in Louisiana (Food Channel Editor, 2011). Reductions of oyster recruitment and harvest were projected to continue over the next 3 years (Martínez et al., 2012). Mortality of the oyster beds and a decline in the oyster catch for 2012 through 2013 resulted in decreased income for oyster fisherman.

By mid May 2010, in an effort to prevent the intrusion of oil from the *DWH* spill into fragile coastal ecosystems, a number of freshwater diversions were in effect, many of them to the east of the Mississippi River. The total measurable flow from these diversions was 29,550 cubic feet per second (cfs) (Louisiana Office of Coastal Protection and Restoration, 2010).

- Bayou Lamoque Diversion: Plaquemines Parish, 7,500 cfs (capacity 12,000)
- Davis Pond Diversion: St. Charles Parish, 10,650 cfs (capacity 10,650)
- Violet Siphon: St. Bernard Parish, 200 cfs (capacity 200)
- Caernarvon Diversion: St. Bernard Parish, 8,000 cfs (capacity 8,800)
- Whites Ditch Siphon: Plaquemines Parish, 200 cfs (capacity 200)
- Naomi Siphon: Plaquemines Parish, 1,500 cfs (capacity 1,500)
- West Pointe a la Hache Siphon: Plaquemines Parish, 1,500 cfs (capacity 1,500)

[8] http://www.lca.gov/Projects/14/Default.aspx.

[9] USGS Davis Pond Freshwater Diversion: http://waterdata.usgs.gov/usa/nwis/uv?295501090190400.

The total water diverted to east of the river was 17,400 cfs, which amounts to less than 1 percent of the water flow down the lower river during flood years 2008–2010 (Allison et al., 2012). Dubravko Justić's FVCOM model grid (personal communication) was applied to Breton Sound, with incorporation of state-designed freshwater diversions to keep oil from entering Louisiana estuaries.

> ***Finding 4.7.*** *The effectiveness of mechanisms for near-shore and onshore protection of shorelines was negligible in terms of berms and freshwater diversions, but somewhat effective in terms of booms. Well-formed scientific advice was provided in advance that sand berms and diversions would be ineffective in keeping the oil from reaching the wetlands. Moving forward, decision analyses may be needed to refine guidelines and to optimize the effectiveness of various technologies and protocols. Additional research is needed to monitor the overall impacts of these response technologies on ecosystems and the services they provide.*

## TREATMENT OF SHORELINE OILED BY THE *DWH* OIL SPILL

Technological efforts offshore do not entirely contain the oil. Oil at sea very often will impact coastal shores if sensitive shorelines are not far away, and if ocean currents are likely to move oil slicks or dispersed oil to those areas. When shorelines and costal habitats are oiled, as they were in the *DWH* oil spill, shoreline cleanup processes are invoked.

### *DWH* Shoreline Cleanup Assessment Technique (SCAT)

In response to the *DWH* oil spill, NOAA formed a shoreline response team, Shoreline Cleanup Assessment Technique (SCAT), which relied upon the creation of a consistent data and knowledge base of the extent of oiling along types of shoreline habitat (Santner et al., 2011). Between 15 and 20 SCAT teams were deployed each day after May 4, 2010, and eventually determined the total length of oiled shoreline to be 1,053 miles. Teams of archaeologists (often as a part of the SCAT teams) also conducted field surveys of the entire impacted shoreline to locate known cultural resources, identify sites of potential cultural concern, and monitor operations to provide direction on minimizing impacts to important ecosystem services related to cultural resources.

The shoreline response team also created multiagency Core Groups, one for Louisiana and one for Mississippi, Alabama, and Florida to (a) set treatment priorities, (b) develop treatment recommendations for the various habitat types, (c) monitor the cleanup process, and (d) determine the effectiveness and impacts of treatment. Technical Working Groups were established within the Core Groups to deal with particular habitat issues and treatment recommendations and to establish "no further treatment" (NFT) criteria for sandy shorelines, coastal marshes and mangroves, and human-constructed shorelines. Their reports serve as appendixes to the Core Groups' final reports, "The Stage III SCAT-Shoreline Treatment Implementation Framework" (Santner et al., 2011). Their findings are described below.

## Shoreline Cleanup Strategies After the *DWH* Oil Spill

Barrier island shoreline treatment following the *DWH* oil spill typically involved manual or mechanical removal, because mixing, sediment relocation, washing, or chemical treatment each requires some form of regulatory approval or permitting (Owens et al., 2011). The differences in oiling conditions (surface and buried oil or oil layers, stained sand, and sunken oil in the adjacent subtidal waters) and sediment characteristics of the central Louisiana coast and the Gulf beaches of Alabama, Florida, and Mississippi precluded the use of a single technique as a panacea (Owens et al., 2011; Santner et al., 2011). Key considerations in choosing the cleanup methods for beaches were minimization of sand removal and therefore waste generation, minimization of restoration time for amenity beaches used for recreation, and maintenance of beach stability against storms.

The removal of bulk oil, mobile oil in intertidal areas that posed a threat to adjacent habitats or resources, from the intertidal areas along the northern GoM was repeated as necessary in areas of high environmental significance such as turtle-nesting areas, high-use tourist beaches, waterfront parks, and local residential areas (Santner et al., 2011). Amenity beaches that suffered recurring oiling from remobilized oil, or reworking of the shoreline by wind and wave action, were also subject to continued oil removal operations through the 2010–2011 winter (Santner et al., 2011).

Cleanup techniques for salt marshes and mangroves were natural attenuation (discussed below), low-pressure ambient-temperature flushing (to float the oil), mobile vacuum systems, securely deployed containment sorbents or snares, manual removal (on sand or shell substrates only), and vegetation cutting from boats for limited access to *Roseau* cane marshes only (see Appendix D in SCAT, 2010). In salt marsh habitats where there was little or no risk of repeated oiling, bulk oil removal was only done once on a limited scale, conducted from floating platforms, skiffs, or shallow-draft barges fitted with flushing and vacuum systems, adapted from concrete pump arms (Santner et al., 2011). These floating crafts were able to reach into oiled fringe wetlands to wash and recover mobile oil. When stranded oil was removed, it was primarily carried out by hand with sorbent material and by cutting oiled vegetation. The preferred oil spill response in salt marshes was natural attenuation (Santner et al., 2011).

## Manual Removal of *DWH* Oil

Human-constructed shorelines of riprap, breakwaters, groins (low walls or timber barriers extending into the sea from a beach to check erosion), and jetties were treated after the *DWH* oil spill through manual removal of bulk oil, followed by washing using a range of temperatures and pressures (Santner et al., 2011). Manual equipment included long-handle hand-mesh and screens, pitchforks with screens, pool nets for surface residue balls along the water line, and mechanical adaptations such as rotary screens and welded mesh screens for extended-reach backhoes working with surface residue balls and patties in water-saturated sand (Owens et al., 2011).

## Challenges to Cleaning Sandy Beaches

Re-oiling during storm events (as observed after Hurricane Isaac in August, 2012; Clement et al., 2012) exposed buried oil, and chronic deposition of new oil occurred from oily sediment mats in the lower intertidal and near-shore subtidal zones, which created challenging issues for cleanup operations (Santner et al., 2011). Oil became stranded in the supratidal zone during storms and buried up to 1.5 m deep in beaches, thus requiring extensive excavation, especially on amenity beaches. Further challenges included delineation of buried oil over large areas, both along-shore and cross-shore. Submerged oil mats were widespread in the lower intertidal zone, but they were difficult to clean because they became covered by high water and clean sand over time (OSAT, 2011b; Santner et al., 2011). The use of heavy equipment was limited because of concerns that mechanical methods would result in increased beach erosion or because of restrictions and prohibitions on the use of mechanical equipment on remote barrier islands. An NFT guideline of no visible oil above background on amenity beaches was a difficult, but required, endpoint.

## Sand Washing

A fixed sand-washing system, constructed with a shaker sieve to remove large surface residue balls and patties along with debris, as well as two heated wash units, was used on Grand Isle, Louisiana, and proved to be successful (Owens et al., 2011). The small amount of residual oil remaining in the treated sediments from this procedure was removed by surf-washing operations.

## Surf Washing

Three field (and several small-scale) demonstrations of surf washing (i.e., enhanced dispersion of oil by the formation of oil-mineral aggregates [OMAs]) were conducted on Grand Isle, Louisiana (Owens et al., 2011). The demonstrations included sediment and water chemistry analysis before and after the event, and sometimes beach profiling, to determine the efficacy of the technique as a sand-polishing step after the removal of bulk oil by sieving or sand washing. Surf washing by relocation of sediment to the lower intertidal zone did not cause significant sediment loss, nor did the technology increase hydrocarbon concentrations in intertidal or subtidal sediments or water. The decision was made to use this technique, and more than 30,000 cubic yards were surf washed within a week. Although one small-scale demonstration in Escambia, Florida, led to an initial decision against surf washing because of strong longshore currents, sediment drift, and little residual oil (Owens et al., 2011), operations were resumed at a later date to treat residual oil in the beach sediments.

***Finding 4.8.*** *The impacts of shoreline cleanup operations have not yet been determined. Cleanup efforts were inconsistently successful because of storms, shifting sediments, vari-*

*able beach habitats, and variable oiling profiles along the northern GoM coasts. Lightly oiled beaches on barrier islands were the easiest to clean and restore.*

## Salt Marsh Impacts from Cleanup Operations

Physical destruction of marsh habitat during cleanup operations is the most common concern, but virtually all options will cause some damage to marshes during cleanup (Aldhous and Hecht, 2010). Fertilizer, such as phosphorus to encourage regrowth of oiled *Spartina alterniflora*, was not applied to encourage marsh recovery, probably because of concerns about root damage as described in studies of nutrient overenrichment of marsh habitats (e.g., Darby and Turner, 2008; Deegan et al., 2012; Turner et al., 2009). Root damage would reduce the power of the marsh sod to hold together, resulting in greater erosion (Aldhous and Hecht, 2010).

However, there is evidence to support the application of nutrients to enhance the remediation of oiled wetlands dominated by *Spartina alterniflora* and *Spartina patens* (Lin and Mendelssohn, 1998). In the case of the *DWH* oil spill, it has been suggested that oiled Gulf Coast marshes will recover by natural attenuation because prior research has demonstrated their intrinsic resilience (DeLaune and Wright, 2011). Natural attenuation was the preferred option in the case of the *DWH* oil spill (Santner et al., 2011).

Aeration from tidal action, along with the addition of nitrogen in the form of ammonia, has been shown to significantly increase oil biodegradation in Louisiana salt marsh sediments (Jackson and Pardue, 1997; Shin and Pardue, 2001a,b; Shin et al., 2000, 2001). Anaerobic biodegradation of oil in marsh sediments can be enhanced in the presence of mixed electron acceptors (sulfate, nitrate). This enhancement occurs in Louisiana marsh sediments where various electron-accepting anaerobes that degrade petroleum hydrocarbons coexist (Boopathy et al., 2012). The recovery rate will depend on the extent of oiling, depth of oil penetration into the sediments, and types of plant species affected. There is already evidence that new shoots are appearing in heavily oiled areas, so the long-term effects may be limited (DeLaune and Wright, 2011). Research is under way evaluating the long-term impacts of the Macondo well oil on coastal ecosystem health and function in the GoM.[10]

## Risk Analysis for Shoreline Cleanup Operations

The current understanding of impacts of residual near-shore oil from the spill on organisms including plants, invertebrates, fish, birds, mammals, and sea turtles has been addressed in the Operational Science Advisory Team report (OSAT, 2011a) and is summarized below. Direct exposure may occur through incidental ingestion when feeding on sand beaches, or through physical contact. Aquatic organisms can be indirectly exposed to toxins by diffusion of dissolved oil fractions across gills and cell membranes, or directly exposed by feeding on suspended oil residues.

[10] http://gulfresearchinitiative.org/.

OSAT considered the risk of acute toxicity to fish and invertebrates to be unlikely, but there is potential for chronic exposure to dibenzothiophenes and dibenzofuran from the water-accommodated fraction. The dynamic nature of the surf zone and subtidal areas is expected to rapidly dilute petroleum concentrations to insignificant levels, and toxicologically significant petroleum concentrations would be expected to be localized in the surface microlayer adjacent to submerged oil mats or surface residue balls.

OSAT considered the risk to sea turtles from oil residues on top of the beach to be low, but it identified a potential difficulty for turtles attempting to nest by digging through subsurface tar mats (OSAT, 2011a). Beach cleanup and spill monitoring activities also impacted sea turtle nesting activities, but to what degree remains unknown. The toxicity of weathered oil to turtle eggs was assumed to be low (OSAT, 2011a). The oil is degrading and becoming less toxic, but media reports have claimed that more than 1,000 turtles, 70 marine mammals, and possibly hundreds more marine mammals (see Chapter 5), and 4,000 birds are thought to have died as a consequence of the *DWH* oil spill (Service, 2010). These numbers may be significantly underestimated. Neither short-term nor long-term risks nor damage to juvenile forms of these animals is known at this time. The release of the damage assessment data is expected to support more accurate quantification of these impacts.

OSAT (2011a) determined that the risks to birds are mostly applicable to their foraging along the beach and inadvertently ingesting oil. Risk appears low for the piping plover, but low to medium for the western sandpiper based on its foraging behavior. Risks to eggs of beach-nesting birds are possible but unknown. Risks from cleanup activities are medium to high. Beach-dwelling mammals may be exposed to surface residue balls by incidental ingestion, or if they build burrows in the vicinity of supratidal buried oil (OSAT, 2011a). OSAT determined the risk of oil exposure for the federally endangered Alabama beach mouse (*Peromyscus polionotus ammobates*) to be low. That said, spill response actions, such as cleanup activities and increased human traffic on the beach, may have altered, reduced, or damaged the mouse's habitat and therefore increased the threat to this species.

Results from the OSAT 2 NEBA analysis indicated that the environmental effects of the residual oil following cleanup are relatively minor, especially considering the pre-spill baseline level of oiling (OSAT, 2011b). Analyses of water and sediment samples along the northwestern Florida coastline have shown that swimming at or visiting the beaches poses no risk to human health from oil spill hydrocarbons (Florida Department of Environmental Protection, 2011). If cleanup were continued, then it would be expected to result in an increasingly greater extent of damage to habitats and associated resources. Consideration must also be given to the use of beaches for recreation in the spring and summer, which would conflict with cleanup operations (Florida Department of Environmental Protection, 2011).

## NATURAL ATTENUATION

Natural attenuation is the "reduction in mass or concentration of a contaminant in the environment over time or distance from its source of release due to naturally occurring physical, chemical and biological processes, such as biodegradation, dispersion, dilution, adsorption, and volatilization" (American Society for Testing and Materials, 2010). The natural attenuation of oil can be defined as the biotic and abiotic degradation and dispersion of oil that results in natural recovery of an oil-impacted environment. When oil enters the marine environment, abiotic weathering processes (evaporation to the air, dissolution in water, emulsification with water, dispersion, and photodegradation) (see Appendix G in SCAT, 2010) alter properties of the oil (density, viscosity, water content, surface and interfacial tensions), which ultimately define its fate.

In the natural environment, oil droplets do not persist intact. Some of the lighter compounds in the oil dissolve to become the water-accommodated fraction. A number of droplets will recombine or coalesce, rising back to the surface. Others undergo colonization by microbes and are biodegraded.

Oil droplets can attract and incorporate suspended particulate matter from the water column (Owens and Lee, 2003), surface water outflows, or stirred up sediment. These OMAs are an important factor controlling the fate and transport of oil spilled in the marine environment (Bassin and Ichiye, 1977; Boehm, 1987; Karickhoff, 1981; Payne et al., 1989). The parameters controlling the quantity, type, and size of OMA include mineral type and surface properties, quantity of mineral fines, oil viscosity and composition, and oil/mineral ratio (Stoffyn-Egli and Lee, 2002). Water turbulence (breaking waves, strong flood currents) greatly enhances OMA formation.

The concept of using enhanced OMA formation as an operational strategy for the cleanup of oil stranded on coastal beaches came about from laboratory studies of oiled sediments from the *Exxon Valdez* oil spill (Bragg and Yang, 1995; Bragg et al., 1994). These studies revealed that micron-sized mineral fines, seawater, and weathered oil interacted to form clay-oil flocculations (a type of OMA) that promoted the release of oil into the water column. This process was hypothesized to be responsible for the natural cleaning in sheltered shorelines of Prince William Sound after the oil spill. Enhanced OMA production by surf-washing operations has now been recognized as a strategy to remove oil stranded within beach sediments (Lee et al., 1997, 1999; Lunel et al., 1996, 1997; Owens, 1999; Owens and Lee, 2003; Swannell et al., 1999; see section, "Treatment of Shoreline Oiled by the *DWH* Oil Spill," subsection "Surf Washing").

## Biodegradation

A large number of microorganisms are capable of biodegrading hydrocarbons (Atlas, 1984; NRC, 2003: ZoBell, 1973), and bacteria are the predominant hydrocarbon degraders in the marine environment (Atlas and Hazen, 2011). Biodegradation by microbial communities is the major process controlling the eventual removal of oil that enters the marine environment from natural seeps (Atlas, 1995; Atlas and Bartha, 1992; Leahy and Colwell, 1990). Although much slower, anaerobic (oxygen absent) biodegradation of oil should not be underestimated as a strategy, because it has been shown to be a major process in anoxic marine sediments, where sulfate, nitrate, manganese (IV), and iron (III) are the primary terminal electron-acceptors (Canfield et al., 2005; Finke et al., 2007; Lovley et al., 1997). Although normally present in small numbers in pristine environments, oil-degrading microbes can multiply rapidly upon the introduction of oil (Atlas, 1995; Horel et al., 2012), as was the situation in the GoM after the *DWH* oil spill (Hazen et al., 2010).

The individual components in crude oil do not biodegrade at the same rate. Biodegradation rates are highest for the lightest components such as alkanes, followed by light aromatics (up to three rings), with high-molecular-weight aromatics (four rings and higher) and polar compounds exhibiting extremely low rates (Prince, 2010). Because of the chemical complexity of crude oil, its biodegradation may require the cooperation of a consortium of microbes or syntrophic interspecies (McInerney et al., 2008). Some microorganisms may also naturally produce extracellular biosurfactants that promote emulsification of hydrophobic hydrocarbons for subsequent transport into the hydrophilic intracellular space for biodegradation (Desai and Banat, 1997; Southam et al., 2001). Microbes are well adapted to their surroundings, including extremely low temperatures (Delille et al., 2009; McFarlin et al., 2011a), and Hazen et al. (2010) confirmed that oil in the low-temperature, deep-water environments of the GoM may be degraded by indigenous microorganisms at higher rates than previously thought.

***Finding 4.9.*** *Abiotic weathering processes and microbial degradation are major processes controlling the eventual removal of oil that enters marine and coastal environments. Thus, natural attenuation of some oil will occur without technological interventions.*

## Natural Attenuation Following the *DWH* Oil Spill

Regardless of the response technique employed, some oil will remain in the environment. For the *DWH* event, because of the volume of oil spilled and the natural rates of oil weathering, oil reached near-shore waters and coastal habitats, to which a range of removal technologies were applied. With the exception of buried oil that was reexposed after storms (e.g., Hurricane Isaac in 2012) and the associated wave surge, to which additional cleanup activities were applied, the fate of this residual oil was left to natural attenuation. The *DWH* oil spill highlighted the significance of natural oil degradation. Marine bacteria may have been primarily responsible for the observed loss of residual oil evident in the GoM (American Society of Microbiology, 2011; Walsh, 2010).

Unlike oil spills occurring at the sea surface, much of the petroleum hydrocarbons erupting from the Macondo wellhead experienced a prolonged, buoyancy-driven ascent through the 1,500-m water column (Hazen et al., 2010). Hazen et al. (2010) reported that the disappearance of residual oil in the GoM water column from the *DWH* oil spill was associated with microbial degradation processes. Genes for chemotaxis, motility, and aliphatic hydrocarbon degradation were found to be elevated in samples taken from inside the *DWH* oil plume compared to samples taken from outside of the plume, and cells of *Oceanospirillales* bacteria isolated from plume samples were found to possess gene sequences for the degradation of *n*-alkane and near-complete oxidation of cyclohexane (Mason et al., 2012). Temperature did not appear to be a major limiting factor because significant rates of oil degradation were observed within the subsurface plume of dispersed oil at a depth of 1,100- to 1,200-m and temperatures below 4°C (Hazen et al., 2010). Some estimates indicate that as much as 30 percent of the oil and gas never reached the sea surface, but instead formed hydrocarbon-rich plumes within the cold waters present at 1,100–1,200 m depth (Hazen et al., 2010; Valentine et al., 2010).

An examination of petroleum hydrocarbons in oil mousse (emulsion of water and oil) collected between May and July 2010 from the seafloor and the sea surface, and as close as 6 km and as distant as 185 km from the wellhead (Liu et al., 2012), showed only light to moderate degradation of oil deposited in sediments near the wellhead (Liu et al., 2012). This contrasted considerably with the extensive weathering of oil observed in mousse samples in the marshes. Trace metal concentrations, particularly of aluminum, iron, and manganese, increased significantly from *DWH* crude oil, to surface mousse, and then to the marsh sediment, suggesting that clay minerals and dissolved metals were aggregated during oil weathering and mousse transportation. The fate of the oil in the subsurface plume remains an active field of research.

> ***Finding 4.10.*** *The fate and transformations of oil from the* DWH *oil spill were and will continue to be influenced by natural attenuation processes. These processes may be considered the best practical methods for oil removal in sensitive habitats, where the application of technology may cause more harm, or in logistically inaccessible habitats. Continued observations, monitoring, and research of these processes, especially in relation to technological countermeasures, are warranted.*

Because biodegradation of all components within crude oil is not a rapid process, reliance on natural attenuation as the sole strategy might result in reduced provisioning of finfish and shellfish because of impacts to juveniles in spawning and nursery grounds, and a corresponding loss of commercial fishery, recreational fishing, and tourism income (discussed further in Chapter 5, "Fisheries" subsection). Thus far, however, the documented losses of ecosystem services after the *DWH* oil spill have been due to fisheries closures and public fears of consuming contaminated seafood, rather than reduced juvenile recruitment (McGill, 2011; Upton, 2011).

The longer-term impacts of selecting natural attenuation over more proactive spill responses to the ecosystem services provided by the deep-sea GoM, such as provisioning of bluefin tuna and royal red shrimp, recreational whale watching, and the provisioning of oil and gas, are uncertain (see Chapter 5). Regulating and cultural services provided by larger animals such as turtles and bottlenose dolphins (see Chapter 5) may also be adversely impacted. In addition, wetland losses due to the inundation and smothering of marsh plants have affected the ecosystem services provided by these habitats (additional discussion found in Chapter 5).

## SUMMARY

A number of spill response technologies and techniques were applied during the *DWH* oil spill, including direct recovery from the wellhead, skimming, in situ burning, chemical dispersion, flushing (river diversion), and erection of physical barriers. In deeper water, enhanced dispersion via the use of chemical dispersants was one of the most effective response methods in terms of impacts on total oil volume. Skimming and burning removed about 7–10 percent of total oil. Once the oil made it to shore, cleanup efforts depended upon the type of shoreline.

To clean oil from beaches, sand washing and surf washing were performed. To clean oil from marshes, physical removal of the oil by manual labor, floating machinery, and manual cropping of oiled vegetation were employed.

Almost all countermeasures and response options have limitations and environmental effects that ultimately will impact ecosystem services. If the goal of response efforts is to protect the environment that sustains the ecosystem services, then we must realize that prioritization among the services involves complex decisions and the selection and application of response technologies involves tradeoffs. Most decisions regarding the application of spill response technologies after the *DWH* oil spill were based on NEBA,[11] and they were made in consideration of the best available information at the time.

The immediate goal of the response technologies applied during the *DWH* oil spill was to prevent oil from reaching sensitive coastal habitats. Although none of the response tools was completely effective, the use of in situ burning, booms, skimmers, and particularly dispersants significantly mitigated the impacts of the spill along the coast. Continued remediation efforts will eventually result in diminishing returns, at which time natural attenuation of the remaining oil may be the best option. To this end, more science needs to be incorporated into the remediation process (both on- and offshore), not only to determine the most effective methods to apply under various conditions, but also to feed into the NEBA process of ascertaining the operational endpoint. Regarding the *DWH* oil spill, two factors raised important research questions: (1) the depth of the spill and (2) the large-scale application of emerging new technologies, such as subsurface injection of dispersants and oceanic in situ burning.

At the time of this writing, the consequences of the technologies applied to contain the oil and to prevent its reach into sensitive coastal and wetland habitats, as well as the response of the GoM ecosystem to the spill, have not been, and may never be, fully documented. Work to understand the fate and effects of the spilled oil is ongoing. Much work to determine the long-term effects remains. Understanding of the potential impacts of response technologies to ecosystem services and of the impacts of oil on multiple layers of ecosystems and their services is limited by the availability of baseline information and of important data collected during the NRDA process. Until this massive and unprecedented collection of data is complete and the results made available, any assessment of the impacts of the response technologies will not be fully informed. These issues will be discussed further in Chapter 6.

***Finding 4-11.*** *Any conclusive assessment of the impacts of response technologies on the GoM marine ecosystem and ecosystem services would be premature at this time, given the amount of key data and analyses that is being held confidentially in the ongoing NRDA process. Nonetheless, the committee believes that the technologies applied offshore mitigated the impacts of the oil spill on sensitive coastal habitats. A further factor for consideration is that only 3 years have passed since the spill started, and many substantial impacts may not become apparent for several more years.*

[11] http://www.crrc.unh.edu/workshops/dwh_researchneeds_11/index.html.

CHAPTER FIVE

# *Ecosystem Services in the Gulf of Mexico*

## INTRODUCTION

In Chapter 2 we outlined the general concept of ecosystem services and the basic principles and challenges of applying an ecosystem services approach to damage assessment for an event of the magnitude and duration of the *Deepwater Horizon* (*DWH*) oil spill. Chapter 3 explored the concept of resilience in the context of ecosystem services and the challenges faced by managers in attempting to restore or increase the resilience of the Gulf of Mexico (GoM) ecosystem. Chapter 4 reviewed the response technologies used during and after the *DWH* oil spill and their impacts on GoM ecosystem services. This chapter brings the discussion of ecosystem services into focus by examining in more detail the specific ecosystem services provided by the GoM. The chapter begins by considering the characterization of GoM ecosystem services within a geospatial context and how ecosystem services vary as a function of scale and in response to changes in the physical and environmental setting.

The remainder of the chapter is dedicated to presentation of four case studies representing each of the primary ecosystem service types—supporting, regulating, provisioning, and cultural—and chosen to capture the opportunities and challenges that emerge when applying the ecosystem services approach to assessing the impact of the *DWH* spill on the GoM. For each of these case studies, the committee identifies key ecosystem services, considers how they may have been impacted by the *DWH* oil spill, examines methods for taking baseline measurements, and explores the adequacy of existing baseline data for the GoM. Additionally, the committee offers suggestions for additional measurements that can enhance an ecosystem services approach to damage assessment (Tables 5.5, 5.6, 5.7, and 5.8).

## ECOSYSTEM SERVICES IN THE GULF OF MEXICO

As a starting point for examining ecosystem services specific to the GoM, the committee utilized a list of GoM ecosystem services that was developed by a panel of regional experts during a workshop convened in Bay St. Louis, Mississippi, in June 2010 (Yoskowitz et al., 2010). The workshop panel, composed of representatives from academic institutions, nongovernmental organizations, the private sector, and state and federal agencies, defined ecosystem services as the "contributions from GoM marine and coastal ecosystems that support, sustain and enrich human life." As shown in Table 5.1, the panel identified 19 ecosystem services provided by the GoM natural infrastructure and grouped them under the four primary types of ecosystem ser-

**TABLE 5.1** Gulf of Mexico Ecosystem Services by Millennium Ecosystem Assessment Category

| |
|---|
| **Supporting services** |
| Nutrient balance |
| Hydrological balance |
| Biological interactions |
| Soil and sediment balance |
| **Regulating services** |
| Pollutant attenuation |
| Water quality |
| Gas regulation |
| Climate regulation |
| Hazard moderation |
| **Provisioning services** |
| Air supply |
| Water quantity |
| Food |
| Raw materials |
| Medicinal resources |
| Ornamental resources |
| **Cultural services** |
| Aesthetics and existence |
| Spiritual and historic |
| Science and education |
| Recreational opportunities |

vices (supporting, regulating, provisioning, and cultural) defined by the Millennium Ecosystem Assessment (MEA, 2005).

Panelists in the Bay St. Louis workshop (Yoskowitz et al., 2010) recognized that the coastal and marine habitats of the GoM also constitute a natural infrastructure that contributes to the provisioning of ecosystem services. When ecological production functions are not well understood, integrated assessments of ecosystem services tend to use natural structures such as habitats to map the complex interactions of different components of the ecosystem. The scale for assessing an ecosystem service must be determined by the threshold at which changes in ecosystem functioning (or its habitats) can be detected (measured) and at which the ecosystem sustains functions that contribute to its resilience (as discussed in Chapter 3). Table 5.2 organizes a number of important GoM ecosystem services by habitat, which could be used to guide efforts in delineating and determining changes in ecosystem services after the *DWH* oil spill.

Ecosystem services can also be classified according to their spatial characteristics (see Table 5.3). Each of the 19 ecosystem services provided by the GoM can be mapped to at least one of the five different spatial classes (global nonproximal, local proximal, directional flow-related, in situ or point of use, and user movement-related) proposed by Costanza (2008). For example,

**TABLE 5.2** Synthesis of Services Provided by the Gulf of Mexico by Service Category and Habitat

| Ecosystem Service Category | Habitat | Example in the GoM |
|---|---|---|
| ***Supporting services*** | | |
| Soil and sediment balance | Brackish marsh | Upper Barataria estuary, Louisiana |
| | Dunes/beaches | Barrier islands, Texas |
| | Forested coastal ridge | Chenier Forest/Woodlands, Louisiana |
| | Intertidal sediments | Mud flats in Laguna Madre, Texas |
| | Subtidal sediments | Widespread throughout the Gulf |
| | Mangroves | Everglades, Florida |
| Nutrient regulation | Brackish marsh | Upper Barataria estuary, Louisiana |
| | Freshwater marsh | Rockefeller State Wildlife Refuge, Louisiana |
| | Macroalgae | Floating and beached Sargasso |
| | Swamp/bottomland hardwood | Maurepas Swamp, Louisiana |
| | Subtidal sediments | Widespread throughout the Gulf |
| ***Regulating services*** | | |
| Water quality | Oyster reef | Mobile Bay, Alabama |
| | Seagrass | Redfish Bay, Texas |
| Hazard moderation | Oyster reef | Barataria Bay, Louisiana |
| | Salt marsh | Mississippi River Delta, Louisiana |
| | Freshwater marsh | Barrier island freshwater marshes, Texas |
| | Swamp/bottomland hardwood | Sabine River floodplain swamp, Texas and Louisiana |
| | Dunes/beaches | South Pacific Island, Texas |
| | Forested coastal ridge | Chenier Forest/Woodlands, Louisiana |
| | Mangroves | Everglades, Florida |
| ***Provisioning services*** | | |
| Food | Oyster reef | Galveston Bay, Texas |
| | Seagrass | Laguna Madre, Texas |
| | Open water | Widespread throughout the Gulf |
| | Offshore shoals and banks | Sabine Bank, Texas and Louisiana |
| | Subtidal sediments | Widespread throughout the Gulf |
| Raw materials | Oil and gas fields/reservoirs | Shelf/slope of central, western planning areas of the Gulf |
| | Offshore shoals and banks | Sabine Bank, Texas and Louisiana |
| ***Cultural services*** | | |
| Aesthetics and existence | Spiritual and historic | Shell middens throughout the Gulf |
| | Coral reefs | Florida Keys National Marine Sanctuary, Florida |
| | Dunes/beaches | St. George Island State Park, Florida |

*continued*

**TABLE 5.2** Continued

| Ecosystem Service Category | Habitat | Example in the GoM |
|---|---|---|
| Recreational opportunities and tourism | Coral reefs | Florida Keys National Marine Sanctuary, Florida |
| | Salt marsh | Northern Barataria Bay, Louisiana |
| | Forested coastal ridge | Grand Isle, Louisiana |
| | Intertidal sediments | Bays and estuaries anywhere in the Gulf |
| | Open water | Widespread throughout the Gulf |
| | Offshore shoals and banks | Florida Middle Grounds, Florida |
| Science and Education | | Widespread throughout the Gulf |

SOURCE: Modified from Yoskowitz et al., 2010.

**TABLE 5.3** Gulf of Mexico Ecosystem Services by Spatial Characteristics

| |
|---|
| **Global nonproximal (does not depend on proximity)** |
| Climate balance |
| Gas balance |
| Air supply |
| Existence |
| Spiritual and historic |
| **Local proximal (depends on proximity)** |
| Hazard moderation |
| Pollutant attenuation |
| Biological interactions |
| **Directional flow-related: flows from point of production to point of use** |
| Water quality |
| Water quantity |
| Sediment balance |
| Nutrient balance |
| Hydrological balance |
| ***In situ* (point of use)** |
| Soil balance |
| Food |
| Raw materials |
| Ornamental resources |
| **User movement-related: flow of people to unique natural features** |
| Medicinal resources |
| Recreational opportunities |
| Aesthetic |
| Science and education |

a service such as gas balance (an intermediate input to climate regulation) is classified as "global nonproximal" because carbon sequestration occurs across the entire GoM and beyond. "Local proximal" services, on the other hand, are dependent on the spatial proximity of the ecosystem service to the human beneficiaries. For example, "hazard moderation" requires that the ecosystem service be proximal to the human settlements and assets being protected. "Directional flow-related" services are dependent on the flow from upstream to downstream, as is the case for water quality and water quantity.

An examination of the spatial relationships of ecosystem services highlights the need to identify and manage natural infrastructure at scales that improve its ability to withstand chronic or acute impacts, such as those associated with the *DWH* oil spill. If the natural infrastructure is damaged, then human communities that benefit from the services provided by these ecosystems will likely become more vulnerable, therefore decreasing their overall well-being and resilience. Any degradation of the ecosystem structure and function may lead to a reduction in the supply of the essential services that will be needed to help communities build the resilience needed to withstand damages and recover after the spill (see discussion of socioeconomic resilience in Chapter 3). Examples of spatially specific benefits in the GoM region that could be severely impacted by an oil spill include storm mitigation by coastal wetlands and food provisioning by commercial fisheries. Consideration of the spatial characteristics of ecosystem services is important not only when assessing damages, but also when deciding which services to restore. Tradeoffs will be necessary, and recognizing which communities or sectors may benefit (or lose) from restoration efforts will be useful when priorities are set by the Trustees and stakeholders.

## CASE STUDIES

The committee conducted four case studies to explore specific GoM ecosystem services in detail, to highlight some of the opportunities and challenges that emerge when applying an ecosystem services approach to damage assessment, and to demonstrate how to apply such an approach under various conditions and across wide levels of understanding regarding the services in question. These conditions and levels include the amount and utility of available data, the value of the service in market and nonmarket terms, and the range of the impacts of the spill on the services. In addition, the selected ecosystem services were considered in the context of the linkages between ecosystem services and the constituents of well-being identified in Chapter 2 (see Figure 2.1).

Coastal wetlands, which cover a large region in the northern GoM, are the subject of the first case study. Half of the nation's coastal wetlands are found along the GoM, and, of these, approximately 40 percent are in Louisiana. Unfortunately, many wetlands were among the closest land points (only about 40 miles) from the Macondo well (NOAA, 2012b). Coastal wetlands, including salt marshes and mangrove plant communities, provide a wealth of supporting, regulating, provisioning, and cultural services that include maintenance of soil and sediment (shoreline stabilization), regulation of nutrients and water quality, provisioning of food, recreational opportunities, and hazard moderation (Barbier et al., 2011; Shepard et al., 2011). This case study focuses primarily on the regulating service of hazard moderation (specifically storm mitigation) to illustrate the opportunity that exists in using the ecosystem services approach when the underlying ecosystem science and the particular ecosystem service are well known and supported by a rich literature. The ecosystem service of storm mitigation also benefits from having been monetized—that is, the costs of storm damage and reductions in losses due to wetland buffers can be quantified.

The second case study focuses on fisheries, a provisioning service with a rich literature about its valuation and assessment. In recent times, this service has been considered to be a good candidate for a holistic integration of management at an ecosystem scale that includes both ecological and human components. Although the field has been developing this integration by promoting an ecosystem approach to fisheries management, it does not yet consider ecosystem services as a guiding principle. Fisheries, however, offer many examples of quantification of human impacts on the ecosystem structure and of ecological and economical productivity. This case study specifically explores the provision of seafood by the GoM and how the ecosystem services approach may help to quantify the possible impacts of oil spills on seafood provision.

Bottlenose dolphins were chosen as the subject for the third case study for numerous reasons, including their role in three of the four types of ecosystem services—regulating, supporting, and particularly cultural. This case study allowed for the exploration of approaches to estimating the value of passive use and existence—a key, but difficult-to-establish, metric for cultural ecosystem services. The stranding of many dolphins in the GoM before, during, and especially after the *DWH* spill has stimulated considerable public concern, which speaks to our cultural needs and sensitivities regarding their value as an ecological resource and ecosystem service.

Bottlenose dolphins are capable of self-recognition, which ranks them highly on a cognitive scale (Reiss and Marino, 2001). As apex predators, a role they share with humans, they play a role in regulating the GoM food web and their health and well-being serve as important indicators of the health of the GoM and oceans in general. More is known about this species than virtually any other cetacean. The world's longest-running study of a wild dolphin population, spanning five generations, focuses on Sarasota Bay in the eastern GoM (Wells, 2003). Their position as the most studied and arguably the most popular and charismatic marine mammal makes them a centerpiece for conservation science, education, and ecotourism.

Finally, the deep GoM was selected as the subject for the last case study in part because of its location with respect to the *DWH* blowout and spill. The deep sea was also selected because of increasing concern about the risk posed by the energy industry's activities as it employs the cutting edge of engineering in the most poorly understood of the impacted habitats. The biota of the surrounding seafloor at 1,500-m depth and of the water column through which a plume of hydrocarbons and dispersants flowed received the most immediate impact of the uncontrolled discharge. Although there has been some habitat mapping of the deep GoM, the data are quite sparse and the ecological consequence of a spill is incompletely understood.

# Wetlands

## Introduction

Coastal wetlands, including salt marshes and mangroves, provide a wealth of supporting, regulating, provisioning, and cultural services that include soil and sediment maintenance (shoreline stabilization), nutrient regulation, water quality regulation, provision of food and other biological resources, recreational opportunities, and hazard moderation (Barbier et al., 2011; Shepard et al., 2011). The marshes of the Mississippi River Delta comprise almost 40 percent of the coastal wetlands in the 48 contiguous states, support 30 percent of the nation's commercial fishery production, and protect important oil and gas reserves and refineries (Mendelssohn et al., 2012). During the *DWH* oil spill, coastal salt marshes were significantly affected, with 1,100 linear miles of wetland impacted at some time during the event (NOAA, 2012b). Crude oil can smother vegetation by coating leaf surfaces and can cause toxic effects, particularly from the light fractions of the oil that are more water soluble.

With wetlands, the values of some of its ecosystem services, such as storm mitigation, can be quantified within an order of magnitude, while for others, such as nursery habitat for commercial marine resources, the values are much more difficult to quantify. The following discussion focuses on hazard mitigation because it represents an example of ready application of the ecosystem services approach using the existing knowledge base.

## Regulating Services

*Hazard Moderation* Several wetland characteristics are positively correlated with the regulating ecosystem services of both wave attenuation and shoreline stabilization, including vegetation density, biomass production, and marsh size (Shepard et al., 2011).

The topography of wetlands (including plant architecture) provides enhanced friction, which tends to decrease wind speed, wave height, storm-driven steady currents, and storm-surge height. Wetlands can also decrease tropical storm intensity by inhibiting the transfer of heat from the ocean to the atmosphere. It is this latter energy transfer that serves as the basic engine that drives a tropical storm.

Reduction of wave energy depends on the structure of the plant canopy, its height and density, and the cross-shore and along-shore extent of the wetland (Koch et al., 2009; Krauss et al., 2009; Massel et al., 1999; Narayan and Kumar, 2006; Shepard et al., 2011; Vosse, 2008). The velocity of water traveling within a plant canopy is relatively lower than above the canopy. Canopy height in relation to water depth is relevant because water flowing through the vegetation encounters a higher friction than does the water above the vegetation. Therefore, the total friction in the water column will change with the depth of vegetated and nonvegetated areas. Because a mangrove canopy is taller and exerts more drag than a salt marsh community, mangroves are more effective at reducing water inflow and waves than are salt marshes. Quartel et al. (2007) suggested that the drag force exerted by a mangrove forest can be approximated by the function $CD = 0.6e^{0.15A}$, where CD is the coefficient of drag and *A* is the projected cross-sectional area of the submerged canopy. For the same muddy surface without mangroves, the drag is a constant 0.6. Mazda et al. (1997) observed that a 100-m-wide strip of mangrove forest was capable of reducing wave energy by 20 percent. Reduction in water levels across a mangrove area in Florida was 9.4 cm/km (Krauss et al., 2009).

Wave energy is also affected by topography. In a modeling study of sea level rise and storm surge across the Louisiana coast, Vosse (2008) found that when the relative land elevation was decreased by 20 cm and 50 cm, wave heights increased 5–10 cm and 10–20 cm, respectively, across the model domain. The conclusion is that friction by the plant canopy dissipates energy and reduces wave heights, but the effect of the wetland surface depends on water depth. Consequently, the relative elevation of a coastal wetland, not simply its presence or absence and structure, is a determinant of its effectiveness in storm hazard mitigation.

Like waves, storm surge can be reduced by the presence of wetlands because of the increased dissipation. The U.S. Federal Emergency Management Agency (FEMA) provides estimates of the impact of various types of wetland vegetation on frictional dissipation (FEMA, 1985). United Research Services (URS) used the FEMA estimates and a well-validated numerical model to examine the impact of vegetation on surge height. It found a 25 percent reduction in the inland surge simply by assuming the marshland was composed of long grass instead of short grass (Ayres Associates, 2008). Along with these reductions in surge height are substantial secondary benefits. For example, the current velocity is reduced and so is the wave height because larger waves require higher surge, all else being equal. Finally, a reduction in wave height reduces the wave setup, which is a contributor to the surge.

Although there is ample evidence that wetlands reduce surges and waves, the evidence is less clear for winds. One problem is that storm winds (as well as surges and waves) can severely alter the vegetation during the course of a storm. Dingler et al. (1995) discuss this issue, using data collected during Hurricane Andrew, which hit the Mississippi River Delta in 1992. They found that wind dissipation was higher over the wetlands than over the ocean, but only when wind speeds exceeded 20 m/s. This effect occurred despite the fact that the wetlands in the study area were composed primarily of *Spartina* and bulrushes, which would have "flattened" during the stronger winds. Somewhat in contrast to Dingler et al. (1995), FEMA (1985) suggested factors that show a substantial increase in wind dissipation over open-ocean values for all types of wetlands and wind speeds. In total contradiction of the other two researchers, Speck (2003) showed a decrease in dissipation with rising wind speed above 1 m/s over 4-m reeds compared to open ocean. In short, no consistent picture emerges from the research regarding wind dissipation, which is no doubt partially due to the facts that the vegetation type in these studies was highly variable and that the dissipation effectiveness varies over the course of a storm as vegetation is damaged by wind, wave, and current.

Wetlands may also have an effect on tropical storms by mitigating storm intensity. As explained by Emanuel (1987), the intensity of a tropical cyclone is driven by a Carnot cycle (a thermodynamic cycle) that requires warm, humid air near the sea surface. Anything that disrupts that supply will reduce storm intensity. Wetlands clearly have that potential, especially if they contain a high percentage of vegetation. Cubukcu et al. (2000) used a numerical model to examine the effects of land and found a significant weakening in the surface winds in large part because the storm entrains much drier air. However, Shen et al. (2002) showed that not much water (0.5 m deep) was required to substantially counter this weakening effect, a finding that supports anecdotal evidence from numerous tropical storms that have cut across the wetlands of southern Florida without losing a great deal of intensity.

In summary, the general trend is for wetlands to reduce the severity of a storm and its associated wind and waves through numerous physical processes, most of them related to the enhanced friction, which serves as an energy sink.

## Changing Baselines

Although the *DWH* oil spill had multiple impacts on wetlands (discussed below), the most serious threat to GoM wetlands is their inability to keep up with relative sea level rise (Boesch et al., 1994). The reasons for this inability have been discussed in numerous publications (Boesch et al., 1984; Britsch and Kemp, 1990; Day et al., 2009; Dokka et al., 2006; Mallman and Zoback, 2007; Meade, 1982; Reed, 2002; Reed and Wilson, 2004; Turner, 1997). One notable problem is a reduction in sediment supply caused by the construction of levees and dams. Periodic overbank flooding supplied large pulses of sediment to marshlands behind natural levees, but these sediment supplies have been all but eliminated by the hardening and extension of the levees for flood control and navigation. GoM coastal wetlands have also been extensively dredged to provide access to oil and gas platforms, which has seriously degraded freshwater wetlands by providing a conduit for saltwater intrusion (Turner et al., 1982). The piling of dredge spoils on the banks of wetlands adjacent to the canals has also disrupted the natural flow of surface water and sediments across the marshes.

The Mississippi River Delta consists of an estimated 25,000 km$^2$ of wetlands, open water, distributaries, and beach ridges. Of the remaining coastal habitat, there has been a net loss of approximately 4,800 km$^2$ over the past century (Day et al., 2005). The loss rate decreased from a high of about 80 km$^2$/year in the 1970s to about 45 km$^2$/year by the turn of the century (Barras et al., 2003; Bernier et al., 2007). It is not clear if the recent decline in loss rate is due to variation in sea level rise (e.g., Bernier et al., 2007; Kolker et al., 2011), subsidence (e.g., Morton et al., 2002), the resolution of GIS technology, the reduction of dredging activities in the marshes, or other factors. The National Oceanic and Atmospheric Administration (NOAA) gauge on Grand Isle, Louisiana, has registered a variable, long-term relative rate of sea level rise of 6.7 mm/year (Figure 5.1). More recently, a 2005 coastwide analysis indicated that more than 4,714 km$^2$ of the pre-storm coastal wetland area experienced a substantial decline in vegetation density and vigor after Hurricane Katrina, with the majority of persistent damage through November 2006 in the western areas (Steyer, 2008). In addition, the background rate of marsh loss is not uniform across the coast, and is especially acute in the region most impacted by the *DWH* oil spill (Figures 5.2 and 5.4).

Current literature supports the conclusion that wave and storm surge attenuation and damage avoidance are related to wetland area, either nonlinearly with diminishing returns

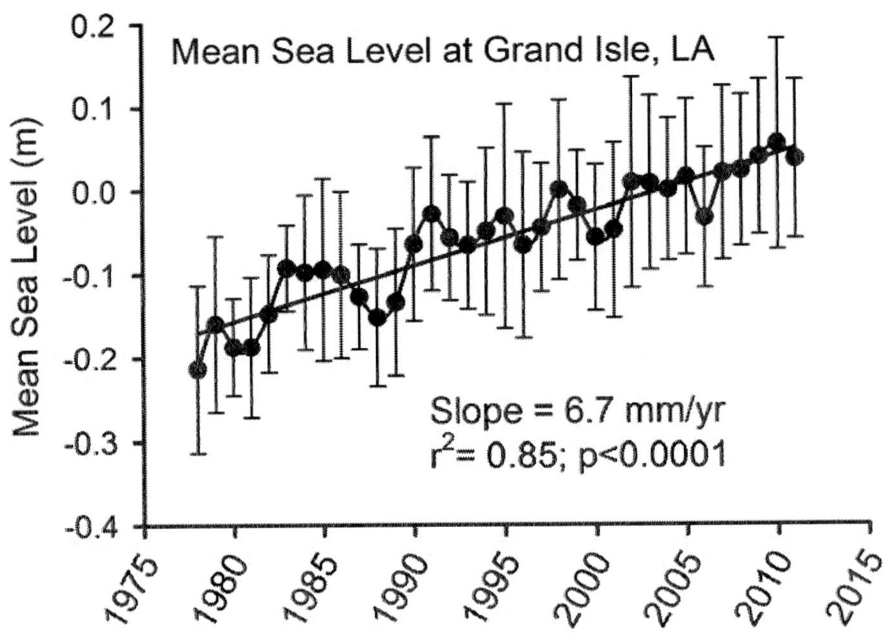


**FIGURE 5.1** Linear regression of mean (±1 SD) annual sea level at Grand Isle, Louisiana (NOAA station 8761724). NOTE: Annual means were computed from monthly means. SOURCE: http://tidesandcurrents.noaa.gov/sltrends/sltrends_station.shtml?stnid=8761724.

to scale (Barbier et al., 2008) or linearly (Costanza et al., 2008). Consequently, change in total wetland area is the most direct and practical measurement of change in ecosystem services in Gulf Coast wetlands. To quantify changes in wetland area, remote sensing is a highly effective method for analyzing estuarine and coastal landscapes (Kelly and Tuxen, 2009; Klemas, 2001; Phinn et al., 2000). Remote sensing is used to efficiently map, monitor, and detect changes in wetlands (Ramsey et al., 2011; Zhang et al., 1997). New satellites carry sensors with spatial resolutions of 1–5 m and spectral resolutions of 200 nm, providing the capability to accurately detect changes in coastal habitat and wetland health (Bourgeau-Chavez et al., 2009; Klemas, 2001, 2011; Ozesmi and Bauer, 2002).

The classification of wetland areas and plant communities is also improving as data from satellites are combined with those collected from fixed-wing aircraft. LiDAR (light detection and ranging), an optical system that can measure the distance to a target and other properties using pulses from a laser, is one of the sensors now commonly used on fixed-wing aircraft. This tool is used to construct digital elevation models and to develop digital profiles of plant canopies (e.g., Omasa et al., 2006). Classification schema based on combinations of these data

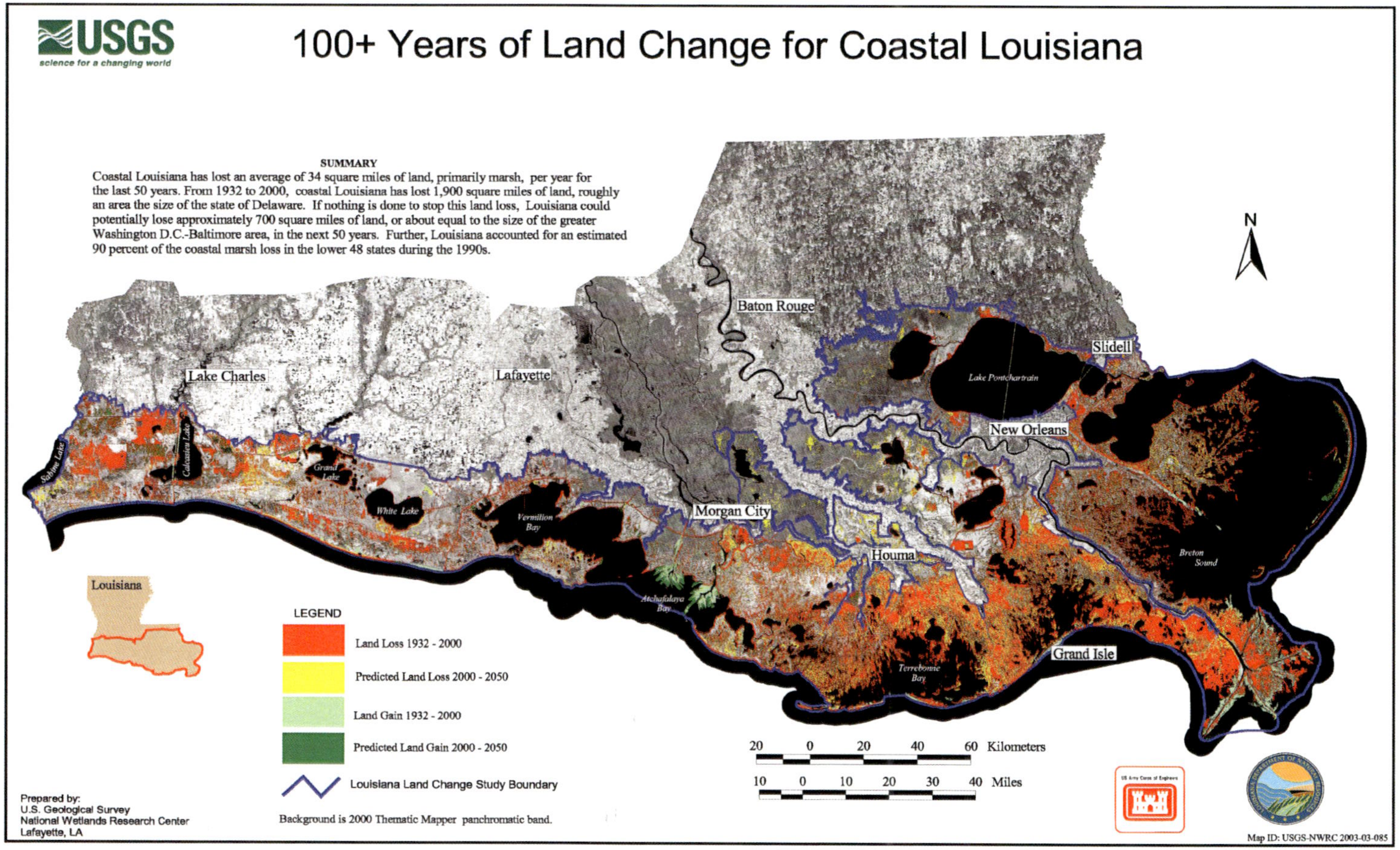


**FIGURE 5.2** Land loss change in coastal Louisiana. SOURCE: http://www.nwrc.usgs.gov/upload/landloss11X17.pdf.

sources can delineate wetland plant communities with a high degree (76–97 percent) of accuracy (Gilmore et al., 2008), and temporal changes can be resolved with repeated measures. The vertical accuracy of airborne LiDAR systems is probably not great enough to detect changes in elevation that are less than 5–10 cm (Hladik and Aber, 2012; Montané and Torres, 2006), but ground-based LiDAR scanners can accurately profile the elevation (millimeter accuracy) of a limited marsh surface area (1- to 10-m scale) (Boehler et al., 2003).

There are also episodic changes due to storms (Steyer et al., 2010) and now possibly to the *DWH* oil spill. Distinguishing the background trends and noise associated with storms, droughts, and other factors from oil-spill-related effects will be challenging (Figure 5.5). Four detectable oil responses are possible, assuming none is positive. One possible response is no response; that is, the wetland area continues along a declining, unstable baseline because of other forces (line A in Figure 5.3). A second possible response is a long-lasting episodic increase in marsh area loss (line B in Figure 5.3), and a third response is a loss of area followed by a recovery or restoration to a previous (lower) baseline followed by subsequent decline or stability (line C in Figure 5.3), depending on the type of restoration. Finally, a hypothetical stimulation of plant growth by light oiling and short-term marsh expansion (still to a lower baseline) is depicted by line D in Figure 5.3. Detection of these alternatives would not be possible without

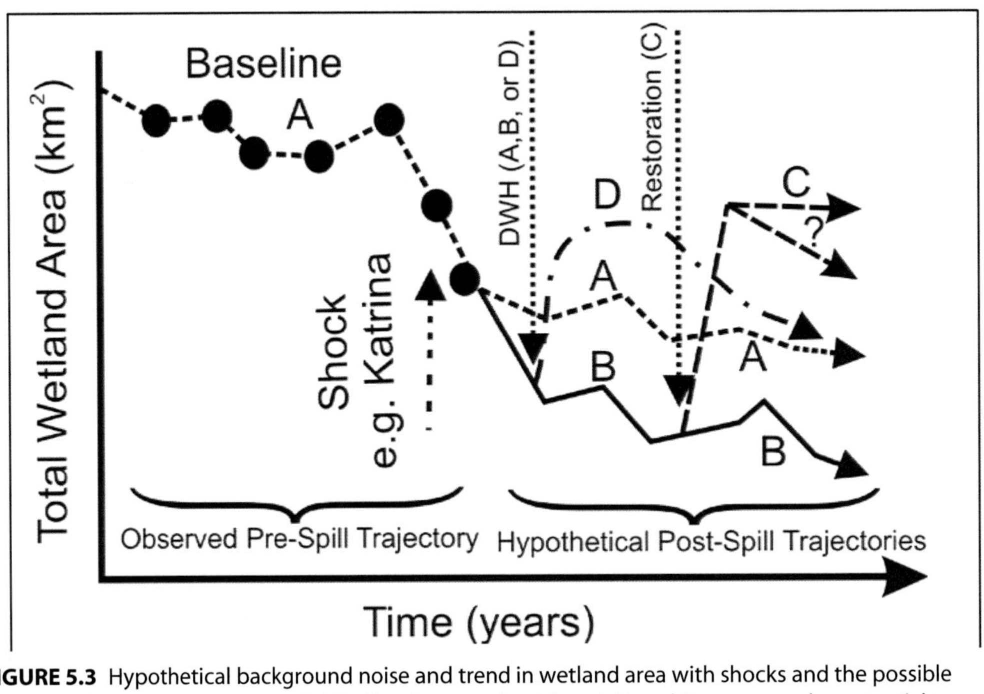


**FIGURE 5.3** Hypothetical background noise and trend in wetland area with shocks and the possible responses following exposure to *DWH* oil and restoration. Lines A, B, and D represent the potential trajectories wetland areas may take because of the impacts of the spill, and line C represents the possible trajectories post-restoration. SOURCE: Committee.

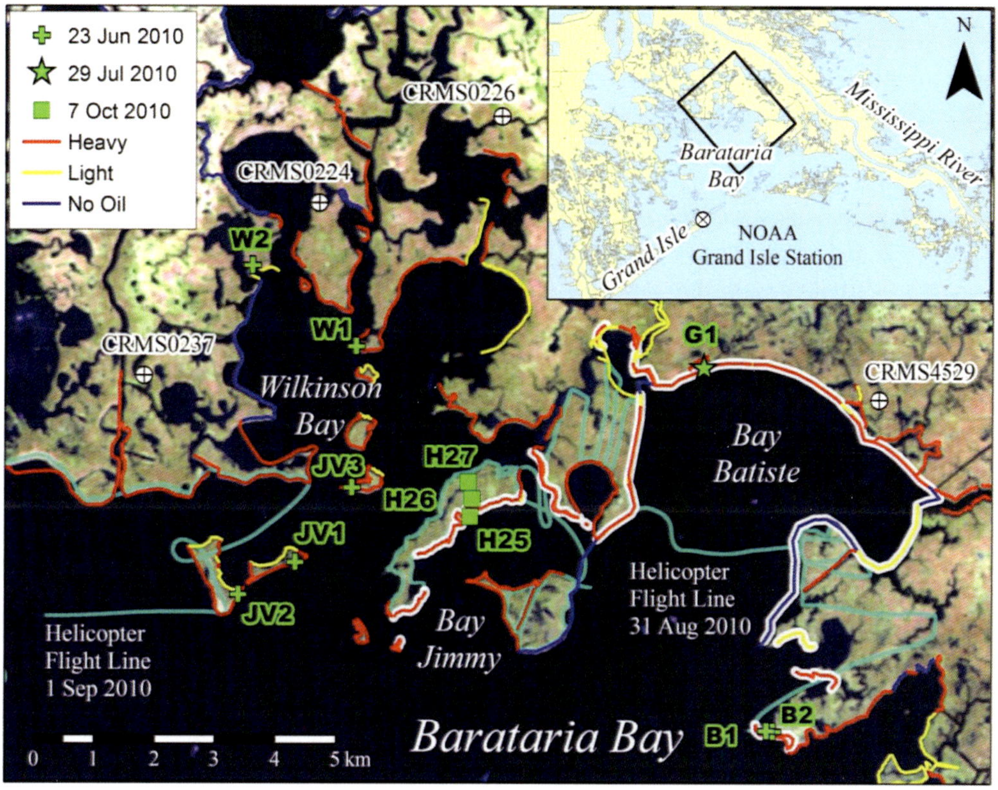


**FIGURE 5.4** A Landsat Thematic Mapper image of study area depicting locations and dates of ground observations of oil contamination from the *DWH* oil spill. SOURCE: Reproduced with permission from Ramsey et al. (2011).

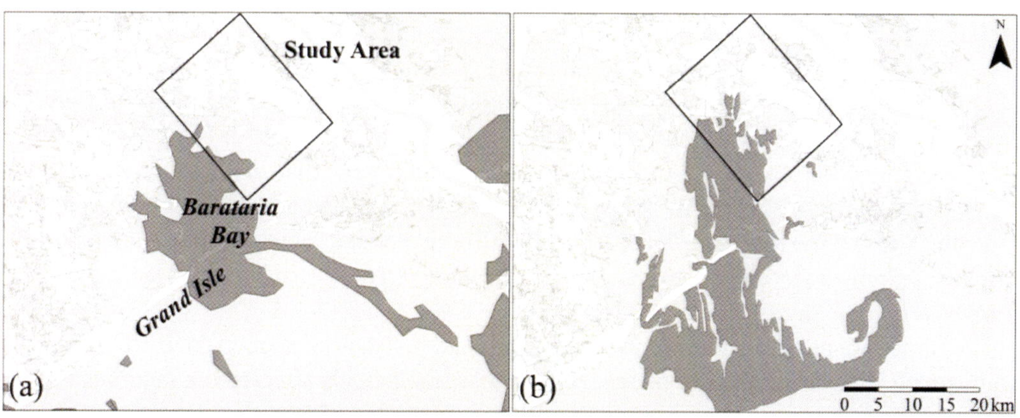


**FIGURE 5.5** Dark grey denotes the extent of the oil slicks in the vicinity of Barataria Bay, Louisiana, based on satellite observations made on (a) May 23, 2010, and (b) June 4, 2010. SOURCE: Reproduced with permission from Ramsey et al., (2011).

a monitoring network and pre-spill data, and subtle impacts will be challenging to detect even with a ground network and remote sensing. The effort will require sampling large areas with sufficient temporal frequency and spatial detail to resolve the changes. Fortunately, the tools exist and the groundwork is in place.

> ***Finding 5.1.*** *Wetlands provide a wealth of ecosystem services, ranging from fishery production to hazard mitigation (particularly storm surge protection). Storm protection by wetlands is achieved through several mechanisms, most of which ultimately trace their origins to the increase in frictional resistance to the storm afforded by vegetation..*

In 1990, the U.S. Congress enacted the Coastal Wetlands Planning, Protection and Restoration Act (CWPPRA) in response to ongoing wetland loss along the Louisiana coast. CWPPRA authorized funding in 2003 for the Louisiana Office of Coastal Protection and Restoration (OCPR) and the U.S. Geological Survey (USGS) to implement a Coastwide Reference Monitoring System (CRMS) as a mechanism to monitor and evaluate the effectiveness of CWPPRA projects across the region (Steyer et al., 2003). As a part of the monitoring program, reference sites were selected to establish the status and trends of existing wetlands. Three hundred and ninety (390) sites with a fixed annual sampling design were approved and secured for CRMS data collection. These sites located within nine coastal basins covering the entire Louisiana coast are sampled annually using a fixed design (Figure 5.4). Sample collection from the ground began in 2005 and will be an important complement to remote sensing techniques. Few of the CRMS sites were impacted by oil (Brady Couvillion, USGS, personal communication, 4/14/2011). Nevertheless, the distribution of sites provides important information about the status and trends of existing wetlands. Data collected at each site include information on sediment accretion and plant community metrics.

## *DWH* Spill Impact on Wetlands

A review of the effects of oil spills on wetlands in general can be found in Mendelssohn et al. (2012). Largely, the sensitivity of vegetation and soil organisms to hydrocarbon exposure depends on the type and concentration of hydrocarbon (Cermak et al., 2010). The lighter fractions and especially the aromatic hydrocarbons are more acutely toxic than are the heavier fractions (Chaîneau et al., 1997; Ziółkowska and Wyszkowski, 2010). Toxicity increases with the number of aromatic rings on the molecule (Baek et al., 2004). There are toxic effects to marsh vegetation from the light hydrocarbons in crude oil and smothering effects from fouling of leaf surfaces from heavy crude (Pezeshki et al., 2000).

A number of field and greenhouse studies of oiled marshes have been conducted. One study found that, although both diesel and heavy oil significantly affected the salt marsh vegetation, 1,000 parts per million (ppm) of diesel alone resulted in >90 percent mortality of above-ground biomass of smooth cordgrass, *Spartina alterniflora*. Plants took much longer to recover from exposure to diesel fuel than from exposure to crude oil (Lin and Mendelssohn, 2003). In contrast, Lippencott (2005) observed 90 percent survival of saltmeadow cordgrass, *Spartina patens*, in plots containing sediments with <12 percent petroleum hydrocarbons, and DeLaune et al. (1984) showed that direct oiling of a Louisiana marsh caused no reduction in macrophyte production or oil-induced mortality for the marsh macrofauna or meiofauna. Similarly, salt marsh vegetation in Texas recovered within 1–2 years after exposure to crude oil (Webb et al., 1985). A greenhouse study of the effect on *Spartina alterniflora* of chronic exposure to aromatic hydrocarbons found that application of 3.3 g C $m^{-2}$ $d^{-1}$ unexpectedly stimulated plant growth, while application of 33 g C $m^{-2}$ $d^{-1}$ inhibited growth (Li et al., 1990).

Addition of heavy oil to a *Spartina* salt marsh in the autumn resulted in high concentrations of the polycyclic aromatic hydrocarbons phenanthrene, chrysene, and fluoranthene in sediment and benthic animals, but hydrocarbon concentrations rapidly decreased over the next 20 weeks (Lee et al., 1981), probably because of degradation and dispersal of the hydrocarbons. Hydrocarbons are degraded primarily by bacteria and fungi, with prior adaptation to hydrocarbons by microbial communities increasing hydrocarbon degradation rates (Atlas, 1981; Leahy and Cowell, 1990), as is likely the case in the Gulf Coast region where navigation and oil seeps are chronic sources of hydrocarbons. Enhancement of hydrocarbon degradation by rhizosphere-associated bacteria also appears to be a significant process (Daane et al., 2001; Lippencott, 2005).

Mendelssohn et al. (2012) and Silliman et al. (2012) discussed impacts specific to the *DWH* oil spill. The expectations for long- and short-term effects of oil fouling of marshes depend on the aerial extent of the exposure and its magnitude. Acute effects of oil fouling of marshes appear to be confined to the margins of bays, canals, and creeks in a limited area (Figures 5.4 and 5.5) (Ramsey et al., 2011; Silliman et al., 2012). Although oil from the spill impacted about 690 km of shoreline vegetation from the Chandeleur Islands to Point Au Fer in Louisiana, the oil was already weathered by the time it reached the shore, so initial considerations were physical coating and hypoxia (Mendelssohn et al., 2012). Wetlands that were oiled include salt marshes, mangroves dominated by the black mangrove, *Avicennia germinans*, and marshes along the Mississippi River Bird's Foot Delta dominated by the common reed, *Phragmites australis*.

Field surveys conducted between July 7 and August 26, 2010, in central Barataria Bay and the Mississippi River Bird's Foot Delta of Louisiana, found that oiling in Barataria Bay marshes ranged from lightly oiled sections of stems of the predominant marsh plant species, *Spartina alterniflora* and *Juncus roemerianus*, to wide zones of oil-damaged canopies and broken stems, penetrating as far as 19 m into the marsh (Kokaly et al., 2011). The oil-damaged zone in Barataria Bay extended an average of 6.7 m into the marsh and usually more than 100 m parallel to the shore. Oil was observed on the marsh sediment at some sites, both above and a few centimeters below the water surface. The common reed, *Phragmites australis*, was the dominant vegetation in oil-damaged zones in the Bird's Foot Delta. Oiling of the leaves and portions of the thick stems of *P. australis* was observed during field surveys. In contrast to the marshes of Barataria Bay, fewer areas of oil-damaged canopy were documented in the Bird's Foot Delta. In both areas, oil was observed to be persistent on the marsh plants from the earliest (July 7) to the latest (August 24) surveys. At sites that were repeatedly visited in Barataria Bay over this time period, oiled plant stems and leaves, laid over by the weight of the oil, broke and were removed from the vegetation canopy, likely because of tidal action. In these areas, a zone of plant stubble 2–5 cm high remained at the edge of the marsh. Further signs of both degradation and recovery were observed and varied by site. Some oil-damaged sites experienced complete reduction of live vegetation cover and erosion of exposed sediments, while other damaged sites experienced regrowth of vegetation (Silliman et al., 2012).

> ***Finding 5.2.*** *Field surveys have been conducted to evaluate the damage to wetlands from the* DWH *oil spill, most of which is thought to be caused by physical coating and hypoxia. Those surveys identified sites where the vegetation had died in heavily oiled areas; sediment erosion often led to the conversion of once productive marshland to open water. However, in view of numerous studies that document a rapid recovery from oiling and a relatively low sensitivity of perennial marsh vegetation to hydrocarbons, marsh vegetation can be expected to suffer little or no long-term impairment from the spill in areas where roots and rhizomes survived the initial impact of oil fouling.*

Seven months after oil from the *DWH* oil spill came ashore in the salt marshes of Bay Jimmy in northern Barataria Bay, Louisiana, the concentration of total petroleum hydrocarbons at a sediment depth of 2 cm was as high as 510 mg/g, causing almost complete mortality of *Spartina alterniflora* and *Juncus roemerianus* (Lin and Mendelssohn, 2012). Moderate oiling had no significant effect on *Spartina* (which was borne out by greenhouse studies confirming its tolerance to oil covering the shoot) but caused a significant decrease in live aboveground biomass and stem density of *Juncus*. *Spartina* was able to recover from complete oil coverage of shoots in 7 months. Penetration of the sediment significantly affected both species in greenhouse tests, which along with shoot coverage and recurring oiling explains the severe impacts to coastal marsh vegetation from the *DWH* oil spill (Lin and Mendelssohn, 2012). Some recovery has been noticed in oiled delta marshes at the mouth of the Mississippi and throughout Louisiana, but as of autumn 2011, most marshes had shown little recovery (Mendelssohn et al., 2012).

As discussed in detail in Chapter 4, cleanup activities after an oil spill can cause indirect effects (Table 5.4), including physical disturbance and compaction of vegetation and soil, which are detrimental to marshes (Pezeshki et al., 2000). Among the cleanup options, the strategies with the least risk and greatest reward are no response, bioremediation, burning, and vegetation cutting, depending on the type of sediment substrate and severity of fouling.

Where the vegetation has died in heavily oiled areas, sediment erosion will likely lead to the conversion of once-productive marshland to open water. However, in view of numerous studies that document a rapid recovery from oiling and a relatively low sensitivity of perennial marsh vegetation to hydrocarbons, marsh vegetation can be expected to suffer little or no long-term impairment from the *DWH* oil spill in areas where roots and rhizomes survived the initial impact of oil fouling.

**TABLE 5.4** Advantages and Disadvantages of Cleanup Techniques Used in Marshes

| Advantages | Disadvantages |
|---|---|
| **No response** | |
| Minimal impact (if oil degrades quickly), no physical impact | Potential oiling of birds or wildlife, oil may impact adjacent areas, heavy oils may degrade slowly or form asphalt |
| **Vacuum/pumping** | |
| Can remove large quantities of oil | Access/deployment of equipment, physical impacts |
| **Low-pressure flushing** | |
| Assists in removal by herding oil, lifts oil off sediment surface | Requires careful monitoring, pressure must be controlled, physical impacts (loss of sediment) |
| **Burning** | |
| Potential to remove oil quickly, can minimize impacts from trampling | Kills or damages oiled (burned) plants and associated fauna |
| **Sediment removal** | |
| May be only remediation possible for heavily oiled sediments | Can destroy the marsh, increased erosion potential, elevation changes may impede regrowth of plants, replanting necessary |
| **Vegetation cutting** | |
| Preserves belowground plant parts, prevents oiling of birds | May kill plant depending on species, potential for increased erosion, must be carefully monitored |
| **Bioremediation** | |
| Great theoretical potential, low impact | Potential for nutrient enrichment, may increase sediment hypoxia |

SOURCE: Modified from Hoff (1995).

## Storm Hazard Moderation: Ecological Production Functions and Valuation

Well-established relationships among plant canopy, plant height, plant density and extent, and topography relative to sea level can moderate the impacts of storms. These features of the plant communities across the wetlands collectively make up part of the ecosystem structure that utilizes the physical processes (ecosystem function) to provide the ecosystem service of storm hazard moderation. The ecological production function in this example specifies the output of ecosystem services (coastal protection) generated by the wetlands given its current condition (the state of the vegetation and topography) relative to its ecosystem function (storm mitigation). Knowledge of these relationships should allow for quantitative modeling of the capacity of wetlands to moderate storms.

The next step in the ecosystem services approach is to attempt to attribute a monetary value for this service. Two methods for doing so include quantifying the costs avoided, as in the case of storm mitigation, or the replacement costs, as in the case of engineered defenses.

Quantifying the avoided costs is probably the most common method used to value coastal protection (Badola and Hussain, 2005; Costanza et al., 2008; Danielsen et al., 2005; Das and Vincent, 2009). The value of coastal protection afforded by coastal wetlands can be estimated by finding the difference in likely damages to coastal communities from a storm event with intact coastal marshes versus an event with degraded or no coastal marshes, considering their relative capacities to absorb wave energy and reduce storm surges (NRC, 2011). Costanza et al. (2008) used this method and concluded that 60 percent of the variation in damage caused by 34 major U.S. hurricanes was explained by wind speed and wetland area in the swath of the storm.

The marginal value of a wetland varies with the level of economic activity in the path of an average storm—the higher the level of economic activity, the higher the costs avoided. A wetland's marginal value is also inversely related to its size—the larger the acreage, the lower the marginal value of an additional acre of wetland. The estimated marginal value of wetlands in the Gulf Coast region after Hurricane Katrina was \$4,363 $ha^{-1}$. However, the marginal value of wetlands in the region varied greatly in an analysis by Costanza et al. (2008), from a low of \$126 $ha^{-1}$ $yr^{-1}$ in Louisiana to a high of \$14,155 $ha^{-1}$ $yr^{-1}$ in Alabama. This large range of values is partially an artifact of the areal extent of marsh versus the economic activity that it can protect.

The marsh's effectiveness in moderating storms is determined not only by its size, but also by its biophysical condition. Fragmented marsh will not be as effective at reducing storm surges as will continuous marsh (Loder et al., 2009), but in current approaches to valuation only areal extent, not degree of fragmentation, is considered.

***Finding 5.3.*** *The ecological production function for storm mitigation from wetlands relates changes in wetland area to changes in coastal protection. The value of storm mitigation by wetlands can be measured using the avoided cost method, which assesses the benefits of maintaining an ecosystem that provides protection against storms, floods, or other natural disasters.*

## Natural Resource Damage Assessment

As discussed in the Interim Report (NRC, 2011) and in Chapter 2, Natural Resource Damage Assessment (NRDA) reports historically utilize habitat equivalency analysis (HEA) to account for wetland losses. HEA provides an analytical framework for estimating how much restoration is needed to compensate for the interim loss due to the spill. The objective of compensatory restoration is to deliver a fair level of compensation for the interim loss of the resource, meaning the value of the increase in the identified resources from the replacement projects should be equivalent to the value of the resources lost due to the injury. Thus, using HEA, the NRDA Trustees (see Chapter 2 for additional details on NRDA and Trustees) select restoration projects that are comparable in type and quality to replace the lost or damaged habitats.

To perform HEA calculations, Trustees must determine how long the injury will persist, the relative service level of the injured and replacement resources, and the lifetime of the replacement project. With this information, the Trustees can calculate the amount of the restoration needed by establishing an equivalency between the quantity, usually measured by acreage, of lost or damaged habitat (and its associated services) and that generated through the compensatory restoration project over time. As noted in earlier discussions (NRC, 2011), HEA can fall short in identifying and valuing the habitat that connect to the public's use of or benefits from the various ecosystem services provided by the wetlands.

Table 5.5 summarizes the state of the present NRDA practice to value the reduction of wetlands due to a spill as well as the metrics necessary to determine a value under the ecosystem services approach.

Because of the complexity of and challenges in the GoM ecosystem, including subsidence, sea level rise, and human activities, the HEA approach may not always offer the full range of restoration options. In contrast, the ecosystem services approach may offer a broader range of options to restore ecosystem services and enhance wetland resilience. As depicted in Table 5.5, application of the ecosystem services approach in the GoM is fairly simple from a concep-

**TABLE 5.5** Provision and Valuation for Wetland Storm Mitigation

| **NRDA Practices** | |
|---|---|
| Resource | Wetlands (salt marsh) |
| Typical approach to assessment | Determine exposure pathways and spatial extent of vegetation oiled; collect and document any dead and oiled wildlife |
| **Valuation Under the Ecosystem Services Approach** | |
| Ecosystem Service | Hazard Moderation |
| Type of data needed for ecological production function | A. Plant type (or species), height, and density.<br>B. Percentage of area likely to experience acute toxicity and die-off.<br>C. Cross-shore and along-shore extent of wetland harmed.<br>D. Estimates of ability of the wetland to reestablish with or without human intervention. |
| Ecological production function | Relationship between plant type, height, density and areal extent of vegetation and reduction of wave energy and storm surge. |
| Type of data needed for valuation | A. Location of structures, infrastructure, agriculture, etc., near the coast.<br>B. Value of structures, infrastructure.<br>C. Areal extent of marsh and biophysical condition. |
| Valuation method | Avoided cost; calculate the expected damages associated with storm surge. The value of the ecosystem service is equal to the reduction in expected damages. |
| Type of data needed for valuation of ecosystem service | 1. Data on (A), (C), and (D) from the ecological production functions. Data on wetland extent and amount oiled would be collected in a standard NRDA, but other data likely would not.<br>2. Building the functional relationships to translate from data on plant type, height, density, and extent to likely height of storm surge; may be done via empirical relationships or modeling.<br>3. Building the functional relationship that translates height of storm surge to expected damage. |

tual standpoint. For the GoM, where the baseline is changing rapidly, the ecosystem services approach first estimates the wetland loss over the next few decades that would have been expected if the spill had not occurred. The baseline must be observed over several decades in order to derive a relatively stable valuation, because episodic events such as storms, which may have both positive and negative effects, and long-term trends affect the trajectory. Next, the approach calculates the expected wetland loss over the same time period with the spill effects considered. This estimate will have high uncertainty, which ideally should be considered, but it will not be considered here for the sake of simplicity.

> ***Finding 5.4.*** *An ecosystem services approach can develop production and valuation functions, which calculate a net present value of the incremental loss of wetlands and human-engineered structures due to storms. This approach allows for identification and comparison of alternative mitigation options to find the preferred restoration option, which may increase resilience more so than simply restoring the damaged wetland to its original state.*

## Opportunities for Restoration

Against a natural backdrop of significant continuing wetland loss due to a variety of circumstances, including subsidence, dredging of canals, salt intrusion, and sediment starvation, are a number of options for wetland restoration. These options include elevating existing wetlands to a range that restores the resilience of the wetland and increases its elevation capital, and creating and introducing genetically modified wetland plants or hybrids that produce massive quantities of roots and rhizomes that more efficiently trap mineral sediment.

The former can be accomplished more broadly by diverting sediment-laden water from the Mississippi River into degrading wetlands or, more locally, by creative uses of dredge spoil. For example, thin layer disposal of spoil, accomplished by spraying sediment slurry into adjacent marshes, is possible in the vicinity of dredging operations (Wilber, 1993).

With respect to bioengineering approaches, natural hybrids of *Spartina* exist, such as those found in San Francisco Bay, California (Ayres et al., 2004; Rosso et al., 2006), with superior capabilities to expand and trap sediment. The introduction of genetically modified plants, or even hybrids, would be controversial, but would be analogous to what has been developed and practiced in agriculture for many centuries.

To accelerate the degradation of oil residues in marsh sediments, the process of bioremediation can be used to exploit the activities of microbial communities to restore oil-polluted environments (see "Near-Shore and Onshore Biodegradation" discussion in Chapter 4). Bioremediation of petroleum pollutants often involves using the enzymatic capabilities of the indigenous hydrocarbon-degrading microbial populations and modifying environmental factors, particularly concentrations of electron acceptors (oxygen, nitrate), fixed forms of nitrogen, and phosphate, to achieve enhanced rates of hydrocarbon biodegradation (Atlas, 1991; Swannell et al., 1996). It has been demonstrated experimentally that degradation of oil can be enhanced by the addition of the inorganic nutrients nitrogen and phosphorus (Röling et al., 2002). Degradation of oil can also be enhanced by the dispersion and emulsification of oil in aquatic systems and by absorption by soil particulates in terrestrial ecosystems (Leahy and Cowell 1990), processes that happen naturally or, in the former case, with the aid of added dispersants. However, the application of dispersants in wetlands is generally prohibited without a case-by-case review and approval from the Environmental Protection Agency (EPA) and the U.S. Coast Guard in their capacities as cochairs of the Regional Response Teams (NRC, 2005a).

## Fisheries

### Introduction

The second case examines the fisheries in the GoM, particularly the provisioning ecosystem service of seafood and fish-based products. Fisheries clearly offer an important cultural ecosystem service through recreational fishing, which will be considered only briefly here. Like for wetlands, substantial baseline data for fisheries exist, and significant efforts have been made to develop ecological production functions through various fisheries models (see Chapter 2). The values of both the provisioning and cultural services offered by fisheries are measurable by traditional economic methods (Eide, 2009). Given their economic importance, fisheries have been the focus of numerous policies (Feral, 2009) and management plans (Die, 2009) aimed at maintaining the flow of their services to provide the greatest benefit possible to society. The importance of understanding and managing marine ecosystems to ensure sustainable provision of the ecosystem services provided by fishing is now enshrined in state laws such as California's Marine Life Protection Act,[1] federal laws such as the Magnuson-Stevens Fishery Conservation and Management Act ("Magnuson-Stevens Act"),[2] which governs most of the major fisheries in U.S. waters of the GoM, and international treaties that call for an ecosystem services approach to the management of marine fishery resources (Garcia and Cochrane, 2009; Garcia et al., 2003). At both the national and international levels, it has been recognized that fishery management decisions should be based on the best available scientific information (Magnuson-Stevens Act, 16 U.S.C. §1851(2); FAO, 1995).

### Provisioning Service

*Food and Fish-Based Products* In 2011, the GoM produced more than 200 million pounds of shrimp valued at $352 million, which accounts for nearly 68 percent of U.S. landings. The GoM also produced 18 million pounds of oyster meat valued at $84 million, accounting for 64 percent of U.S. landings. Although not used for food, menhaden landings were another regional top earner, with 1.6 billion pounds (66 percent of U.S. landings) earning $110 million (NMFS, 2012). Sport fishing is also a major economic engine. In 2011, an average of 23 million fishing

[1] Cal. Fish & Game Code § § 2850–2863.

[2] 16 U.S.C. §1801 et seq.

trips took place in the Gulf States (NOAA, 2013a), generating sales of $9.9 billion and income of $5.1 billion.

Seafood harvesting and recreational fishing take place in a variety of ecosystems within the GoM, from the coastal bays to the edge of the exclusive economic zone (NMFS, 2012). Seafood includes not only sessile organisms such as oysters, but also highly migratory species such as tunas. Many demersal organisms (associated with the bottom of the sea as adults), such as shrimp and some species of fish, often have pelagic larvae that connect populations over ocean basin scales (Cowen et al., 2006). As a result, a wide array of GoM fishery resources may have come in contact with substrates or waters polluted by the *DWH* oil spill or the spill responses.

### Additional Ecosystem Services

Although fisheries aim to provide two main ecosystem services, provisioning (food) and cultural (recreation), fishing affects our ability to extract other services from the ecosystem. For example, fishing practices may cause changes to benthic communities (Wells et al., 2008) or remove fish predators (Ruttenberg et al., 2011), and thereby affect regulating and supporting services, or degrade reefs (Dulvy et al., 2004). The presence of such interactions between fishing and ecosystem services other than those related to seafood and recreation means that disruption of fishing activities caused by an oil spill will have consequences for other ecosystem services.

### Quantifying Changes in Baselines

Recognition of the value of ecosystem services derived through fishing activities and of the need to base resource management decisions on science has meant that some of the monitoring and assessments required for understanding the provision of seafood have been in place for a number of decades (Beverton, 1998). Therefore, in comparison to other ecosystem services, those related to the provision of seafood, and to a lesser extent recreational fishing, have a long history of research, monitoring, and quantitative evaluation of the impacts of human activities (Hilborn, 2012).

Most of these research and monitoring efforts have focused on how to manage fishery productivity by controlling harvests or fishery inputs in a single fishery population context (Pope, 2009). Some researchers assert that fisheries management has failed (Jackson et al., 2001; Myers and Worm, 2003) because it has not followed a holistic ecosystem approach, while others suggest that this approach is the reason for management failures (Hilborn, 2004; Hilborn et al., 2004; Mace, 2004), while still others suggest that the problems are often more related to the failure of fishery institutions (Mesnil, 2012) and limitations of fishery laws (Dell'Apa et al., 2012). What is clear, however, is that much knowledge about marine ecosystem processes has been gained because fisheries have functioned as large-scale ecological experiments (Jensen et al., 2012).

Unfortunately, the impacts on fishery productivity from coastal development (Mora et al., 2011), pollution from freshwater catchments (Caddy and Bakun, 1995), and dumping of debris (Ribic et al., 2012; Wei et al., 2012b) and contaminants (Brodie and Waterhouse, 2012), including oil spills (Barron, 2012; Garza-Gil et al., 2006; Penela-Arenaz et al., 2009), are not well understood. The impacts of ocean acidification (Kaplan et al., 2010) and global warming (Eide, 2008; Sherman and Adams, 2010) have only recently been studied.

Market-based baseline information collected by state and federal agencies routinely includes data on landings by weight, by dollar value at state, regional, and national scales, by species, and by distance from shore (NMFS, 2012). These data can provide some initial estimates of impacts from events such as hurricanes; for example, in Figure 5.6 the observed reduction in value for Louisiana's landings in 2005–2006 may reflect the impact of Hurricane Katrina, which hit the northern GoM coast in August 2005 and destroyed a large part of the fishery infrastructure (Impact Assessment, Inc., 2007), while the reduction in value for 2010 may be attributed to the *DWH* oil spill.

Most fishery models describe the productivity of a fish population as a function of population dynamics (Quinn and Deriso, 1999). The amount of seafood that can be sustainably harvested is thought to be a function of historical catch rates, the natural mortality suffered by the population, and the population's capacity to reproduce. The resulting balance between these processes is then reflected in the relationship between harvested fish and the abundance remaining in the population, most often measured indirectly through observation of standardized catch rates (Maunder and Punt, 2004). Ecological production functions for fish stocks can then be derived by modifying fishery models that consider economics (Sanchirico and Springborn, 2010).

Even for the best-studied and most valuable fish populations, such as bluefin tuna (Fromentin and Powers, 2005) or Gulf menhaden (Smith and Vaughan, 2011), estimates of fluctua-

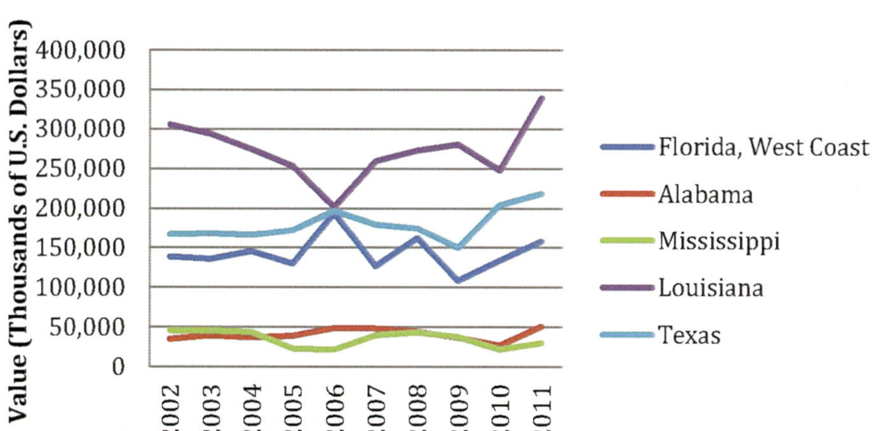


**FIGURE 5.6** Value of commercial landings of fish by the Gulf states. SOURCE: Data from NMFS 2002–2011. Fisheries: Ecological Production Functions and Valuation.

tions in productivity cannot be precisely characterized (Figure 5.7). Long-lived species such as bluefin tuna have shown production cycles ranging from 20 to 100 years, which may be related to the dynamics of the tuna population or to changes in the ecosystem (Ravier and Fromentin, 2001). Because high uncertainty is present in the estimation of historical abundance, predicting future population trends for fish populations that are confined within the GoM (Walter and Porch, 2008); NOAA, 2011b,c,d,e) and those that migrate through it (ICCAT, 2010, 2011, 2012) is a difficult endeavor.

One of the most serious limitations with estimating stock size is that, for most species, current assessments of GoM fishery resources are not spatially explicit—that is, they only provide estimates of abundance for populations of large migratory fish that encompass the entire GoM or the entire Atlantic. Recently, Carruthers et al. (2011) estimated spatially explicit abundance for migratory fish; however, their model still does not resolve abundance at scales smaller than

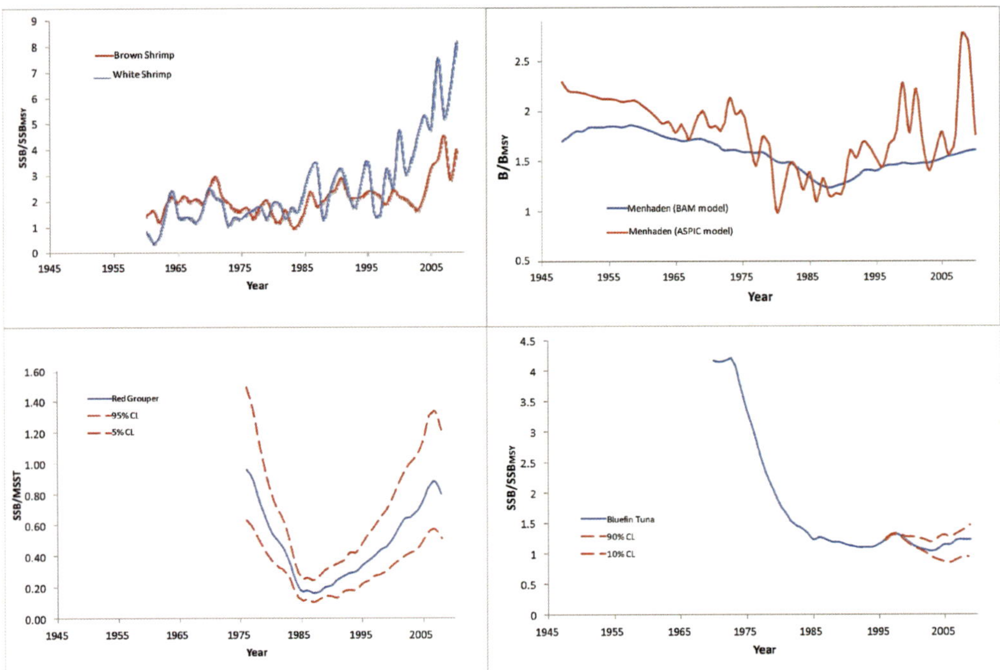


**FIGURE 5.7** Historical sustainability indices for some of the most important fish stocks in the Gulf of Mexico that may have been affected by the oil spill or by the fishing closures imposed after the oil spill. NOTES: Indices for (a) brown and white shrimp (Nance, 2009), (b) Gulf menhaden (NOAA, 2011c), (c) red grouper (SEDAR, 2010), and (d) western stock of Atlantic bluefin tuna (ICCAT, 2011) are provided in the four line graphs above. Indices at or above 1.0 indicate sustainable exploitation. All biomass indices are relative to overfishing reference points, either the biomass at maximum sustainable yield ($B_{msy}$) or the minimum stock size threshold (MSST). Some of the common sources of uncertainty in estimates are highlighted: data and estimation uncertainty (red grouper and Atlantic bluefin tuna) and assessment model uncertainty (menhaden).

the entire GoM. This lack of resolution challenges assessment of the spill's impact on these resources because the direct impact of the spill is most likely to be restricted in space to the areas closest to the oil.

Although population assessments do not estimate spatially explicit population abundance, there are indices of relative abundance for some species that have some measure of spatial resolution. Many of the indices of relative abundance for fish populations in the GoM use the National Marine Fisheries Service (NMFS) statistical areas as explanatory areal factors. These statistical areas (Figure 5.8) encompass vast areas of the GoM (Scott-Denton et al., 2011), and they may provide enough spatial resolution to test hypotheses about the impact of a large spill such as the *DWH*, but are clearly inadequate for smaller spills. More recent developments in fishery data collection, such as the introduction of vessel monitoring systems in the reef fish fishery of the GoM, can provide more precise geolocation data that could be used to obtain finer-scale indices (Saul, 2012).

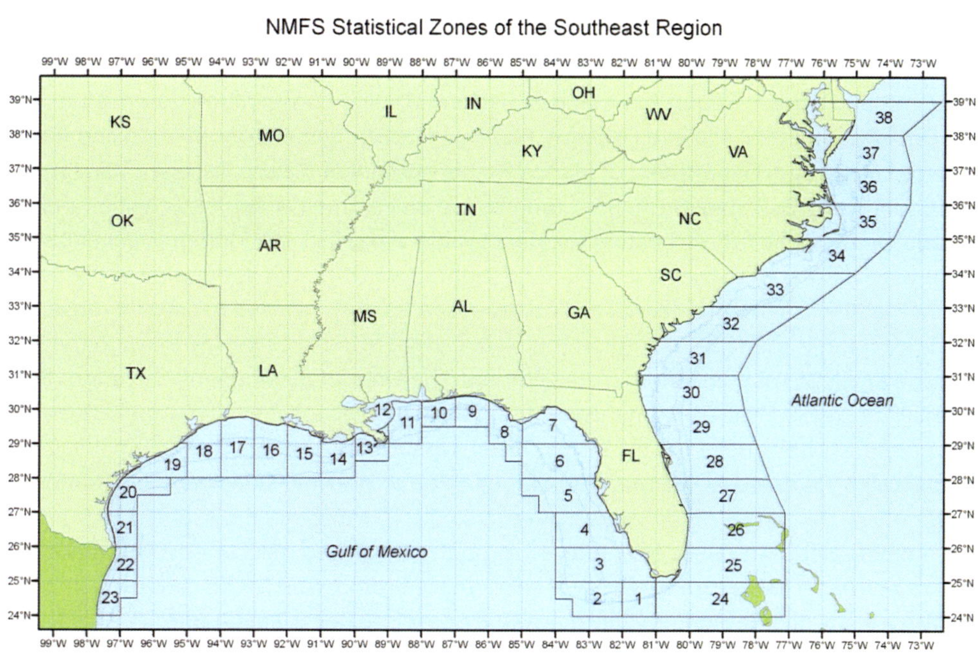


**FIGURE 5.8** Spatial scales for which fishery data are available in the Gulf of Mexico. NOTE: Each area of the ocean represents a statistical area used by NMFS in the analysis of fishery data. Areas represent the smallest scale at which representative and comprehensive historical data on catch and fishing effort are available. SOURCE: NMFS, 2010a.

***Finding 5.5.** Fisheries provide one of most important and lucrative provisioning services in the GoM via seafood production and cultural services via recreational fishing. Despite many long-term studies and ongoing development of fisheries models, the ability to detect spatial and temporal differences in fishery productivity of the GoM fish populations is often limited.*

Given the difficulty in separating the effects of environmental forcing on ecosystems from those related to population processes, fishery management often assumes that the environment has no effect on productivity, or conversely that population productivity is stationary. By assuming that changes in productivity are related to harvest, managers can take the precaution of curtailing fishery removals. Kell and Fromentin (2007) have shown that it is possible to design robust management strategies despite the lack of knowledge about all of the factors that affect the productivity of species such as bluefin tuna. In theory, it should be possible to design and implement precautionary management measures that acknowledge the uncertainty of the impacts of the *DWH* oil spill on fishery resources and therefore ensure sustainable provision of seafood into the future. However, adopting such measures would likely meet some resistance from stakeholder communities that rely explicitly on harvesting of these resources for their livelihood and well-being, as has often been seen when fishery managers seek to implement precautionary management. In fact, the difficulties of implementing precautionary management may be even greater when its scope is at the ecosystem level rather than at the level of a specific species or fishery (Gerrodette et al., 2002).

Given the extensive data collected by various federal and state agencies (NOAA NMFS, Texas Parks and Wildlife Department, Florida Fish and Wildlife Conservation Commission, Louisiana Department of Wildlife and Fisheries, Alabama Department of Conservation and Natural Resources, Mississippi Department of Marine Resources, Gulf of Mexico Fisheries Management Council, and Gulf States Fisheries Commission) on commercial fishing activities and the dockside values of the fish, estimating the direct value of the fish and food provisioning service of commercial fisheries to fishermen is fairly straightforward: the dockside value of the fish (the amount that the fishermen receives) minus any expenses they incurred to capture those fish. What is less clear is the value as the commercially caught fish makes its way through the supply chain and finally to the consumer's plate. Although the amount of commercially caught fish in the Gulf decreased in 2010 from the previous year (544 metric tons, down from 635 metric tons), potentially because of the extensive closures associated with the *DWH* oil spill, in 2011 the total catch increased to 862 metric tons, for a value of $797 million (Figure 5.9) (NMFS, 2012).

Commercial fishing values are captured through established mechanisms, but there is no similar mechanism to capture the value of subsistence fishing, which is an extremely important activity for many people along the northern Gulf Coast (see Box 2.2 in Chapter 2). An effective measurement of the food provisioning service would include not only commercial fishing, but also subsistence fishing.

The methodology for evaluating the economic effects of an oil spill, like any other effect of pollution, is relatively simple in concept (Lipton and Strand, 1997). In commercial fisheries, the costs of economic pollution are derived from either reduced production (due to die-offs

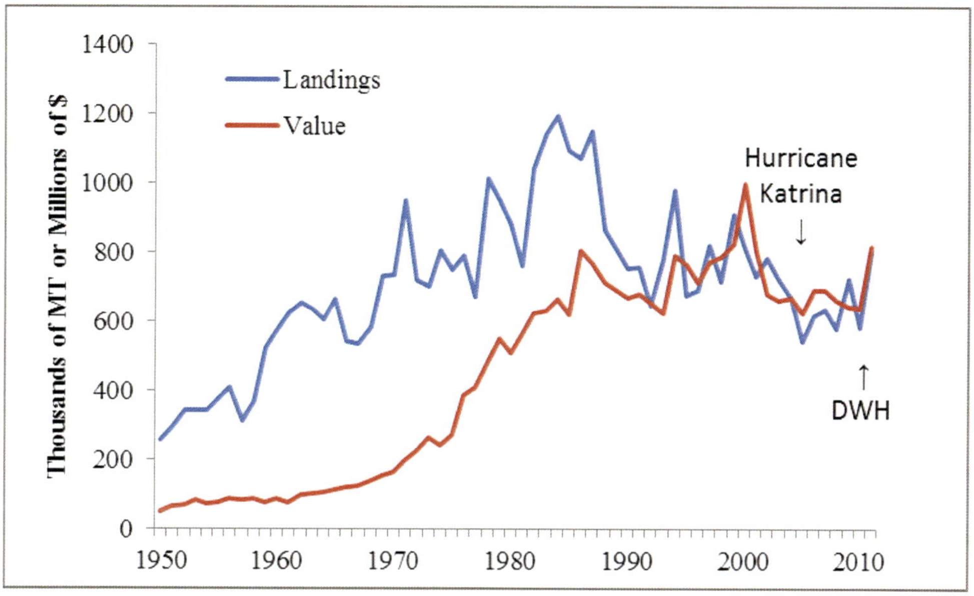


**FIGURE 5.9** Gulf of Mexico commercial fishery landings by weight (thousands of metric tons) and value (millions of $) for 1950–2011. SOURCE: NMFS, 2012.

or fishery closures) or by a drop in consumer demand due to the perception of reduced fish quality or safety (Lipton and Strand, 1997). However, the application of such methods provides few insights into the ecosystem's functioning. Attempts at making long-term predictions of fish abundance with traditional fishery management models have been relatively unsuccessful, making it difficult to believe that we can do better with long-term predictions of impacts of the oil spill. Furthermore, predicting indirect impacts, such as those mediated through the food chain, will be even more challenging. Recent advances in the development of more comprehensive ecosystem models provide some hope; however, these methods can, so far, only provide deterministic predictions (Field and Francis, 2006; Kaplan and Leonard, 2012).

## *DWH* Spill Impact on Fisheries

As already discussed in this report, the impacts of the *DWH* oil spill can be far-ranging and long-lasting. However, only a few studies have started to present evidence of these impacts. In this section we discuss two categories of impacts: those that directly affect animal or plant fitness and those that affect fishers or consumers of seafood.

The most direct impact of oil on living organisms relates to the toxicity of oil compounds. Only a few studies have so far provided quantitative data on the toxicity of the *DWH* oil spill on fish. Analysis of a data set consisting of 853 trawl samples taken between 2006 and 2010 did not show any loss in seagrass-associated juvenile fish abundances after the *DWH* oil spill

(Fodrie and Heck, 2011). In fact, for more than half of 20 commonly collected species, the catch rate was significantly higher in 2010 ($\alpha = 0.05$), which the researchers suggest might have occurred as a result of the fishery closures (see below). Catch-per-unit-effort of the most abundant species, pinfish (*Lagodon rhomboides*), was higher in 2010 (Fodrie and Heck, 2011). Post-spill community-level shifts in seagrass-associated fish assemblages were not observed.

Nonnative zebrafish, *Danio rerio*, were used in models of responses of embryonic developmental to exposure to water-accommodated fractions of *DWH* MC-252 crude oil to assess their potential to induce toxic effects (de Soysa et al., 2012). Results indicated that the oil was capable of causing significant defects in embryogenesis, resulting in cell death (apoptosis) and abnormalities in locomotor behavior, sensory and motor axon pathfinding, somitogenesis, and muscle growth. In addition, heavily weathered oil from the *DWH* oil spill has also been implicated in reproductive impairment and aberrant protein expression in gill tissues (physiological impairment) of larval and adult Gulf killifish (*Fundulus grandis*) that are common to Louisiana and Alabama coastal marshes (Whitehead et al., 2011). Such effects on fish reproduction are consistent with previous reports of direct effects of benzo[*a*]pyrene on the reproduction of fathead minnows (*Pimephales promelas*) (White et al., 1999). Despite the fact that only trace hydrocarbon concentrations were detected in the water column, effects characteristic of polycyclic aromatic hydrocarbon (PAH) exposure were apparent for up to 2 months after the initial exposure at Grande Terre, Louisiana, where oil came ashore (Whitehead et al., 2011).

Indirect effects of the toxicity of oil compounds include those related to the population-level and ecosystem-level responses to fitness change. Whenever the fitness of a given life history stage, such as juvenile fish, is affected by oil toxicity, we may expect population-level responses in subsequent life history stages even if they are free from the direct effects of exposure to oil. This is especially relevant for fish larvae and juveniles that inhabit the coastal areas and the surface of the open ocean, where concentration of oil is often the greatest after a spill. Such indirect effects on fish populations may take years to reveal themselves for long-lived species and thus affect seafood provision for decades.

Some measure of the potential disruption to the provision of seafood caused by the spill can be visualized by examining the spatial extent of the fishery closures (Figure 5.10) imposed by NOAA in its aftermath. These closures were intended to limit the risk of harvested seafood (generally adult fish and shellfish) from reaching the human consumer, not to protect the organisms from the effects of oil (NOAA, 2010). The fishery closures alone may have decreased fishery landings in the GoM by as much as 20 percent by preventing fishermen from accessing resources (McCrea-Strub et al., 2011).

The closures, however, underestimate the spatial extent of the possible indirect impacts of the spill on fishery resources because some fish can migrate through the spill area without being caught (Galuardi et al., 2010) and larvae present in the spill area may survive and end up recruiting as juvenile fish elsewhere in the GoM or other areas of the Atlantic (Muhling et al., 2012).

According to news reports and presentations to the committee by local GoM community spokespeople, a lack of demand caused seafood sales to drop after the spill (McGill, 2011). A study commissioned by the Louisiana Seafood Promotion Board and conducted by the market

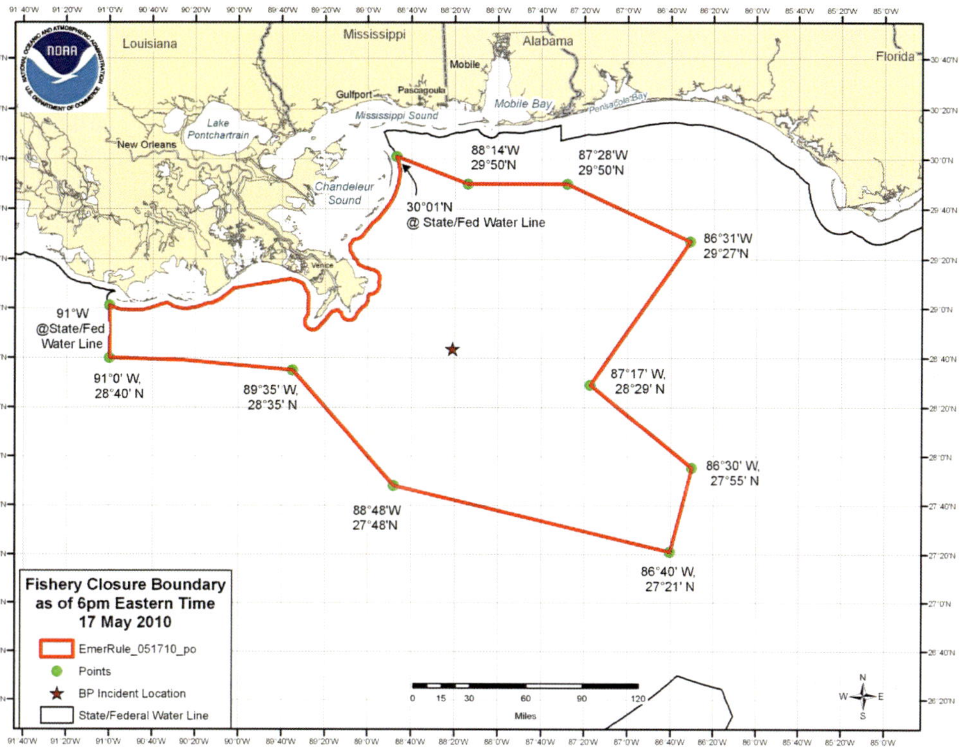


**FIGURE 5.10** Fishery closures imposed by NOAA to ensure seafood safety, effective May 17, 2010. The closed areas changed daily as NOAA sampled seafood in the area and the geographical extent of the detected oil changed. SOURCE: NOAA.

research company, MRops, found that greater than 70 percent of survey participants expressed some degree of concern regarding the safety of GoM seafood after the spill. The study also found that 23 percent of participants claimed to have reduced their seafood consumption after the spill. These results, which demonstrate the potential impacts on the GoM seafood market, have been cited frequently in the media and other reports (McGill, 2011; Upton, 2011).

***Finding 5.6.*** *Fishery closures in response to the* DWH *oil spill serve as one limited measure of the impacts to fisheries. Another impact of the spill and the subsequent closures was the public's concern about the safety of seafood from the GoM, despite efforts by federal and state governments to establish a comprehensive protocol to ensure the safety of GoM seafood. The* DWH *oil spill may result in longer-term consequences to seafood production beyond those caused by the cessation of harvest from oil spill–related fishing closures, including impacts to productivity or fitness, which may take years or decades to determine.*

Further, any mortality or reduction in individual fitness caused by the spill directly or indirectly through trophic interactions may determine the productivity of fishery resources into the future. Potential changes in productivity due to an oil spill, and the corresponding consequences to fishery harvest, may take years or, for some species, decades to transfer through the ecosystem, as shown by the time taken for fishing effects to cascade through marine food chains (Carscadden et al., 2001; Estes et al., 2009; Salomon et al., 2010; Sommer, 2008).

Trophic cascades may result from the indirect impacts of pollution (Jackson et al., 1989; Sala et al., 1998). The ripple effects of such disturbances on an ecosystem can last a long time (Peterson et al., 2003) and may ultimately lead to persistent changes in the ecosystem structure and its functioning. Fish can show detectable levels of contaminants long after a spill (Jewett et al., 2002), and sublethal impacts of these contaminants may be more significant than previously suspected (Peterson et al., 2003).

Prior to the *DWH* oil spill, oil spills have typically been local events that affect mostly the vicinity of the incident where oil was released. Therefore, impacts of oil spills have a strong spatial signal, with effects being stronger on ecosystem components that are closest to the spill site (Roth et al., 2009) and weaker on those that are more distant (Bustamante et al., 2010; Diez et al., 2009). A recently completed survey (Murawski et al., in preparation) of approximately 7,500 offshore fish in 2011 and 2012 reveals a similar signal pattern in the frequency of abnormal skin lesions coupled with elevated PAH levels found in bile of fish caught in the northern GoM. Fortunately, the frequency of lesions dropped significantly by 2012, indicating that the exposure was more episodic than chronic in nature. PAH concentrations are typically indicative of oil-related pollution, and the composition of the PAH compounds in red snapper was similar to those found in fish collected near the *DWH* wellhead. It is prudent to acknowledge that most fish can metabolize PAHs, and liver and muscle tissue samples in these fish had PAH levels falling well below the Food and Drug Administration's "levels of concern" (Murawski et al., in preparation).

## Natural Resource Damage Assessment

As discussed in previous sections of this report, understanding of fisheries is dominated by the economic valuation metrics of the commercial and recreational fish stocks that many of the state and federal resource management agencies have been collecting for many years. These data, coupled with the estimates of stock size and recruitment (which carry varying degrees of uncertainty and predictability), represent some of the metrics used in the traditional NRDA process (see Table 5.6). By contrast, the ecosystem services approach to valuation allows for a more comprehensive integration of environmental variables, such as the condition of relevant habitats, that influence the condition of fisheries. This integration can also include the impacts of fishing on other ecosystem services, so that, in the case of a fisheries closure, for example, changes (including positive changes such as reduced fishing pressure and subsequent cascades for the food web) can also be identified.

**TABLE 5.6** Provision and Valuation for Seafood Provision

| | |
|---|---|
| | **NRDA Practices** |
| Resource | Commercial Fisheries |
| Typical approach to assessment | 1. Measures of fishery landings.<br>2. Measures of fishery stock and recruitment.<br>3. Estimates of losses due to environmental injury/toxic effects. |
| | **Valuation Under the Ecosystem Services Approach** |
| Ecosystem service | Food and Fish-Based Products (commercial and subsistence) |
| Type of data needed for ecological production function | A. Measures of fishery landings.<br>B. Measures of fishery stock and recruitment.<br>C. Indices of environmental parameters affecting recruitment of harvested species.<br>D. Indicators of harvest pressure (catch, fishing effort). |
| Ecological production function | 1. Relationship between ecosystem and habitat conditions and fishery productivity. |
| Type of data needed for valuation | 1. Market price of commercial fish.<br>2. Fishing cost per unit effort (capital, labor, fuel).<br>3. Subsistence fishing activity level and species caught. |
| Valuation method | Market valuation: Calculate profit from fishing. Use market price and harvest data to calculate revenue. Use cost data along with revenue calculation to calculate profit. Calculate the value of subsistence-caught species based on the market equivalent price. |
| Type of data needed for valuation of ecosystem service | 1. Data on industry costs, dockside prices, and catch from subsistence fishermen.<br>2. Building the functional relationship between ecosystem and habitat condition and fishery productivity. This may be done via empirical relationships and/or modeling. |

***Finding 5.7.*** *Relative to the NRDA process, valuation efforts utilizing the ecosystem services approach can integrate more of the environmental variables that influence the productivity and recruitment of the fisheries, particularly the condition of the habitat that could be directly affected by an event such as the* DWH *oil spill.*

## Opportunities for Restoration

The ability to manage possible impacts on the provision of seafood and fish-based products as an ecosystem service also relies on our knowledge of the natural resilience of marine ecosystems. As discussed in Chapter 3, resilience is measured by the rate and extent to which recovery can happen after disturbances, such as an oil spill. Harwell and Gentile (2006) report that it took less than 20 years for Prince William Sound to recover from the *Exxon Valdez* oil spill, and Deriso et al. (2008) partially confirm their conclusion when they dismiss the argument that the *Exxon Valdez* oil spill led to the local collapse of Pacific herring.

Other researchers, however, dispute such assessments (Matkin et al., 2008; Peterson et al., 2003). Furthermore, in marine ecosystems, internal feedback mechanisms can promote the persistence of disturbances. For example, it seems that adult herring, "released" from predator control because of the depletion of cod from overfishing, can prevent the rebuilding of cod after the cessation of overfishing by consuming the early life stages of cod (Bakun and Weeks, 2006). Examples such as this are reminders that marine ecosystems can behave counterintuitively to what is observed in terrestrial systems and have routinely frustrated fishery rebuilding efforts by modelers and managers alike.

In the GoM, restoration efforts are likely to be widespread as resources become available from the responsible parties, but if early efforts are indicative, then coastal habitats including wetlands and oyster and seagrass beds are likely targets of restoration. This is, in large part, because habitat restoration can simultaneously enhance multiple ecosystem services.

Direct evidence of effects of habitat restoration on marine fisheries productivity is mostly limited to the effects observed from restoration of oyster reefs (Kim et al., 2012; Nestlerode et al., 2008). However, a recent analysis of habitat restoration efforts for Columbia River salmon identifies major errors in the conceptual foundation of many aquatic habitat restoration efforts (Williams, 2006). It is therefore essential that habitat restoration efforts are documented with a proper and standard set of habitat project-level metrics so that decision makers have the information they need to decide whether projects are likely to mitigate the effects of the *DWH* oil spill on fisheries (Barnas and Katz, 2010).

> ***Finding 5.8.*** *Aside from application of precautionary catch limits to mitigate the potential impacts of the* DWH *oil spill on fisheries, habitat restoration, because of its potential to enhance a suite of ecosystem services, may prove to be one of the more effective ways to restore fishery productivity in the GoM. It is also essential that habitat restoration efforts are documented with a proper and standard set of habitat project-level metrics so that decision makers have the information they need to decide whether projects are likely to mitigate the effects of the* DWH *oil spill on fisheries.*

## Marine Mammals

### Introduction

The third case study focuses on marine mammals in the GoM, particularly bottlenose dolphins. From February 28, 2010, to December 9, 2012, a period of approximately 33 months, at least 817 dolphins were found on the shores and in the marshes of the northern GoM; all but 5 percent were dead. As a rough comparison, the number of dolphins stranded in that region between 2002 and 2009 was approximately 100 per year.[3] Because some unknown portion of

[3] See running tally at http:// www.nmfs.noaa.gov/pr/health/mmume/cetacean_gulfofmexico2010.htm.

dead dolphins are neither found nor reported, these figures underestimate actual mortality (Williams et al., 2011). NMFS has officially designated an "unusual mortality event" (UME) in the northern GoM based on the unexpectedly high levels of bottlenose dolphin mortality along the coast.[4] Such a designation calls for intensive data and sample collection and a rigorous, coordinated study into the cause. Investigation of the mortality event has been subsumed into the NRDA process related to the *DWH* oil spill. Because it is not yet known if this aspect of the NRDA process will be litigated among the various parties, scientific findings from the investigation of the mortality event have not been made public.

The uncertainty surrounding the UME invites an examination of the range and extent of ecosystem services that would be altered, diminished, or lost if dolphins were affected by the spill. The exercise also has application for other marine mammals that could have been affected by the spill. Although applicable to marine mammals in general, this case study focuses on the scientific and educational, aesthetic, spiritual, and recreational benefits derived from dolphins in the GoM ecosystem.

## Cultural Services

*Scientific and Educational* Because of their proximity to the shoreline, certain populations of bottlenose dolphins that reside in GoM coastal waters, bays, and sounds are ideal subjects for observational studies from the beach or small watercraft. This research provides invaluable scientific information on dolphin populations, and much of what we know about their natural history was derived from such observations. A single day-long vessel excursion can yield a wealth of data—for example, size, age, and composition of the group, social structure and dynamics, feeding behavior, and mother-calf interactions. Gathering the same kind of data on a pelagic species is difficult, costly, and can take decades. Capture-and-release studies on coastal dolphins have become the only safe and reliable way to examine live specimens, from snout to flukes, and to obtain repetitive blood and other samples to determine the health status of the animals and, over time, their population (Wells, 2009).

These studies also provide an opportunity to broaden our understanding of the links among the health of these animals, the quality of their environment, and the potential impact of human activities on their well-being. For example, bottlenose dolphins along the GoM coast from Florida to Texas are commonly exposed to toxins produced by certain harmful algal blooms (HABs), and growing evidence implicates such exposure with dolphin stranding events (Fire and Van Dolah, 2012). Worldwide, HABs are spreading geographically and increasing in frequency for reasons that appear to include nutrient inputs from land sources (Olascoaga et al., 2007). Another issue is that dolphins are continually exposed to infectious agents, such as morbilliviruses (Duignan et al., 1996), and a wide variety of potentially harmful microbes, such as *Brucella* spp., that affect their reproductive and other internal systems, and plausibly affect their productivity (NOAA, 2011f). Dolphins in coastal waters are further exposed to potentially toxic chemical compounds, such as tributyltin (Kannan et al., 1996) and polychlorinated biphe-

[4] Note that the UME was declared more than 5 weeks before the *DWH* oil spill began.

nyls (Houde et al., 2006), which accumulate in fat and other tissues and have the potential to reduce their immunological capacity to fight disease (Lahvis et al., 1995).

> ***Finding 5.9.*** *As with many other species and ecosystems in the GoM, dolphins are often subject to multiple stressors that are linked to environmental conditions and human activities.*

Bottlenose dolphins in experimental settings can detect oil films on the surface, day and night, using vision and to some extent echolocation (Geraci et al., 1983). A dolphin's contact with oil elicits a "startle" response normally associated with stress or annoyance and, given a choice, thereafter avoidance (Smith et al., 1983; St. Aubin et al., 1985). These studies suggest that dolphins found amidst oil may have overriding reasons to be there, like feeding, social cohesion, reproduction, or simply migration through the spill area. Exposure to fresh volatile oil carries the danger of acute, possibly fatal, toxicity (Geraci, 1990). And while mere contact with older oil and sheens is not necessarily harmful to dolphins (Geraci, 1990), such environments may exert pressure on their social order, abundance, and quality of suitable prey, or other features of their life history that are less tolerant of change (Wursig, 1990). The net effect would plausibly add more stress to animals already exposed to contaminants, biotoxins, pathogens, and other stressors that combine to threaten the health of individuals and the vitality of populations. Dependent calves would be of particular concern. Studies under the NRDA process may shed light on whether the *DWH* oil spill has affected the health of dolphins in the GoM and any links between the spill and the mortality event.

Information gathered through these studies, together with fundamental physical, chemical, and biological descriptions of their habitat, is needed to identify the ecological production functions for coastal dolphin populations. As noted in Chapter 2, the cumulative information, in turn, forms the basis for valuing the ecosystem services that GoM dolphins provide. It is a long, multifaceted process that may be somewhat streamlined because dolphins are such desirable subjects to study.

## Existence and Spiritual Services

In the United States, all marine mammals, including the bottlenose dolphin, have special standing. Completely protected under the Marine Mammal Protection Act of 1972, they cannot be harvested for sport or food except in limited circumstances (e.g., Native American subsistence hunting). Therefore, most of the public's association with these species is through knowledge of their existence. Whether it be for their large brains, social behavior (nursing calves, tending their ill), apparent "smile," large eyes, or wide media exposure, dolphins resonate with humans in ways that transcend biology, appearance, or place in the ecosystem. From early history, dolphins have been featured in Greek mythology, as religious symbols, and as icons in art and architecture. Today dolphins are bulwarks for advancing conservation policy regionally and globally. The United States invests significant public resources, including millions of appropriated dollars, into programs designed to protect marine mammal health and well-being, as well as to assure that our use of ocean resources is compatible with the marine mammals' long-term welfare. Much of this attention is tied to conserving a natural resource and the greater ecological good that it confers. Still, dolphins stand out among many species that profit by such protection.

On a personal scale, many people simply want to know that dolphins exist. The passion of that commitment can be seen during any stranding event, where volunteers show up by the hundreds in challenging, often harsh weather to lift an animal back to sea and watch it swim away. These types of events illustrate the high personal value that individual humans place on these animals and highlight how dolphins help humans to realize a sense of well-being (see Figure 2.1 in Chapter 2).

## Recreation

Dolphin ecotourism is a growing industry nationally and globally. Apart from the financial benefits to operators and communities, dolphin-watching tours offer patrons unique noncontact recreational and educational opportunities. All five GoM states have tour operators specializing in dolphin watching. Expenditures that individuals make for these experiences provide some measure of their minimum value.

Revenues from these operations are uncertain, but relevant information is available from other parts of the country and could be used to help model/extrapolate values for this GoM ecosystem service. Direct annual revenues attributable to whale watching in Hawaii in 1999 were \$11 million to \$16 million (Utech, 2000) and roughly \$21 million in New England. Globally, the whale- and dolphin-watching industry in 2008 generated revenues of \$2.1 billion and is growing at a rate close to 4 percent per year (O'Connor et al., 2009). Beyond the economic value to the host communities, a case can be made that activities of this kind increase public awareness of the animals, invite dialogue on larger issues of the ocean environment, and are a natural platform for conservation education.

## Supporting and Regulating Services

Bottlenose dolphins are apex predators in the food web of the GoM and thereby affect the flow of energy and nutrients throughout an ecosystem. What would be the consequences to that ecosystem if a dolphin population was gradually or suddenly reduced or eliminated? Would other large predators, like sharks, fill the vacated apex niche? Would there be increased productivity among their prey species, including commercially important species? Lessons from other dynamic natural systems help to illustrate the difficulty of forecasting the cascade of consequences that such a change might initiate and to highlight the importance of understanding complex trophic links.

Top predators, such as sharks and groupers, are required for a healthy coral reef, and their removal dramatically reduces coral growth and biomass because it leaves coral grazers unchecked (Friedlander and DeMartini, 2002). The decimation of apex shark populations from northwest Atlantic ecosystems is enabling their former prey populations, including smaller sharks, skates, and rays, to expand unabated, similar to the herring-cod feedback loop discussed earlier in the fisheries case study. No longer regulated by shark predation, the cownose ray, *Rhinoptera bonasus*, has contributed to the collapse of local scallop fisheries (Myers et al., 2007). The digging activity of the rays also destroys eelgrass (*Zostera marina*), which otherwise provides habitat for communities with high faunal diversity and density. Removal of the eelgrass leaves unstable, less vital sand habitat (Orth, 1975), demonstrating that trophic cascades of this kind can have significant and long-term consequences. This example of the increase in the cownose ray, coupled with the decline of the local scallop harvest, lends itself to further analysis through the ecosystem services approach. Scallops are valued by humans as food (a provisioning ecosystem service), and the decline in the harvest can be quantified and results translated into a financial metric that would be readily understood by the public and by policymakers.

> ***Finding 5.10.*** *Bottlenose dolphins have a unique role in the GoM as both an apex predator and a charismatic species. Researchers have consequently studied this species extensively for decades, which allows for exploration of their role in providing a number of ecosystem services.*

## Changing Baselines

Estimating changes in the ecosystem services provided by dolphins will ultimately require an understanding of the underlying ecological production functions for dolphins and of the impacts of the *DWH* oil spill on the population dynamics of GoM dolphins. Determination of the causes of the strandings will be an important piece of the puzzle, because they may provide a link between the spill and mortality or loss of fitness for individual dolphins. The ability to relate the fitness loss to the status of GoM dolphin populations will depend on our knowledge of the size of these populations and the processes that may affect their dynamics. Evaluating the long-term effects of the spill will be especially difficult, requiring diligence and substantial commitment on the part of all parties involved.

In the Eastern Pacific, decades of research have yet to determine unequivocally whether certain dolphin populations have failed to recover because of persistent negative effects of a well-known stressor, tuna fishing, or because of yet-to-be-identified changes in ecosystem productivity (Gerrodette and Forcada, 2005; Reilly et al., 2005). The reason for these competing hypotheses relates, in part, to the simplistic assumptions made when trying to explain historical changes in marine mammal abundance (Gerrodette and Forcada, 2005) and in part to the difficulty of obtaining estimates of abundance that are precise enough to detect changes in population size of only a few percent per year, the maximum rates expected for long-lived mammals such as dolphins (Reilly et al., 2005). It is also possible that multiple stressors can be affecting the dynamics of these populations (Gerrodette and Forcada, 2005). The *DWH* oil spill highlights the critical need for long-term studies into population estimates and how they may correlate with conditions at all trophic levels, the roles of natural and human-induced factors underlying health and disease, and the influence of climate change. This kind of information is needed to inform any discussion of the resilience of GoM dolphin populations.

***Finding 5.11.*** *Substantial uncertainty regarding the abundance of GoM dolphins and of the range of stressors that affect them will complicate the assessment of the true impact of the* DWH *oil spill on their populations and on the ecosystem services that they provide.*

The challenge is further complicated by the stock structure of bottlenose dolphins in the GoM. The NMFS recognizes six major stocks: (1) northern bay, sound, and estuarine; (2) eastern coastal; (3) northern coastal; (4) western coastal; (5) northern continental shelf; and (6) northern oceanic. Given the distribution of these stocks and that of the spilled *DWH* oil, it is possible that all suffered some impacts from the spill. If so, the northern bay, sound, and estuarine stocks, northern coastal stock, and the northern continental shelf stock likely were the most affected. Prior to the *DWH* oil spill, the bay, sound, and estuarine dolphin stocks were the least understood with regard to life history information (Merrick et al., 2004).

## *DWH* Oil Spill Impacts on Bottlenose Dolphins

Scientific findings from the dolphins associated with the UME are currently withheld from public review by the NRDA process. However, the scale of the event and timing of the spill demand attention. Because dolphins are an essential part of the GoM biotic community, it is logical to hypothesize that altering any component of that community, particularly with something on the scale of the spill, could affect dolphins at some point in their life history. Such effects may not be immediately obvious because (1) the animals adapt quickly with no observable population-level effects; (2) current research methods and technology are not sufficient to detect or measure certain types of changes that may occur; or (3) the primary insult becomes masked by complicating factors such as contaminants, pathogens, algal toxins, and the pressures of a degraded environment. Any effects of oil, whether by contact, ingestion through prey, or the extensive disturbances associated with containment and cleanup activities, would have to be considered in the context of other stressors the animals already face and would ultimately be expressed in ways that are subtle or removed in time, making it difficult to explicitly link health insults or mortality with the spill. Is it possible the future will reveal some unforeseeable connections as remote in time and space as post-war whaling and damaged kelp forest communities? Once a disturbance starts, cascades can progress for years, decades, or longer, affect multiple trophic levels, and lead to ecological shifts that would not have been predicted (Gerrodette and Forcada, 2005). We should be alert to all factors, natural or otherwise, that place even local dolphin populations at risk, and, when analyzing plausible causes of mortality events, we should take a deep look back in time.

## Cultural Services: Ecological Production Functions and Valuation

As mentioned previously, dolphins provide services that are based in recreation and tourism, as well as aesthetics and spiritual well-being for many humans. The value for recreation and tourism can be assessed in market terms. One method for doing so is the travel cost method. As described in a National Research Council report (NRC, 2011, p. 105),

> *travel cost studies use information on trips that individuals make to recreational sites and the expenditures of time and money involved in making the trip" (Freeman, 2003). Travel cost studies typically use "random utility models" in which the value (utility) of a visit to a given recreational site is a function of distance from sites, site access fees, observable site characteristics (environmental quality, site facilities, etc.), observable characteristics of individuals (income, education, etc.), as well as unobservable characteristics of individuals (idiosyncratic preferences). Travel cost methods allow an analyst to trace out a demand function for site visits by varying the implicit price of a visit (travel cost plus access fees) faced by individuals and observing the number of trips taken."*

Further, by looking at sites of varying environmental quality, the value of improved environmental quality can be estimated or the damage to a site can be quantified. If damage to dolphins occurs and there are fewer dolphins to view, then presumably there will be fewer visitations and therefore lower expenditures. Linking of the changes in dolphin populations to changes in visitations to dolphin-viewing destinations would be required for an effective assessment of the impact on recreational values from this specific activity.

An additional method to estimate the value of services provided by dolphins is passive-use valuation, which is often used when considering spiritual and aesthetic services. Carson et al. (2003) conducted one of the most extensive studies to measure the impact of an oil spill on passive-use values, with regard to the *Exxon Valdez* oil spill of 1989. They assessed the spill's impact on murres and bald eagles rather than marine mammals, but their results demonstrated a significant willingness-to-pay to avoid similar damages in the future of $4.87 billion in passive-use values. A subsequent study (Loureiro et al., 2009) after the *Prestige* oil spill in Spain, which built upon the valuation work conducted on the *Exxon Valdez* spill, estimated passive-use losses with respect to marine birds and mammals and produced similar results. Summing from the household level, total willingness-to-pay was approximately $750 million, a rather substantial amount.

Certain regulations (e.g., under the National Environmental Policy Act of 1969) may require evaluation of passive-use values before any assessments of the dolphins' value are undertaken. In many cases, federal agencies have neither the time nor the resources to conduct original valuation studies, and, in such cases, benefit transfer approaches might be practical. Loomis (2006) demonstrated that a complete quantification of value must include not only the costs of an action, but also the benefits. He identified and monetized several important benefits of sea otters along the California coast that were missing from the draft supplemental environmental impact statement prepared by the U.S. Fish and Wildlife Service on sea otter range expansion (Loomis, 2006). The passive-use/existence values were significantly greater than the direct-use values. Similarly, a national estimation of the willingness-to-pay for expanded protection of the Steller sea lion in Alaska yielded a benefit of at least $5.8 billion annually (Giraud et al., 2002), demonstrating a strongly positive existence value for a protected species.

A recent effort to assess the economic value associated with recovery of marine mammal populations in Canada (Boxall et al., 2012) cites the need to establish marine protected areas to ensure recovery of the valued species. The results indicate that average willingness-to-pay per household ranges from $77 to $229 annually. As a nation, Canada would be willing to support recovery programs that improve the current at-risk status of the beluga whale and harbor seal at a level of $962 million–$2,850 million per year. These results are particularly relevant to the recovery efforts linked to the *DWH* oil spill. If the NRDA process determines that GoM bottlenose dolphins have been significantly impacted, then a part of the required restoration may involve protecting habitats that are important to them. Having an understanding of their value to the public provides a valuable context for any type of tradeoff analysis.

## Natural Resource Damage Assessment

The Interim Report (NRC, 2011) and Chapter 2 of this report discussed the habitat and resource equivalency approaches (HEA and REA) to NRDA that are commonly used by the Trustees to assess injuries and services lost because of damages to habitats and wildlife. With respect to marine mammals, and in particular, the bottlenose dolphin, REA is likely to be one of the methods used by the Trustees for the *DWH* oil spill. Table 5.7 describes the kinds of data that would likely be collected by the Trustees for such an analysis, as well as the data needed to incorporate an ecosystem services approach into the existing NRDA framework. Data limitations may impede the ability to fully develop ecological production functions for dolphins, in contrast to the wetlands and fisheries. Nevertheless, the approaches outlined in the Interim Report (NRC, 2011) can provide a useful framework for the Trustees and other decision makers who are undertaking an ecosystem services approach to evaluate the impacts to marine mammals, while maintaining concordance with the ongoing NRDA process.

## Opportunities for Restoration

Under the current Marine Mammal Protection Act in the United States, there are no specific restoration plans in place for bottlenose dolphins. If the NRDA findings establish a linkage between the recent mortality event of dolphins in the GoM and the *DWH* oil spill, then a major opportunity will exist to establish a plan to protect and restore dolphin habitat and to reduce mortality caused by human activities, such as bycatch from fishing. Strategies to reduce dolphin bycatch have already been implemented in the Bottlenose Dolphin Take Reduction Plan (BDTRP) for neighboring Eastern Atlantic bottlenose dolphin populations (NOAA, 2012c). A similar plan could be developed, if necessary, for the GoM.

**TABLE 5.7** Provision and Valuation for Marine Mammals—Recreation

| **NRDA Practices** | |
|---|---|
| Resource | Marine Mammals (dolphins) |
| Typical approach to assessment | Determine exposure pathway: Collect and document any dead or sick dolphins. Quantify the numbers dead or dying and, where needed, apply models from similar cases to estimate/extrapolate the number of dead or dying that were not directly observable. |
| **Valuation Under the Ecosystem Services Approach** | |
| Ecosystem Service | Recreation |
| Type of data needed for ecological production function | 1. Numbers of marine mammal–watching operations and visitation rate.<br>2. Measures of dolphin standing stocks, reproduction, and recruitment to near-shore GoM.<br>3. Estimates of the ability of populations to recover (resilience) with and without human intervention. |
| Ecological production function | Relationship between dolphin condition and quality of water and food sources. Understanding of the direct impact from exposure to oil, as well as response actions taken to alleviate the spread of the (surface/substrate) plume. |
| Type of data needed for valuation | Survey information on commercial marine mammal–watching trips in the GoM. |
| Valuation method | Travel cost. Use information on recreation trips, time and resource costs of trips to calculate willingness-to-pay for recreational mammal-watching trips. |
| Type of data needed for valuation of ecosystem service | 1. Collecting data on (2) and (3) from above.<br>2. Building the functional relationship between water quality, food or prey quality, and dolphin condition; may be done via empirical relationships or modeling.<br>3. Estimation of value using travel cost (random utility model). |

***Finding 5.12.*** *If a determination is made that the recent dolphin mortality event is linked to the* DWH *oil spill, then an opportunity may exist to establish a plan that includes the protection and restoration of dolphin habitat as well as the reduction of dolphin mortality from human activities.*

## The Deep Gulf of Mexico

### Introduction

The final case study focuses on the GoM deep sea, defined by this committee as ocean depths of 200 m and greater (effectively beyond the continental shelf). The GoM deep sea is a vast and relatively poorly understood subregion of the GoM. It is a region for which baseline and ecological production function data are relatively limited, but for which an ecosystem services approach may be particularly important. There is sufficient knowledge of deep-sea processes and availability of GoM-specific information to begin to identify ecosystem services of this deep-sea habitat, consider how these services may have been impacted by the *DWH* oil spill, examine methods of measuring baselines, determine what baseline data already exist for the GoM ecosystem, and identify gaps in those data. At present, the gaps in knowledge of the GoM deep-sea inhibit the ability to apply an ecosystem services approach in a quantitative way, leaving development of ecological production functions and the most effective use of valuation tools (Figure 2.1 in Chapter 2) for the future.

> ***Finding 5.13.*** *The deep sea is the largest yet least understood and quantitatively characterized subregion of the GoM. Because the* DWH *oil spill occurred in the deep sea, an assessment of the impacts of the spill cannot be complete without considering this vast subregion of the GoM.*

Delineating what we do and do not know about this extensive subregion could be helpful in identifying relevant processes and uncertainties and thus directions for future investigation. The European research community, moving in this direction, highlighted the *DWH* oil spill as evidence of the need to develop an ecosystem services approach to evaluate impacts to the deep sea (Armstrong et al., 2011). The committee recognizes it as a unique opportunity to clarify ecosystem services of the deep sea, to identify meaningful adjustments that could be made to ongoing data collection strategies and decision making, and to highlight areas of future research. To establish a context for our discussion of the deep-sea environment in the GoM, Box 5.1 provides a brief summary of ongoing federal environmental management activities in the deep portions of the U.S. Exclusive Economic Zone.

According to our current understanding, the primary ecosystem services of the deep sea fall in the category of supporting services. In the case of the deep GoM, nutrient resupply is particularly important for overlying primary production that supports the marine ecosystem. This case study focuses on supporting services, but it also identifies ecosystem services from the other categories. The regulatory service of pollution attenuation, which is inextricably linked to primary supporting services, receives particular attention.

When considering the services of the deep GoM, it is important to identify those aspects of this system that make it unique when compared to the entire deep ocean. These distinguishing features include the flow of an oceanic gyre through silled straits (discussed below under "Supporting Services") and the presence of extensive natural petroleum seeps and geological structures that have sequestered massive amounts of carbon as hydrocarbon and carbonate over long periods of geological time (discussed below under "Regulating Services").

**BOX 5.1**
**U.S. Environmental Protection of the Deepest Sea**

When considering the utility of an ecosystem services approach to managing, protecting, and restoring the deep Gulf of Mexico, it is informative to first consider the approaches taken by various agencies that have a mandate to manage such a vast and poorly known part of the ocean system. By recognizing the 200-nautical-mile Exclusive Economic Zone (EEZ) in 1983 (Presidential Proclamation 5030), the United States extended all then-existing environmental regulations to the deepest parts of the world ocean. The U.S. EEZ is both the largest by area and the deepest of any nation, for it contains the Marianas Trench, with its world maximum depth of 10,994 m. In the deep portion of the U.S. EEZ, efforts to address particular habitats or resource issues, including (1) oil and gas development, (2) dumping (waste attenuation), (3) mining, and (4) fisheries, have been and continue to be largely disjointed. The U.S. portion of the Gulf of Mexico (GoM) is a relatively modest portion of the entire deep EEZ, with a maximum depth of only 3,600 m. Its extensive hydrocarbon deposits, however, have made it one of the most exploited and studied deep-ocean areas. Achieving effective science-based management is challenging due to the unique characteristics of the deep GoM, including novel geochemistry, slope instabilities, and steep escarpments.

*Environmental Protection During Petroleum Development*

The Outer Continental Lands Act of 1953, amended in 2000, has a primary intent of developing offshore hydrocarbon resources, but also includes sections requiring that the natural and human environment be protected. The Bureau of Ocean Energy Management (BOEM), in both past and current organizational configurations, carries out the mandates of the Act. Strategies for protection were initiated on the continental shelf and included restricting development where there is conflict of use, such as fishing grounds or sensitive/rare habitats. In response to industry movement into regions deeper than the continental shelf, BOEM, then known as the U.S. Minerals Management Service (MMS), established a deep-water program first outlined by Cranswick and Regg (1997) and described in more detail in a series of later reports (Baud, 2002; French et al., 2006; Richardson et al., 2004, 2008). The Environmental Studies Program (ESP) had initiated deep-water environmental studies in the mid-1980s. Based upon the benthic sampling, especially the discovery of hydrocarbon-seep chemosynthetic communities, MMS initiated requirements that industry consider habitat protection when filing exploration and development plans. Requirements undergo revision as additional information is considered by BOEM; they are communicated to the offshore industry as Notices to Lessees (NTLs).[a] Of special concern is the protection of historical shipwrecks (cultural artifacts), chemosynthetic communities, and deep-coral assemblages. More detail is provided in the discussion of background data for the deep ocean.

*Protection from Unregulated Dumping*

Prior to 1972, the extensive dumping of chemical wastes into the Gulf and the incineration of such wastes on vessels sailing the Gulf were practices endorsed by the U.S. Environmental Protection Agency (EPA) (Ditz, 1988). They were brought to an end when the U.S. Congress adopted the Marine Protection, Research, and Sanctuaries Act of 1972 (MPRSA), which applied the restrictions of the London Dumping Convention to the U.S. EEZ. The MPRSA prohibits disposal activities that would unreasonably degrade or endanger human health or the marine environment, with responsibilities assigned to EPA, the U.S. Army Corps of Engineers, U.S. National Oceanic and Atmospheric Administration (NOAA), and the U.S. Coast Guard. The offshore oil and gas industry are excluded from MPRSA coverage. The small amount of dumping that still takes place, shallow or deep, is allowed by permits requiring minimization of environmental impact. The primary dumping activity in the U.S. EEZ is dumping of dredged material created by dredging in support of navigation. All dumping of dredged material in the GoM is into shallow water.

*Deep-Ocean Mining*

The seafloor metallic oxide nodules found primarily in regions of the central Pacific Ocean, where sedimentation rates are extremely slow, were recognized as a potential ore for strategic metals during the cold war (Mero, 1952). The extensive habitat destruction caused by the planned strip mining gave rise to environmental concerns (Barkenbus, 1979; Gerard, 1976; NRC, 1975). Wanting to protect U.S. mining claims in international waters, yet not having ratified the Law of the Sea Treaty (LOS), the U.S. Congress passed the Deep Seabed Hard Minerals Resources Act (DSHMRA) of 1980 (P.L. 96-283, P.L. 99-507), assigning licensing and environmental regulation to NOAA. Remarkably, both the LOS and DSHMRA included the very progressive environmental policy of allowing mining to proceed only if equal areas of similar habitat, referred to as Stable Reference Areas, were preserved (NRC, 1984; Post, 1983). The LOS assigned environmental monitoring and study of these sites to the license holder, while DSHMA gave similar responsibility to NOAA. While nodule mining interest and activity within the United States has declined, the DSHMRA remains in the U.S. Code as Title 30, Chapter 26, Section 1401. This act serves primarily as a placeholder to protect a single remaining claim filed by Lockheed Martin under the Act's provisions (National Ocean Service, 2012). Internationally, interest in seafloor mining and associated environmental protection remains high, especially for polymetallic sulfides (Hannington et al. 2008).

*Deep-Ocean Fisheries and Fish Habitats*

Commercial fishing for deep-ocean bottom fish is very limited in the United States, but may increase as shallow-water stocks are depleted. The Magnuson-Stevens Act (1976 and amended in 1996 and 2006) is the primary source of authority for regulation of fisheries in the United States. Initially adopted in part to bring U.S. fisheries regulation the same EEZ coverage included in the LOS, the provisions of the Act (amended in 2006) potentially cover deep fisheries. Although the Act contains no specific instructions for deep-ocean fish stocks, Title IV, Section 408 recognizes the importance of deep coral habitat. NOAA has implemented the Deep Coral Research and Technology Program to address the implementation of the Act (Lumsden et al., 2007; NOAA, 2013a). The South Atlantic Fishery Management Council has designated eight deep marine protected areas and deep-water coral habitats of particular concern. For the most part, these areas are along the shelf edge at depths as shallow as 60 m, but no deeper than 400 m (Ross and Nizinski, 2007).

*On the International Front*

In contrast to the United States' fragmented program of deep-ocean management is the European Union's highly organized approach, which adds an ecosystem services component to traditional conservation and protection (Armstrong et al., 2011; Tinch et al., 2012). Regulation of deep-sea living and mineral resources is a very active area of planning, research, and application (Gjerde, 2006). Especially noteworthy is the Convention for the Protection of the Marine Environment of the North-East Atlantic (the OSPAR Convention) in effect since 1992 (Ardron, 2008). The region of concern is far more extensive than the EEZ of Western European states, extending from the North Pole through Greenland to the Straits of Gibraltar. By 2010, 181 marine protected areas (MPAs) had been established within this large area, with six on deep seafloor in international waters (OSPAR Commission, 2010).

[a] http://www.boem.gov/Regulations/Notices-To-Lessees/Notices-to-Lessees-and-Operators.aspx.

## Supporting Services

The supporting services of the deep GoM are greatly influenced by the semi-enclosed nature of the basin, which produces locally unique conditions. At the same time, the GoM is tightly coupled to the larger Atlantic basin by the powerful surface currents that flow through its two straits. Processes of the deep GoM thus provide supporting services on a local scale and on a much larger proximal scale. The following discussion first considers water circulation and exchange in the GoM as critical to understanding oxygen and nutrient resupply (in support of primary productivity) and then addresses nutrient resupply and natural and oil-impacted microbial processes that determine or influence these services.

## Water Circulation and Exchange

As reviewed in the Interim Report (NRC, 2011), water circulation in the GoM consists of a relatively high-velocity, rapidly mixed upper layer and a low-velocity, slowly mixed lower layer. Transfer of oxygen, dissolved nutrients, heat, and momentum between the upper and deeper layers is limited by density stratification. The circulation and exchange of the GoM has two special aspects. First, the GoM is an important segment of the intensified western boundary current of the North Atlantic Gyre, receiving strong flow from the Caribbean and on through to the Atlantic. This Gulf Loop Current links the surface water to the larger Atlantic and causes local processes to have far-distant effects. Presumably, much more will be learned about transport and mixing from the extensive studies following the *DWH* oil spill.

Second, the deep thermohaline circulation of the GoM is influenced by the sill depths of the basin, connecting to the Caribbean in the south through the straits of Yucatan, with a sill depth of approximately 2,040 m, and to the Atlantic to the east through the Florida Straits, with a sill depth of approximately 740 m. Water flows out of the GoM only through the Florida Straits, where the depth of the sill restricts the flow. The 1,300-m difference between the depth of the Yucatan sill and the Florida sill sets up a deep-flow regime that is not fully understood. Sturges (2005) review of deep-flow data for the Caribbean and GoM supported previous suggestions that flow through the straits of Yucatan consists of cold flow into the Gulf balanced by an equal but somewhat warmer outflow created by mixing enhanced by the topography of the sill. These two-way deep flows traverse the Caribbean and connect to the open Atlantic via the Windward and Anegada-Jungfern passages.

These circulation patterns may contribute to unusually high levels of oxygen in depths of the GoM below 500 m. The water column of the deep GoM has a moderate oxygen minimum of approximately 2.5 ml/L at a mid-water depth of about 500 m, compared to levels greater than 5.0 ml/L at greater depths (Rivas et al., 2005). High levels of oxygen below the oxygen minimum in the deep GoM (despite active biological consumption and topographic restriction) can only be explained by frequent "ventilation," the inflow of oxygenated water from the Caribbean matched by an outflow of older deep water. Such deep exchange is only possible through the Yucatan Channel because of the shallower sill depth of the Straits of Florida and the upper-water outflow there.

The Yucatan Channel is now recognized to be a two-way passage, with deep water leaving and returning after some residence period in the GoM (Rivas et al., 2005, 2008; Sturges, 2005). Estimates of deep-water residence times vary. Shiller (1999) reviewed estimates and set residence time in the eastern GoM at 50 years and the western GoM at 270 years. This range is consistent with a model-based rate of 100 years (Welsh and Inoue, 2002). A 100-year period for the turnover of water in the semi-enclosed GoM is short, relative to open deep-sea basins, and helps to explain maintenance of the oxygenated and nutrient-rich state of the deep GoM, yet long compared to the generation times of marine organisms and their ability to be resilient in the face of stressors or disturbances that lead to oxygen or nutrient depletion in deep waters of the GoM.

## Nutrient Resupply in Support of Primary Productivity

The most extensive studies of primary productivity in the GoM have been carried out in shallow coastal waters, especially associated with the discharge of the Mississippi River. High levels of productivity have been observed over the deep GoM using sea-surface color imagery (Biggs et al., 2008; Müller-Karger et al., 1991). Regions of transient high productivity are associated with the Gulf Loop system, which can cause upwelling (and thus nutrient resupply) when impinging on the shelf break and at the junction of gyres. Primary productivity in waters over the deep GoM supports the larger marine ecosystem; upwelling or other mechanisms that resupply nutrients (e.g., inorganic sources of nitrogen and phosphorus) depleted during photosynthetic activity are essential to continuing that support. The fraction of surface production not transferred to higher trophic levels in surface waters (or recycled there) is exported to the deep GoM, where it supports the extensive benthic heterotrophic communities that dwell there (Rowe et al., 2008).

The oxygenated state of the deep waters in the GoM, coupled with the availability of methane and hydrocarbon-associated hydrogen sulfide emerging from natural seeps, supports another form of (nonphotosynthetic) primary productivity: chemosynthesis by prokaryotic microbes, particularly symbiotic chemosynthetic bacteria that support whole communities of metazoans such as tubeworms and mussels (Cordes et al., 2009; Orcutt et al., 2005). Seafloor seepage from extensive hydrocarbon reservoirs through an unknown number of sites, even if individually of limited spatial scale, makes the deep GoM one of the most chemosynthetically productive regions of the world's oceans. These seeps clearly support a dense mix of endemic and nonendemic fauna (and microflora) that exploit the emerging resources. Actual productivity rates are unknown, but the substantial age of some metazoans indicates that the biomass pools have accumulated very slowly (Cordes et al., 2009). Trophic web tracing using stable isotopes has identified some transfer of chemosynthetic carbon and nitrogen to the surrounding normal (heterotrophic) benthic and pelagic animals, but this export is localized and few species have been documented to participate in these cross-habitat links (Carney, 2010; Cordes et al., 2010; Demopoulos et al., 2010; Macavoy et al., 2005).

Lacking photosynthesis entirely, and despite the localized bottom areas of chemosynthesis, the deep GoM is primarily a region of respiration, the process by which heterotrophic organisms (from aerobic bacteria to all higher organisms) consume oxygen and release carbon dioxide from the organic matter on which they feed. The primary source of organic matter in the deep sea is photosynthetic detritus that sinks from surface waters; respiration of this material not only releases carbon dioxide, but also inorganic nutrients essential to resupplying and thus supporting photosynthetic primary production in surface waters. To a presumably lesser (but unknown) extent, seeping hydrocarbons are also respired; importantly, given the nitrogen-poor status of hydrocarbon compounds, they are respired without an appreciable release of inorganic nitrogen to support primary production. Indeed, oil-degrading bacteria, whether responding to naturally seeping or "spilled" hydrocarbons, must take their required nitrogen from the resupply pool in order to consume and respire the hydrocarbons.

Comparative measurements on samples collected from inside and outside of the deep oil plume emanating from the *Macondo* well site provide clear evidence of early and significant nitrate depletion within the plume coincident with hydrocarbon degradation, with nearly twice as much nitrate consumed as oxygen (Hazen et al., 2010). No evidence for a compensatory response by nitrifying bacteria or Archaea (microorganisms responsible for generating new nitrate from reduced forms of nitrogen, particularly ammonia, in the deep ocean) was observed, nor was the concentration of ammonia inside the deep oil plume statistically distinguishable from that in surrounding deep water (Hazen et al., 2010).

An increase in the baseline of hydrocarbon respiration in the deep GoM may thus translate to a reduction in the resupply of nitrogen (particularly nitrate) to the surface ecosystem, yet estimates of that baseline, whether static or fluctuating, are lacking, as are evaluations of any long-term or persistent impacts of the *DWH* deep oil plume on the available nitrate pool. Suitable models to identify the linkages and balances between these various processes are also lacking. Some existing models, targeting other goals such as trophic transfers of carbon and nitrogen (Rowe and Deming, 2011; Rowe et al., 2008; Wei et al., 2012a), may provide starting points for developing the more complex systemwide models ultimately required to evaluate supporting ecosystem services of the deep GoM.

## Biological Interactions: Resilience and Habitat Diversity

The deep GoM is a very large and low-turnover subsystem that provides the whole ecosystem with greater resilience (see Chapter 3 for definitions and a discussion of resilience). This role is especially obvious with respect to nutrient recycling. In systems where plants have direct access to nutrient uptake, the available supply of nutrients may fluctuate in an unstable manner. Being aphotic and vastly larger than the photic zone, the deep GoM provides a resupply of nutrients at a rate totally independent of the photic zone's biological demand and is thus a very stable source of nutrients to overlying life. Although the *DWH* blowout triggered more research on hydrocarbon-degrading bacteria and their reactions to dispersants used in the deep GoM (see Chapter 4), investigations of nutrient-regenerating microbial processes (e.g., the production of nitrate from ammonia) essential to stable long-term support of overlying productivity appear to be lacking.

When considering ecosystem services, it is important to note that the deep GoM is not homogeneous, but has several special seafloor habitat types—hydrocarbon seeps, massive carbonate platforms, and complex mud bottoms—which figure in the discussion of other ecosystem services (below). The seep habitats are best studied and comprise a suite of chemically and morphologically distinct habitats associated with hydrocarbon and brine seepage. These habitats are widespread in the northern GoM and can be categorized as ranging from fluid- to mineral-prone (Roberts and Carney, 1997). The fluid-prone habitats have ongoing effluxes of methane, liquid hydrocarbon, and/or brines. The mineral-prone habitats have accreted massive carbonate hardgrounds over a long period of microbial anaerobic hydrocarbon oxidation. The chemosynthetic ecology of the fluid-prone systems has been studied extensively since discovery (reviewed by Fisher et al., 2007). Research on the heterotrophic ecology of the carbonate hardgrounds is under way in conjunction with efforts to protect deep coral and relate potential impacts from oil and gas development (Becker et al., 2009; Sulak et al., 2008).

The full complexity of the extensive mud bottom is only slowly being recognized because of its vast size and the difficulty in detecting spatial or temporal differences in its biological processes. Bathymetric zonation of biota is the most obvious form of habitat heterogeneity on a regional scale (Wei et al., 2010b). On a more local scale, geological processes may create ecologically unique habitats. In the northern GoM, the continental slope is topographically complex, with many small basins and ridges. The ecological consequences of this topography have yet to be effectively examined. The slope is made additionally complex by extensive submarine landslides and megaslides that create seafloor scars where geochemically older sediments are exposed to seawater and biological colonization. The same slides also produce large deposits of slide material that entomb geochemically young sediments. Because submarine landslides pose both a hazard to oil drilling and a tsunami risk, their geophysics are being actively explored, but the ecological and ecosystem services aspects of deep-water slides are virtually unknown (Mienert et al., 2003; ten Brink et al., 2009). Massive carbonate habitats, unrelated to sites of hydrothermal activity, are comprised of the nearly vertical walls of the carbonate platforms of west Florida and the Yucatan. These platforms represent one of the most extensive deep hardground habitats in the world's oceans (Poag, 1991). Study of the ecology and biodiversity of these deep habitats is in its infancy.

Its location within the larger North Atlantic circulation and its abundance of special habitats make the deep GoM a likely source—and center of distribution—for seep and hardground fauna. This conjecture is now under examination.

***Finding 5.14.*** *The deep Gulf of Mexico's supporting services of water circulation and stable nutrient resupply to the photic zone and regulating service of gas and climate regulation are of broad benefit to all people. These provide large-scale and long-term benefits that involve the entire global ocean. Additionally, the deep biota may play a critical role in maintaining the diversity and resilience of the larger Atlantic Ocean.*

## Regulating Services

*Pollution Attenuation* The ecosystem service of pollution attenuation as it pertains to hydrocarbon degradation (breakdown to smaller organic compounds) is well documented for shallow marine and intertidal environments (see Chapter 4), but was unknown for the deep-sea water column prior to the *DWH* oil spill. The release of crude oil from the spill site, creating a plume of hydrocarbons in the deep GoM, provided unique impetus and resources for a scientific evaluation of the ability of natural communities of marine bacteria to attenuate hydrocarbons in situ in deep water. Several groups of researchers, working on different ships and at different times from the start of the oil spill, documented the indigenous microbial response to this deep-sea plume by comparing microbial abundance, diversity, and selected activities inside the plume with similar attributes of the surrounding deep water (Camilli et al., 2010; Hazen et al., 2010; Mason et al., 2012; Valentine et al., 2010).

Within several weeks of the initial accident, the total number of bacteria inside the plume was twice as high as in surrounding waters ($5.51 \times 10^4$ cells $ml^{-1}$ compared to $2.73 \times 10^4$) and bacterial community diversity had dropped (Hazen et al., 2010), with known hydrocarbon degraders becoming dominant (Abbriano et al., 2011; Lu et al., 2011: Valentine et al., 2010). Of particular importance were species of *Oceanospirillales*, a genus known to degrade simple aliphatic hydrocarbons; *Cycloclasticus* ("ring-breaker"), a genus first discovered in marine sediments, named for its ability to degrade complex aromatic hydrocarbons (Dyksterhouse et al., 1995), and known to occur in GoM waters prior to the *DWH* oil spill (Geiselbrecht et al., 1998); and *Colwellia*, a genus first described from nearby deep Caribbean waters for adaptation to the low temperature and high pressure of the deep sea (Deming et al., 1988). The latter two groups were confirmed experimentally to degrade hydrocarbons in seawater samples collected directly from the oil plume in the deep GoM (Redmond and Valentine, 2012). Microbial community succession was thus deduced from available studies conducted during the spill (Abbriano et al., 2011) to proceed from dominance by *Oceanospirillales* in the first weeks, to *Cycloclasticus* and *Colwellia* by June, and to methanotrophic (methane-oxidizing) genera, including *Methylococcaceae*, *Methylophaga*, and *Methylophilaceae*, by September. This succession implied that the sequential process for pollution attenuation was degradation of simple hydrocarbons, followed by more complex hydrocarbons, and finally methane gas (Kessler et al., 2011). Within-plume measurements of oxygen consumption, inherent to all of this microbial study, fell short of a depletion level that would threaten fisheries (Camilli et al., 2010). Indeed, as oxygen (and nitrogen) is depleted in the water column, the causative microbial activity itself will slow.

> ***Finding 5.15.*** *Attenuation of the* DWH *oil spill by natural populations of deep-sea oil-degrading bacteria represents a novel interpretation of this category of ecosystem service, and one of potentially high value in an overall cost-benefit analysis of the impacts of the* DWH *oil spill.*

The deep waters of the GoM clearly provide the ecosystem service of pollution attenuation that can foster or enhance hydrocarbon bioremediation. A valuation of this service could be attempted based on an estimate of the amounts of hydrocarbons degraded in the plume and the costs of other means of attenuating that amount. The valuation would also require some understanding of the known costs of other remediation efforts and where they were applied (Chapter 4). Complicating such a valuation is that other attenuation means are not typically compound specific, just as in situ bioremediation can leave many components of crude oil unaltered.

Marine sediments in the deep GoM, as hosts to natural hydrocarbon seeps, harbor marine bacteria capable of degrading hydrocarbons (Geiselbrecht et al., 1998; Orcutt et al., 2005). Until recently, this capability was understood to be dependent on the availability of dissolved oxygen, supplied from overlying oxygenated water, as it is in other marine environments. Oxygen depletion events would slow or halt hydrocarbon attenuation. If indeed restricted to oxygenated settings, then the tradeoff between pollution attenuation and oxygen consumption would place limits on the value of this ecosystem service in marine sediments, where only upper sediments near the sediment-water interface remain oxygenated. Recent research on hydrocarbon degradation within anoxic strata of marine sediments, including from the deep GoM, clarifies that degradation can proceed in the absence of free oxygen through the activities of newly discovered sulfate-reducing bacteria (Kniemeyer et al., 2007). Severe oxygen depletion would be required in the water column (e.g., in the oil plume) to allow the activities of these anaerobic bacteria to dominate, but natural communities of heterotrophic sulfate-reducing bacteria perform near continuously within the oxygen-depleted strata of the sediments that blanket the GoM floor. Unknown is whether the specific types of hydrocarbon-degrading sulfate-reducing bacteria discovered at natural oil seeps in the GoM (Kniemeyer et al., 2007) also occur throughout the other anoxic sediments of the GoM, including those impacted by the *DWH* oil spill.

Just as indigenous hydrocarbon-degrading bacteria responded to an injection of oil into deep waters (Abbriano et al., 2011), indigenous hydrocarbon-degrading bacteria in oxygenated surface sediments, and potentially within anoxic strata, may be attenuating pollution in the formation of deposited hydrocarbons, even if slowly or intermittently over time. Deposits of undegraded hydrocarbons from the *DWH* oil spill onto sediments (Joye, 2013; Joye et al., 2011) will be buried rapidly by bioturbation (the oil sequestered in mud-dwelling animal burrows) and over time by natural sedimentation processes. Buried oil will continue to be subjected to long-term attenuation by bacteria dwelling within the sediments. The limits on hydrocarbon bioavailability over time (as described in Chapter 4) may be partially alleviated by surfactants produced naturally by sediment-ingesting biota (Mayer et al., 1996).

Deep muds of the GoM thus provide an ongoing ecosystem service of pollutant attenuation that is currently understood only in limited, largely qualitative ways. Baseline information on the magnitude of in situ hydrocarbon degradation by marine bacteria inhabiting the vast seafloor sediments of the deep GoM is not available. Obtaining measurements of in situ rates of hydrocarbon degradation in GoM sediments and scaling the impacts accordingly would facilitate a quantitative valuation of this ecosystem service. Methods to estimate in situ rates exist, as applied to evaluations of the fate of pollutants more toxic and recalcitrant than light-crude hydrocarbons (e.g., phenanthrene) in coastal marine sediments elsewhere (Tang et al., 2006).

## Pollutant Attenuation Valuation

Valuation of pollutant attenuation (or waste treatment) by natural systems in theory is not difficult and in practice has been conducted primarily in fresh- and saltwater wetlands for the coastal and marine environment (Breaux et al., 1995; Kazmierczak, 2001; Su and Zhang, 2007). Studies focusing on pollutant attenuation/waste regulation primarily rely on the replacement cost or treatment cost approach. As described in the Interim Report (NRC, 2011), this approach calculates the cost of providing the ecosystem services, such as pollutant attenuation, in an alternative way, such as replacing the service provided by ecosystems with a human-engineered approach. This approach only focuses on cost and therefore is not a direct measure of benefits, except in very rare circumstances: (a) when there is no difference in the quantity and quality of the service; (b) the engineered alternative is the least costly; and (c) society demands these services (Shabman and Batie, 1978).

Determination of the suitability of the replacement cost or treatment cost approach to quantify the value of pollution attenuation, specifically hydrocarbons, requires confirmation that the GoM can perform this service and that society needs this service.

As described earlier in this case study, the GoM is well equipped to handle the breakdown of hydrocarbons into smaller organic compounds. Hydrocarbon-degrading bacteria, present in the plethora of natural seeps in the deep GoM, responded rapidly when the plume was formed, which demonstrated that the GoM environment does have the ability to attenuate pollution, specifically hydrocarbons. Has society expressed a need for this service? Given the legal requirement under NRDA to "make the public whole" and the spill responses detailed in Chapter 4, society expects that the pollution will be "cleaned up." The remaining question is: What would be the engineered equivalent of the services provided by the interaction of indigenous microbes and nitrogen recycling metazoans at depth? The techniques utilized to clean up surface oiling are well understood (see Chapter 4), but at depth, where the plume formed, it is unclear what current technology could be dispatched to "clean" the deep. However, in this case, damage to the service did not take place, nor was evidence of a limiting factor observed (nitrate was consumed during the service but not depleted). The service became manifest given the presence of the hydrocarbons. Therefore, the value to be calculated would not be for an NRDA assessment. Rather, it would be the benefit that society receives from the presence of hydrocarbon-degrading bacteria as characterized by the ecosystem services approach in Table 5.8, below.

The benefits derived from GoM pollutant attenuation have changed over the past several decades. For many years, coastal communities and inland chemical industries benefited from low-cost ocean disposal of wastes under the assumption that dilution of the pollutants by the marine waters provided a degree of protection from the toxicities of those pollutants. More recent recognition of environmental damage has since resulted in additional restrictions on

**TABLE 5.8** Provision of and Valuation for Deep Sea Pollution Attenuation

| | |
|---|---|
| | **NRDA Practices** |
| Resource | Deep Sea |
| Typical approach to assessment | Not previously assessed |
| | **Valuation Under the Ecosystem Services Approach** |
| Ecosystem Service | Pollutant Attenuation |
| Type of data needed for ecological production function | A. Circulation and mixing models.<br>B. Rates of microbial attenuation of pollutants. |
| Ecological production function | 1. Relationship between physics, geomorphology, and microbial degradation to estimate the rate of natural removal of pollutants. |
| Type of data needed for valuation | 1. Engineered cost for pollution removal.<br>2. Amount of hydrocarbons released and available for attenuation.<br>3. Attenuation rates. |
| Valuation method | Replacement cost: Calculate the cost of engineering a similar pollutant removal process that is provided by the deep ocean. |
| Type of data needed for valuation of ecosystem service | 1. Data on (A) and (B) would be required to understand the potential pollution attenuation rates.<br>2. Cost data and rates of engineered attenuation equivalent to what is being provided by the deep ocean.<br><br>This approach, while theoretically possible, might be difficult to employ given the scale of the deep ocean and the ability to effectively measure the amount (and specific components) of pollutant that was attenuated. |

polluting activities, but ocean discharge of processed wastes is still a critical component of water management plans in the GoM region. In the specific case of partial attenuation of deep-water oil spills, stakeholders that are able to continue exploiting deep oil sources after a major accident will still be able to benefit.

## Greenhouse Gas and Climate Regulation

The climatic influence of the GoM sea surface on the eastern two-thirds of North America is profound, yet identifying special roles of the deep GoM in greenhouse gas and climate regulation is challenging. Certainly the amount of hydrocarbons commercially extracted from the deep GoM and subsequently burned throughout North America contributes to greenhouse gases well above natural levels of seepage. The unusual geomorphology and physical oceanography of the deep GoM may give it important roles in nutrient resupply, a critical supporting service of the deep sea, and thus in carbon dioxide fixation and carbon sequestration.

The important question of whether the deep Gulf is a net repository or source of carbon to the ocean and atmosphere, however, is unresolved. Several lines of evidence suggest that the GoM sequesters more than the average amount of carbon in its sediments. First, the GoM lacks the great depth needed to dissolve biogenic carbonates, thus entombing these planktonic particles in the thickening sedimentary column. Second, the high sedimentation rates on the continental slope near the Mississippi River (Santschi and Rowe, 2008) allow for greater preservation of organic detritus. Third, the massive deep carbonates of the West Florida and Yucatan escarpments indicate a high rate of sequestration over geological time. Last, the prevalence of microbially deposited carbonates within the sedimentary environment (mediated by methane-consuming bacteria in sulfide-rich strata) illustrates a long-term conversion and sequestration of seeping hydrocarbons. The extent to which an oil spill of the magnitude of the *DWH* oil spill may impact seafloor carbon sequestration (or hydrocarbon attenuation) in the deep GoM is unknown.

The argument that the GoM is a net exporter of carbon would seem to be supported by the increasing information on the spatial extent and seepage rates of natural hydrocarbon seeps. Actual rates remain speculative, but surface slicks caused by liquid hydrocarbons can be effectively mapped using synthetic aperture radar (SAR) (MacDonald et al., 1996) and are being found to be prevalent over offshore oil deposits. Detection of methane bubble plumes rising from the seafloor has now become much more effective with multibeam sonar systems (Weber et al., 2011).

Another potentially important greenhouse gas–regulating role may be performed by members of the deep GoM microbial ecosystem. Based on the strength of the methane-consuming response of methanotrophic microorganisms to the *DWH* oil spill, and the likelihood that methanotrophic microorganisms are prevalent throughout the deep GoM due to natural seepages, Kessler et al. (2011) suggested that the deep GoM provides a "dynamic biofilter" to large-scale methane releases in the deep, whether natural (e.g., from methane hydrates) or by accident. Methane ranks among the most potent of greenhouse gases. Evaluation of this regulating ecosystem service requires further investigation into the prevalence and rates of microbial methanotrophic activity in the deep GoM. Tradeoffs with the supporting services of the deep GoM, in particular nitrogen resupply to primary productivity, also need to be examined specifically in relation to methanotrophy.

## Provisioning Services

*Finfish, Shellfish, Marine Mammals* Provisioning services of the deep GoM seafloor are currently minimal, although deep-water finfish species, commercially fished in the Atlantic Ocean, are present. There is a limited and intermittent fishery for deep shrimp marketed as "royal reds." The surface waters overlying the deep GoM provide commercial and recreational harvests of large migratory finfish such as the bluefin tuna. The question of how the deep GoM interacts with the rich fisheries of the shallow GoM is open and in need of investigation. The old presumption that the shallow and deep are largely isolated from one another is not necessarily correct.

An example of an unexpected link between shallow and deep can be seen with deep-foraging marine mammals. Offshore upper waters support diverse populations of cetaceans that are generally distinct in species composition from those in continental shelf waters. Modern tagging technology is providing insight into the extent of deep foraging by predatory whales (Arranz et al., 2011). The deep scattering layer in the water column and the benthic boundary layer are both concentrating features for prey organisms exploited by cetaceans diving from the surface to depths as great as 1,500 m. Gouge marks in the sediments of mud volcanoes as deep as 2,100 m in the eastern Mediterranean have been attributed to Cuvier's beaked whales (Woodside et al., 2006); similar marks characteristic of Cuvier's and other beaked whales have been reported for the GoM. Cetaceans have not been commercially harvested in the GoM since the early 1900s.

As discussed in the fisheries case study, impacts on the lower levels of the trophic web have the potential to cascade up to the surface-dwelling species, which exploit the deep aggregations of living biomass at the deep scattering layer and the sediment-water interface (Abbriano et al., 2011; Kaltenberg et al., 2007). Such impacts, if they occur, might be detected by a combination of short-term and long-term population monitoring.

> ***Finding 5.16.*** *The Macondo well blowout had the potential for creating both lethal and nonlethal impacts on deep-sea species of potential future commercial value and upper-ocean species now commercially harvested.*

## Energy Oil and Gas

The U.S. EEZ in the GoM is the major offshore oil and gas production region in the United States. Historically, the continental shelf provided most of these hydrocarbons, but a transition is now under way, with the greatest contribution to energy production currently coming from depths greater than 200 m. The *DWH* blowout and persistent spill caused a loss of commercially valuable hydrocarbons, resulting in a decreased value of the formation where the drilling occurred. Relative to the combined value of all deep GoM reservoirs, the loss of commodity can be seen as small. The blowout did, however, alter the regulatory environment and cause temporary delays in new production through drilling moratoria and extended permitting times (NRC, 2011). The extent to which better-regulated and safer exploitation negatively or positively impacts the provisioning service remains to be seen.

Of the provisioning services, oil and gas extraction and production contribute significantly to the region's economic productivity. Socioeconomic impacts of offshore oil in general and deep-water oil in particular have been assessed by BOEM research programs and are monitored on a 5-year basis (Petterson et al., 2008; Tolbert, 2006). Each stage of deep-water development, from exploration to production, employs a wide range of experts and results in the contracting and hiring of salaried and wage workers both within oil companies and in the supportive service industry. The populations that benefit from these provisioning services includes geologists, petroleum engineers, and structural engineers as well as offshore workers who operate drilling rigs, production platforms, and support vessels.

## Chemical Production

The GoM was once a major source of elemental sulfur extracted as a molten liquid by super-heated steam (the Frasch process), but the last offshore facility ceased operation in 2000 because sulfur could be recovered more cheaply as a byproduct of ore processing (Kyle, 2002). Fossil-water brines extracted along with oil are a potential source of halides, but they are currently discharged into the ocean and regulated as toxic waste. The prevalence of the petrochemical industry in the northern GoM coastal zone is not due solely to the proximity of hydrocarbons. Salt domes within the zone are mined for halogens, primarily chlorine, by means of hydraulic dissolution. Chlorinated hydrocarbons are produced as solvents and feedstock for plastics manufacturing. The GoM is so geologically complex because of salt tectonics that the presence of igneous and hydrothermal ore bodies cannot be discounted. With the exception of a few xenoliths, exotic basement rock carried upward by salt movement (Stern et al., 2011), no deposits have been found to date. The *DWH* blowout did not impact the existing chemical industry or potential deep GoM chemical industry, but an increased regulatory environment for oil and gas exploration will eventually impact these industries.

## Cultural Services

### *Aesthetics and Existence*

> *When you stare into the abyss the abyss stares back at you.*
>
> *Friedrich Nietzsche*

> *I asked them to look into the Abyss, and, both dutifully and gladly, they have looked into the Abyss, and the Abyss has greeted them with grave courtesy of all objects of serious study, saying: "Interesting, am I not? And exciting, if you consider how deep I am and what dread beasts lie at my bottom. Have it well in mind that a knowledge of me contributes materially to your being whole, of well-rounded, men."*
>
> *Lionel Trilling, 1961 essay "On the Teaching of Modern Literature"*

As Lionel Trilling's essay describes, the Abyss has become a powerful metaphor in recent literary history. Vast, dark, and inhabited by seemingly terrible beasts, to Nietzsche it stood for total meaninglessness, but to the more modern writer with a nod to oceanography, it is fascinating. The public at large retains a fascination with the deep ocean and its exploration by pioneers such as William Beebe and modern visionaries such as James Cameron. There is a spiritual or cerebral satisfaction in knowing deep-sea animals or ecosystems exist, evidenced by the unwillingness to use marine mammals as a source of food and measurable as the willingness to pay for conservation.

It can be argued that the deep GoM, as with the entire vast and remote deep ocean floor, has a high existence value, but this argument requires an examination of how high that value is perceived to be. The usual pattern in society is to ascribe a high existence value to habitats that are both desirable in some manner as well as dwindling in supply. Thus, there is often an implied connection between perceived value and rarity.

Preservation of scenic wilderness otherwise threatened by overly exploitative encroachment in the National Park System is a prime example. Protection of a watershed to ensure the survival of a rare endangered species in either a technical or legal sense is another. Rare species abound in the deep ocean, and they are endangered by ongoing and planned resource exploitation that includes destructive trawling and habitat removal similar to strip mining (Carney, 1995; Ramirez-Llodra et al., 2011). The deep ocean is certainly wilderness, and human experience will be restricted to telepresence, except for scientists working with government support and explorers with access to substantial wealth. However, the vastness of the deep sea, the largest ecosystem on the planet, combined with the low level of human awareness of the system have the effect of making the existence value of the system unrecognized. Public interest and curiosity have somewhat effectively been focused on special subsystems such as hydrothermal vents, hydrocarbon seeps, and coral-supporting deep hardgrounds, but these are a tiny fraction of the total area of the deep seafloor. For management of the vast mud bottoms, the assumption of homogeneity has been clearly proven wrong by existing information on zoogeography and biodiversity (Menot et al., 2010; Rex and Etter, 2010).

## Cultural Artifacts

Submerged cultural artifacts in the deep GoM provide an aesthetic benefit to people. The lifetime of such artifacts is prolonged by conditions in the deep sea (e.g., colder temperatures, higher pressure, reduced rates of biological activity) compared to those in shallow waters, making the extended preservation of submerged cultural artifacts an ecosystem service of the deep sea. MMS/BOEM includes in its responsibility the protection of cultural artifacts under the provisions of Section 106 of the National Historic Preservation Act of 1966. The initiation and early implementation of the program in the GoM and decisions about necessary technology were fully explained by Irions (2002).[5] In the deep GoM, cultural artifacts consist primarily of shipwrecks that occur at all ocean depths, although usually clustering on the continental shelf along navigation routes between ports. More than 400 shipwrecks dating from 1625 to 1951 have been verified (BOEM, 2010) in the GoM. The deep GoM shipwrecks represent a broad historical span from colonial times to World War II, including the sinking of a German U boat.

Especially noteworthy cultural artifact studies supported by MMS/BOEM are Church et al. (2007) and Church et al. (2009), which were carried out by multi-institutional partners, including oil companies and the offshore service industry. These studies have led to the formulation of more effective archaeological survey techniques employing high-resolution seismics and the implementation of spatial restrictions intended to provide protection for the artifact. In most cases, industry submits the required surveys to BOEM as part of the permitting process,

[5] Current information about the program is available online at http://www.boem.gov/Environmental-Stewardship/Archaeology/Gulf-of-Mexico-Archaeological-Information.aspx. This link provides access to relevant documents such as survey requirements placed on industry in 2005 and revised in 2011.

although BOEM does support surveys for its own use. Industry-collected survey data are considered proprietary, but they have been shared with independent researchers on occasion. The coordinates of shipwrecks are not shared with the public because of concerns about site destruction. The BOEM Gulf of Mexico Archaeology program continues to be proactive in outreach to make the public aware of the deep artifacts. This effort is carried out through the agency's websites and through partnerships with the NOAA Ocean Exploration Program.

Deep-water oil and gas activities have definitely increased knowledge of submerged artifacts due to discovery and investigation using the advanced tools of the industry. BOEM specifications for survey and safe distances during development afford a high degree of protection. Impacts to artifacts caused by a seafloor blowout are probably limited to damage of physical structures in proximity to the well. No report of damage to a submerged cultural artifact has been reported as a result of the *DWH* oil spill.

## Impacts of the *DWH* Spill on the Deep Sea

Simplistically and under ideal circumstances, environmental impacts are assessed through comparison of baseline data for the before-impact state to data for the after-impact state (Schmitt and Osenberg, 1996). The degree to which pre-spill data constitute an adequate baseline requires a more complex examination than can be undertaken here. The data for possible impacts are considered first, followed by the data that constitute the available baseline. Much of the study of the impacts of the *DWH* oil spill on the deep sea has been conducted under the auspices of the NRDA process, and, at the time of this writing, the full NRDA results for the deep GoM have not been released. A general account of deep-sea activities is available in a status report of April 2012 (NOAA, 2012b). More detailed findings from 2010 are included in Operational Science Advisory Team's *17 December 2010 Summary Report for Sub-Sea and Sub-Surface Oil and Dispersant Detection: Sampling and Monitoring* (OSAT, 2010).[6] Research cruise activities have overlapped, but generally it can be said that water column sampling with the intent of detecting oil and understanding its trajectories began in early May 2010, with a later inclusion of faunal studies that might allow for the assessment of impacts. Taking advantage of ongoing deep-coral studies, a work plan was accepted in July 2010 for NRDA investigation of that habitat based primarily on imaging and observation of hardgrounds in the vicinity of the spill. Related investigations of hardgrounds have continued into 2012.

During phase I, May–August 2010, approximately 4,000 water and sediment samples, taken at depths greater than 200 m, were collected offshore with the primary purpose of detecting and quantifying the presence of spilled oil. These samples provided the basis of findings in OSAT (2010), in which hydrocarbon levels were consistent with those from submerged-plume models and rarely exceeded levels associated with injury except within 3 km of the blowout site.

Plans for investigation of soft-bottom impacts were put forward in July 2011 (Deepwater Benthic Communities Technical Working Group, 2011). These involved new sampling and

[6] Partially redacted study plans are available at http://www.gulfspillrestoration.noaa.gov/oil-spill/gulf-spill-data.

analysis of 65 sites selected from previous studies: 17 within 3 km of the spill, 23 within 25 km, 15 sites more than 25 km away along the modeled route of a deep hydrocarbon plume, 2 sites more than 25 km under the known route of the surface slick, and 8 reference sites previously sampled in the Deep Gulf of Mexico Benthos (DGoMB) study (Rowe and Kennicutt, 2009). The sampling resembled traditional infaunal surveys with supporting geological and chemical analysis. A multicorer was proposed as the primary sampling device. Supplementing these traditional deep-sea approaches has been the use of a sediment-profiling camera, a device useful for assessing the thickness of deposited layers. Macrofauna would be examined at each site in three cores with a surface area of 0.03 $m^2$, and meiofauna from a single core with an area of 0.01 $m^2$. A workplan to investigate large animals living on the sediment surface at 10 sites using remotely operated vehicle (ROV) imaging surveys of bottom and water column was approved in October 2011.

Plankton studies were initiated offshore in November 2010, but they seldom included samples deeper than the 200-m upper limit of the deep ocean. A work plan to sample deep mesopelagic and bathypelagic fish using midwater trawls was submitted on November 30, 2010.

In terms of the supporting and regulating ecosystem services, the primary impacts to the deep system will include reductions of dissolved oxygen and nitrate and toxic effects on biota. A blowout at the seafloor of a deep-water oil well has been recognized by regulatory agencies as a worst-case scenario since the onset of deep drilling (Carney, 1998), with the subsurface behavior of plumes examined during an experimental release (Johansen, 2000). Study of the acute effects on the deep-water column, deep bottom, and surface were initiated as part of the NRDA process as well as by separately funded independent scientists. As noted earlier, NRDA findings were not public as of this writing, but publicly available, peer-reviewed results provide a basic scenario. Because the hydrocarbons forcefully jetting into the bottom waters consisted of a wide range of particle sizes, a subsurface plume of slowly rising droplets as well as a soluble fraction was expected. Camilli et al. (2010) confirmed and documented this expectation, with little early indication of microbial consumption and associated oxygen reduction. An extensive amount of soluble component was subsequently reported (Reddy et al., 2011), as well as microbial consumption accompanied by drawdown of oxygen (Camilli et al., 2010; Valentine et al., 2012) and nitrate (Hazen et al., 2010). Thus, the *DWH* oil spill is highly likely to have impacted regional gas exchange and nutrient dynamics. Hydrocarbons, unlike natural detritus, lack nutrients such as nitrogen and phosphorus. Microbial shifts to consuming hydrocarbons will thus alter the cycling of these essential nutrients. Larger-scale mixing will reduce these impacts over time. Such an event was anticipated and modeled by BOEM-supported researchers prior to the spill using the smaller *Ixtoc 1* oil spill from 1979 as an input example, along with mixing models of the GoM (Jochens et al., 2005).

Among the damage assessments for the water column, mud bottoms, and hardgrounds, only results for the latter two have been reported. At the time of the blowout, the MMS (now BOEM) was supporting exploration and study of deep coral aggregations in the northern GoM in conjunction with regulation development. That effort was redirected to impact assessment. Dead and dying corals with brown, flocculent material on them were reported, which was reasonably attributed to exposure to hydrocarbon/dispersant plumes (White et al., 2012). Three months after the Macondo well was capped, nine sites of deep-sea coral communities, located more than 20 km from the well and at depths of 290 to 2,600 m, were investigated using a ROV, the *Jason II* (White et al., 2012). These sites included seven that had been observed in 2009, and all were healthy, showing no impact from the spill. However, one site at 11 km southwest of the well and 1,370 m deep, which would have been in the path of the hydrocarbon plume, was covered by brown flocs and had coral colonies exhibiting obvious signs of stress such as bare skeleton, tissue loss, sclerite enlargement, excess mucous production, and abnormally colored or malformed commensal ophiuroids (White et al., 2012; WHOI, 2012). Of 43 corals, 46 percent showed impacts extending to more than half of the colony, and 25 percent showed impacts exceeding 90 percent of the colony at this one site only. Strongly reinforcing the interpretation of a spill impact was the analysis of hopanoid petroleum biomarkers isolated from the floc collected from the impacted corals: comprehensive two-dimensional gas chromatography revealed a high degree of similarity with Macondo well oil (Boehm and Carragher, 2012). Some experts contend that these observed effects on the coral colony are due to nearby natural seeps or submarine landslides (Boehm and Carragher, 2012) rather than to the *DWH* oil spill, but the possibility of these alternatives is low, given the extensive surveys of the area in question (White et al., 2012).

Information indicating impacts on the large fauna on the extensive mud bottom are based on analysis of ROV video recordings (Valentine and Benfield, 2013). The bottom was surveyed at four locations north, south, east, and west 2,000 m from the *DWH* blowout preventer (BOP); a fifth site was 500 m north of the BOP. Faunal density, species richness, and species composition varied spatially in a manner consistent with impact at the 500-m site and at the western and southern 2,000-m sites. The conclusion that impacts had occurred was reinforced by observation of low numbers of apparently dead holothuroids and sea pens at those locations. Impact to mid-water zooplankton was indicated by the presence of dead pyrosomes and salps on the seafloor at all locations. Unlike the observations of impacted corals on deep hardgrounds (White et al., 2012), no collected specimens or sediments were taken for hydrocarbon analysis.

The reported impact on corals and soft-bottom biota are consistent with these observations. Effects might be sufficiently great so as to influence potential commercial upper-ocean pelagic species, which are somewhat dependent on the biomass aggregations of the deep-scattering layer and on the seafloor. Spill impact on the nonliving resources of the deep GoM may come through increased regulation and prohibitions intended to prevent environmental damage.

> ***Finding 5.17.*** *The generally low level of understanding about the deep GoM makes it very difficult to assess the full impact of the* DWH *oil spill on ecosystem services. There are few, if any, ways in which the spill will have altered the larger-scale physics of the deep GoM, leaving only the biological and biogeochemical processes subject to major effects. It is, however, possible to consider the likely impact scenarios on supporting, regulating, and provisioning services. The cultural impacts of the spill are more nebulous, but they include what might best be called a loss of wilderness.*

## Baseline Data for the Deep GoM

The U.S. Department of the Interior has the authority to manage seafloor mineral resources in federal submerged lands to the extent of the EEZ. Since receiving the initial mandate, the required tasks have been carried out under different organizational arrangements, beginning with the Bureau of Land Management and then followed by the Minerals Management Service (MMS) and the Bureau of Ocean Energy Management, Regulation and Enforcement (BOEMRE). Following the *DWH* oil spill, BOEMRE was divided into the Bureau of Ocean Energy Management (BOEM) and the Bureau of Safety and Environmental Enforcement. The critical task of identifying information needed to ensure protection of the natural and human environment has been the primary responsibility of the Environmental Studies Program (ESP), which was initiated in 1973 (NRC, 1992) and is currently a component of BOEM. ESP has a headquarters division and regional programs in the GoM, the Pacific, and Alaska.[7]

ESP has a scientifically trained staff, but it lacks a dedicated research component and, in most instances, supports information gathering by means of the Department of Interior procurement authority, which includes competitive bidding, cooperative agreements, and memoranda of understanding. Contracts to carry out studies in deep water have traditionally gone to various combinations of academic institutions and the offshore service industry. Partnerships within government have made use of research platforms and staffs of NOAA and the Biological Resources Division of the U.S. Geological Survey.

As a result of the continuing efforts of ESP and its partners and contractors, the GoM is one of the most extensively studied regions of the world's oceans on a relative basis (volumetrically, most of the global ocean remains unexplored). As will be discussed later, data gaps and inadequately addressed phenomena prevent the establishment of ecological production functions for the deep GoM. However, a basic of understanding of species inventory and the physical environment has been established for the continental shelf and the deep basin. Until the late 1980s, the primary sources of information about the deep GoM came from the sampling programs directed by Willis Pequegnat at Texas A&M University between 1964 and 1973 and were supported without regard to oil and gas development primarily by the U.S. Navy's Office of Naval Research. With the intention to bring these results into the decision-making process for pioneering deep oil leases, BOEM supported an initial synthesis of results from 111 sampling stations above 1,000 m in the U.S. EEZ between Brownsville, Texas, and Desoto Canyon (Pequegnat et al., 1976). Subsequently, a Gulf-wide synthesis of 246 stations was supported (Pequegnat, 1983). The stations were sampled primarily for megafauna on the continental slope to a basin depth of 3,658 m. Extensive hardgrounds on the West Florida Escarpment and Yucatan Peninsula prevented investigation of these eastern and southeastern regions.

The Pequegnat et al. (1976) sampling efforts collected some hydrographic data, but it was primarily a zoogeographic study and biodiversity inventory. As such, it lacked the more comprehensive suite of measurements that became standard in BOEM-initiated baseline studies on the continental shelf. These more comprehensive investigations included the South

[7] http://www.boem.gov/About-BOEM/BOEM-Regions/Index.aspx.

Texas Outer Continental Shelf Study, the Mississippi Alabama Florida study, and the Southwest Florida Shelf study, which included sediment hydrocarbon and metal contaminant measurements along with biological surveys (Carney, 1995). These studies were ecologically comprehensive in that they examined both the seafloor (benthos) and water column (pelagos). A BOEM deep-sea baseline-type study was initiated and carried out on five cruises between 1983 and 1985. The primary results of the Northern Gulf of Mexico Continental Shelf Study (NGMCS) were presented in its third- and fourth-year annual reports (Gallaway, 1986, 1987). The sampling design was not hypothesis-driven in a formal sense; rather, it incorporated Pequegnat's system of faunal zones and the need to compare eastern, central, and western planning regions. In addition to benthic fauna, alkanes were analyzed for indication of petroleum hydrocarbon, along with metals. Pelagic work was limited to water column hydrology and chemistry.

The 10 years following the completion of the NGMCS study saw the initiation of sufficient deep-oil drilling that a second comprehensive benthic study was initiated in 1999, the DGoMB project. Sampling was carried out between 2000 and 2002. Major results were published in a special volume of *Deep-Sea Research* (Rowe and Kennicutt, 2008) and a final report issued by Rowe and Kennicutt (2009).

Although lacking a substantial analysis of the water column, the DGoMB study is noteworthy with respect to future transition from habitat characterization to an ecosystem services approach. It included ecosystem function and model development in addition to the more traditional biological surveys. Sampling design was structured around specific hypotheses and included stations in close proximity to the *DWH* site. Primary productivity and carbon flux to the bottom, although not directly measured, were estimated from surface chlorophyll data. Benthic microorganisms were included as well as animal taxa.

BOEM has contracted three major studies of seep communities and has afforded this special habitat protection through issuance of a Notice to Lessees (BOEM, 2012; National Ocean Service, 2012). When protection of deep coral habitat became a major issue in the European Union, BOEM contracted a series of studies to map and assess similar systems in the GoM and elsewhere within the U.S. EEZ. The second large study of deep corals, "Lophelia II," was under way at the time of the *DWH* blowout and was converted into an NRDA task.

## Role of Industry in Baseline Development

People generally familiar with environmental permitting may assume that the offshore industry carries out substantial environmental surveys prior to any drilling activity. Were that the case, a baseline would have been determined for the thousands of wells already drilled. In fact, the Department of the Interior, in the role of manager of nationally owned offshore lands, carries out studies and prepares reports that meet National Environmental Policy Act requirements. The offshore industry has relatively minimal requirements to contribute to the overall understanding of the ecology of either the deep or shallow GoM. Much of the data developed by industry and submitted as a part of the permitting and planning process are proprietary and not available for review. The primary requirement in deep water is that the proposed drilling must avoid sensitive habitats such as seep or hardgrounds. As evidence that such systems are being avoided, BOEM accepts geophysical or video recordings of the bottom taken by ROVs. The industry does from time to time exceed BOEM's minimal baseline requirements. Any ecological data gathered, however, are considered proprietary or otherwise are not made available for independent scientific analysis.

A particularly noteworthy program in which industry makes a major contribution to data gathering and understanding of deep circulation is the BOEM Deepwater Current Monitoring on Floating Facilities program initiated in 2005. With few exceptions, floating facilities in water deeper than 400 m must acoustically monitor water movement through most of the water column. The data are publicly available online from the National Data Buoy Center (Bender and DiMarco, 2008).

### The Heavily Studied But So-Poorly-Known Contradiction

Given that BOEM studies have provided so much information about the deep GoM, it is critically important to understand why the region has often been characterized as "poorly known" during and after the *DWH* spill. Beyond a lack of familiarity with GoM studies on the part of some spill responders, four causes can be suggested and corrected: (1) sparse data, (2) data less suited to serving as a baseline, (3) limited ability to access data collected over three decades, and (4) lack of conceptual frameworks that produce useful syntheses of existing data.

With respect to sparse data, a major challenge to managing the deep GoM or any other portion of the United States' deep EEZ is the problem of obtaining adequate data density to support management decisions and to provide a baseline from which to assess any damage due to exploitation. The deep GoM is a very large area that must be sampled by very small devices for understanding ecology and ecosystem services. This is especially the case for biotic inventories, sediment analyses, and benthic metabolism. How little of the actual deep habitat is actually sampled can be seen from an examination of the two major biological surveys of the deep GoM bottom: the NGMCS study (Pequegnat et al., 1990) and the DGoMB study (Rowe and Kennicutt, 2008). In the relatively recent DGoMB study, the sedimentary biota was sampled with a corer 271 times for a total area sampled of only 46 $m^2$ (Wei et al., 2010b). The older NGMCS study took 324 cores for similar analysis, but these were of a smaller size and the total area sampled was only about 20 $m^2$. Thus the major portion of deep GoM sampling that has contributed to habitat classification and that might be used as a baseline for the deep sedimentary environment has covered a seafloor area of less than 70 $m^2$. Given that the deep GoM is not a homogeneous region, it remains grossly undersampled. The deep water column biota has largely been ignored: there are few pre-spill data sets to either characterize the habitat or serve as a baseline.

With respect to collecting relevant data, management agencies such as BOEM are faced with the challenge of balancing regulatory obligations and limited budgets with evolving strategies of natural systems management and an incomplete understanding of what parts of natural systems are most important. Based on prior recommendations to increase data density and gather data relevant to ecosystem functioning (NRC, 1992), it might be useful to assess the progress and limitations of deep GoM understanding of the past 20 years. As part of that assessment, the data requirements of the ecosystem services approach can be determined.

With respect to data access, all reports produced by ESP since its inception are available online at the Environmental Studies Program Information System.[8] Studies completed prior to the use of digital media are available as scanned images. The lack of a central topic index and the lack of geospatial information other than regional seriously limit the utility of these reports. The extent to which the reports include the critically important raw and processed data is highly variable. BOEM partners with the National Oceanographic Data Center (NODC) of NOAA for data archival. Much oceanographic data collected before 1995, however, remain in the Multi-Discipline Archives Retrieval System format, which is no longer supported. In addition to ESP studies, BOEM obtains information about the deep environment from industry during the process of approving exploration and development plans. This information includes video surveys to determine if protected habitats are present and seismic data for multiple purposes. Although of considerable ecological relevance, these data are proprietary and unavailable for independent analysis.

Whether or not we can obtain and facilitate access to more data of greater relevance depends upon how the data are to be utilized. As more and more data are being collected, there is a growing need for an integrative process for search, retrieval, and analyses.

> ***Finding 5.18.*** *As discussed throughout this report, contemporary management of deep-sea resources will benefit from the adoption of new perspectives such as ecosystem-based management, ecosystem services approach, and management for resilience. For each of these perspectives, new ideas are being proposed and critically examined. Meshing this process of idea evaluation with appropriate and adequate field data is a critical activity that should engage the participation of the academic community, federal agencies, and the offshore industries.*

BOEM's deep-water program initiated a limited geospatial ecological synthesis effort in the form of Grid Programmatic Environmental Assessments (GPEAs). The deep GoM was divided initially into 17 and later 18 regions (Richardson et al., 2008). Depending primarily upon proprietary data submitted as part of exploration and development plans, BOEM would determine the adequacy of existing information. For example, *DWH*'s Mississippi Canyon lease block 252 is in grid cell 16. MMS issued an environmental assessment for this lease block in 2002 (MMS, 2002), indicating that the data were judged to be adequate—a conclusion that might profitably be reexamined in light of events. The effectiveness of GPEA executions prior to the *DWH* oil spill can be argued, but with more careful examination of criteria and with all data available and accessible, a similar approach may be useful going forward.

[8] http://www.data.boem.gov/homepg/data_center/other/espis/espismaster.asp?appid=1.

## Deep-Sea GoM Conclusions

The *DWH* oil spill occurred within the deep habitats of the GoM, producing a sustained buoyant plume that rose to the surface, crossing the density gradient (pycnocline) and contaminating both the slowly mixed lower and more rapidly mixed upper water column. In addition, nonbuoyant plumes remained in the lower volume, contaminating water and bottom areas as these plumes intersected the continental slope. The environmental impacts of these contaminations are being investigated as part of the NRDA process, but few results had been released at the time of this writing. That work requires the time-consuming processing of samples of small fauna. In the case of deep corals that are more easily observed in the field, however, areas of injury consistent with exposure to deep plumes have been documented. Because of the increasing exploitation of deep-water areas, including the GoM, deep habitats will be increasingly at risk.

With respect to the deep-sea environment of the GoM, an ecosystem services approach affords the potential to greatly improve the effectiveness of environmental management, especially as the ecosystem services approach is refined to better consider the ecological functions of this very large but poorly understood system. In addition, this approach is directed at an understanding of ecosystem interactions rather than the promotion of species inventory and habitat classification. It is likely that microbial and faunal contributions to hydrocarbon attenuation, carbon sequestration, and nutrient recycling are critical ecosystem services that may require careful management. Actual assessment of ecosystem services and their distribution across the deep GoM will, however, require substantial innovation both conceptually and technologically. Work needs to be done from a thoughtfully developed conceptual base, guiding careful sampling, with access to integrative tools for analysis and synthesis.

Of the ecosystem services considered, the linked biological, geochemical, geological, and physical systems of the GoM interact to provide critical supporting and regulating services. The exact nature, rates, and distribution of the interactions and resulting services present many critical questions. How does the complex circulation across two sills and the Caribbean prevent deep hypoxia within the GoM basin under normal and spill-impacted conditions? How does the complex deep circulation interact with the Loop Current to determine the distribution of recycled nutrients within the GoM and the larger Atlantic? From a comprehensive perspective, what is the carbon balance of the GoM: is the deep a net sink or a net source?

Of very special interest is the ability of the microbial system, closely linked to the physical system, to consume naturally seeping liquid and gas hydrocarbons. How does this attenuation capacity vary within the GoM, as well as other regions of the global ocean? How does it impact nutrient and gas balance in slowly mixed deep water? Can this capacity be decreased by industrial accidents or other mismanagement? Through the development of broad-based knowledge of the ecosystem dynamics of the deep GoM, required for an ecosystem services approach, we can hope to answer many of these fundamental questions.

## SUMMARY

The four case studies presented in this chapter (wetlands, fisheries, marine mammals, and the deep sea) were chosen to provide examples of how an ecosystem services approach may be applied to assess the impact of the *DWH* oil spill on several key ecosystem services in the GoM. They represent a range of conditions with respect to the amount and utility of available data, our fundamental understanding of the functioning of the ecosystem subcomponents, the values of the services in market and nonmarket terms, and the range of the impacts of the spill on the services. As such, they serve as exemplars of how an ecosystem services approach can add to the ability to capture the full impact of an event such as the *DWH* oil spill and, at the same time, illustrate the challenges faced when attempting this approach.

The case studies should make it clear that, within the GoM, some ecosystem services (e.g., storm mitigation from wetlands) are associated with years of research and baseline measurements, which creates a situation in which adequate ecological production functions and valuation processes exist to carry out an ecosystem services approach to damage assessment, with a high likelihood that the result will provide a more holistic view of the impact of the *DWH* oil spill and a wider range of restoration options. In the case of the ecosystem services provided by fisheries, valuation techniques are well established (at least for the provisional services) and a significant amount of baseline data exist, but these data suffer from a lack of spatial specificity, which affects the ability to assess impacts on the current and future productivity of the fisheries. The final two examples (marine mammals and the deep GoM) highlight the difficulties in estimating the full range of impacts when the current database, level of understanding of ecosystem interactions, and approaches to valuation are clearly inadequate. Nonetheless, in each case, the potential benefits of an ecosystem services approach are outlined. The next chapter discusses the research efforts that are needed to realize these benefits.

CHAPTER SIX

# *Research Needs in Support of Understanding Ecosystem Services in the Gulf of Mexico*

## INTRODUCTION

As has been discussed throughout this report, the concept of ecosystem services offers the potential to view human-ecosystem interactions in a holistic fashion that may provide decision makers with broadened opportunities to understand the impacts of natural and human-caused events in complex systems such as the Gulf of Mexico (GoM). An ecosystem services approach to damage assessment may also offer managers a broader array of restoration options as well as strategies to ensure long-term resilience. Although an ecosystem services approach to damage assessment offers many advantages, its application in real-world situations faces a number of challenges, many of them revolving around a limited mechanistic understanding of the detailed linkages and interactions in an ecosystem as complex as the GoM. Other fundamental challenges involve how to relate changes in the ecosystem directly to benefits (or deficits) in human well-being and how to manage the potentially complex tradeoffs among the benefits (or deficits) when selecting from various restoration strategies.

Some ecosystem services are well defined and well documented—for example, wetland mitigation of storm impacts or contribution to fisheries production (see Chapter 5). For some of these well-defined ecosystem services, methods exist to establish a monetary value of the service, which provides one measure of benefits. For less-well-defined ecosystem services, the understanding of relationships between ecosystems and the value humans derive from them is still in a nascent stage. Considerations of ecosystem services are particularly important in the GoM region, where the livelihoods and lifestyles of many coastal communities have been closely tied to access to and quality of GoM ecosystems, often for multiple generations.

Previous chapters have outlined many of the challenges to application of an ecosystem services approach to understand the impact of an event such as the *Deepwater Horizon* (*DWH*) oil spill on the GoM ecosystem. This final chapter focuses on the question posed in Statement of Task 8: "What long-term research activities and observational systems are needed to understand, monitor, and value trends and variations in ecosystem services and to allow the calculation of indices to compare with benchmark levels as recovery goals for ecosystem services in the Gulf of Mexico?" In the fleshing out of these research and observational needs, it is hoped that the groundwork will be laid for establishing methodologies that will facilitate the application of an ecosystem services approach in the GoM and other ecologically sensitive regions.

## CONTEXT OF RESEARCH IN THE GULF OF MEXICO IN LIGHT OF THE *DWH* OIL SPILL

### Directives and Funding for Restoration

Long-term research and data collection activities to support an ecosystem services approach to GoM management will be conducted in a framework of legislative directives and focused funding that stem from government actions and monetary settlements resulting from the *DWH* oil spill.

To supplement the traditional Natural Resource Damage Assessment (NRDA) process (see Chapter 2), which began almost immediately after the spill, President Obama directed, through Executive Order 13554, the establishment of the Gulf Coast Ecosystem Restoration Task Force and charged it with developing a "Gulf of Mexico Regional Ecosystem Restoration Strategy" within a year of the order.[1] This strategy (Gulf Coast Ecosystem Restoration Task Force, 2011) proposed a science-based Gulf Coast ecosystem restoration agenda, which included goals for ecosystem restoration, development of a set of performance indicators to track progress, and a means of coordinating intergovernmental restoration efforts guided by shared priorities. Federal efforts were to be efficiently integrated with those of local stakeholders, and a particular focus was to be given to innovative solutions and complex, large-scale restoration projects. The strategy also identified the monitoring, research, and scientific assessments needed to support decision making for ecosystem restoration and to evaluate existing monitoring programs and gaps in current data collection.

The U.S. Congress passed the Resources and Ecosystem Sustainability, Tourist Opportunities, and Revived Economies of the Gulf Coast States Act of 2012 (RESTORE Act) on June 29, 2012, and President Obama signed it into law on July 6, 2012. The RESTORE Act allocates 80 percent of the administrative and civil penalties levied under the Clean Water Act (CWA) to the Gulf Restoration Trust Fund within the Treasury Department (Figure 6.1). The remaining 20 percent of the CWA fines will go to the Oil Spill Liability Trust Fund. The amount of civil penalties that may be recovered, as well as the timing of recovery, is currently unknown, but it is very likely that the amount will involve billions of dollars.

Thirty percent of the RESTORE Act funds will be managed by the Gulf Coast Ecosystem Restoration Council, an entity that replaced the Gulf Coast Ecosystem Restoration Task Force through an Executive Order dated September 10, 2012.[2] The Council is charged with developing a comprehensive plan for ecosystem restoration in the Gulf Coast (Comprehensive Plan), that includes "identifying projects and programs aimed at restoring and protecting the natural resources and ecosystems of the Gulf Coast region, to be funded from a portion of the Trust Fund; establishing such other advisory committees as may be necessary to assist the Gulf Restoration Council, including a scientific advisory committee and a committee to advise the Gulf Restoration Council on public policy issues; gathering information relevant to Gulf Coast resto-

[1] Exec. Order No. 13554, 75 Fed. Reg. 62313 (Oct. 8, 2010). Available at http://www.epa.gov/gcertf/pdfs/GulfCoastReport_Full_12-04_508-1.pdf.

[2] Exec. Order No. 13626, 77 Fed. Reg. 56749 (Sept. 10, 2012).

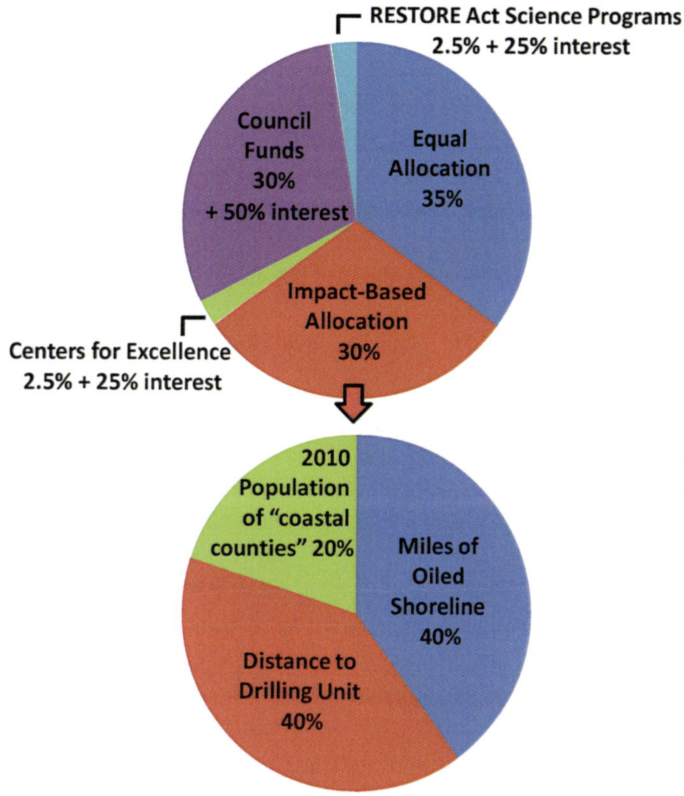


**FIGURE 6.1** Distribution of RESTORE Act funds. SOURCE: Based on Tulane Institute on Water Resources Law and Policy (2013).

ration, including through research, modeling, and monitoring; and providing an annual report to the Congress on implementation progress." Thirty-five percent of the RESTORE Act funds will be divided equally among the five impacted states for ecological restoration, economic development, and tourism promotion; 30 percent will be divided among the states according to a formula to implement individual state expenditure plans as approved by the Council; 2.5 percent will be given to the National Oceanic and Atmospheric Administration (NOAA) for a monitoring, observation, and technology program; and the final 2.5 percent will be used by the five impacted states for the establishment of research centers of excellence. The CWA is only one of the potential grounds for statutory liability for damage caused by the spill for the responsible parties; the Oil Pollution Act, discussed below, is another.

In May 2010, BP Production and Exploration committed $500 million over a 10-year period to create a broad, independent research program to be conducted at research institutions primarily in the Gulf Coast states.[3] BP initially granted year-one block grants in June 2010 to Gulf

[3] http://research.gulfresearchinitiative.org/.

Coast state institutions and the National Institutes of Health to establish critical baseline data as the foundation for subsequent research as well as to support study of the health of the oil spill workers and volunteers. Smaller "bridge" grants, totaling $1.5 million, were awarded in the summer of 2011, allowing researchers with National Science Foundation (NSF) Rapid Response Research (RAPID) grants to complete their research. The first major competition for research consortia was completed in fall 2011, with the awarding of $112.5 million in grants to eight research consortia. These consortia completed their first year of activities in December 2012. Smaller investigator grants totaling $22.5 million were awarded in fall 2012.[4] A new request for proposals for research directed to larger overall research questions will entertain proposals in the early part of 2014 for the next 3-year period.[5]

Additionally, a November 5, 2012, settlement between the U.S. Department of Justice and BP Exploration and Production, for the latter's guilty plea for a number of violations related to the Macondo blowout, provides several billion dollars for restoration, research, education, outreach, and monitoring.[6] An additional settlement between the U.S. Department of Justice and Transocean Deepwater Inc. was reached on February 14, 2013, regarding Transocean's violation of the CWA via the *DWH* oil spill.[7] Included in these two settlements were awards of nearly $2.5 billion dollars to the National Fish and Wildlife Foundation to remedy harm and eliminate or reduce the risk of future harm to Gulf Coast natural resources and $500 million dollars to the National Academy of Sciences for a program focused on human health and environmental protection, including issues relating to offshore drilling and hydrocarbon production and transportation in the GoM and on the Outer Continental Shelf.

Additional research funds are being provided by federal entities, states, nongovernmental organizations, and private foundations. For instance, almost immediately following the spill, the NSF made available significant funds from its programs to support *DWH* oil spill research. More than 150 RAPID grants and Major Research Instrumentation (MRI) RAPID grants were awarded to ocean scientists for research directed to certain aspects of the spill. Subsequent NSF funding has been awarded through the competitive grants system. Thus, from an unprecedented environmental event, significant funding has become available to conduct research across a broad spectrum of issues.

The final amount of funds to be directed to various activities through potentially numerous avenues depends on the outcome of legal proceedings between the Department of Justice and the responsible parties. The NRDA process and fines determined under the Oil Pollution Act, the establishment and collection of penalties under the CWA, and plans for Gulf Coast restoration create unprecedented opportunities for research and restoration, to address not only the damage caused by the spill but also the longer-term impacts of a number of human activities on the GoM ecosystem. Restoration will directly respond to damages from the spill, but it offers the opportunity for a system recovery that considers human and natural systems resilience.

---

[4] http://www.unols.org/meetings/2010/201010anu/201010anuap09.pdf.

[5] http://gulfresearchinitiative.org/2013/save-the-date-gomri-rfp-for-2015-2017-research-consortia.

[6] http://www.justice.gov/opa/pr/2012/November/12-ag-1369.html.

[7] http://yosemite.epa.gov/opa/admpress.nsf/0/B80C763F3BFC11A385257B1200788616.

***Finding 6.1.*** *The funding from the settlements stemming from the* DWH *oil spill, coupled with public- and private-sector investment, present an unprecedented opportunity to create research and restoration programs that could establish a level of understanding and baseline information about the GoM that may overcome the challenges faced in fully evaluating ecosystem services in this vulnerable region.*

## Research Activities

The early recognition of the *DWH* oil spill as an environmental disaster of unprecedented proportion led to a rapid response by researchers from many sectors. As outlined in the Interim Report (NRC, 2011), immediately after the spill began, the federal government initiated a large-scale sampling program to fulfill its obligation to assess the injury to the public under the NRDA process. The data collected in support of NRDA represent some of the most comprehensive sampling that has ever been done in the GoM. As discussed in the Interim Report and in Chapter 2, these data are intended to support traditional damage assessment approaches (i.e., losses are generally measured in simple ecological terms rather than in terms of reductions in the value of ecosystem services), and therefore may not be directly useful for enhancing our understanding of ecological production functions and ecosystem services. Data collection and analyses continue under NRDA, and some of the data have been made public,[8] but many of the results will remain confidential until litigation is complete. Nonetheless, once fully available to the public and the research community, the NRDA data sets will inevitably be an invaluable addition to the overall database and understanding of the GoM.

Along with the federal government, researchers from private industry, universities, research institutions, and nongovernmental organizations (NGOs) moved as quickly as possible to begin the scientific assessment of the oceanic and onshore processes and impacts. Methodologies included collection of biota census data (with multiple variables), the Before and After Control Impact (BACI) approach (in those situations where scientists were able to assess habitats before and after exposure to the oil), comparisons of assorted environmental variables to long-term data, if available, and deployment of sensitive and state-of-the-art instrumented arrays in the deep ocean and atmosphere. Although the process is still young, the research results, unless confidential, are being published with unprecedented speed in highly respected peer-reviewed journals. The research will continue, and findings about the impacts of the spill and the fate of the GoM ecosystem will be communicated for years to come. However, at some point in time, the mandate for the liable entities and the resource managers to conduct oil-spill-related research and interpret the findings will most likely be removed.

As outlined above, the Gulf of Mexico Research Initiative (GoMRI) has funded eight regionally led consortia to conduct leading-edge research on the GoM. GoMRI, in cooperation with a Research Board, administers the research funds to ensure high scientific standards and to isolate the research programs from the funding source. Research projects focused on the following tasks:

[8] See http://www.gulfspillrestoration.noaa.gov/.

1. Physical distribution, dispersion, and dilution of petroleum (oil and gas), its constituents, and associated contaminants (e.g., dispersants) under the action of physical oceanographic processes, air-sea interactions, and tropical storms.
2. Chemical evolution and biological degradation of the petroleum/dispersant systems and subsequent interaction with coastal, open-ocean, and deep-water ecosystems.
3. Environmental effects of the petroleum/dispersant system on the seafloor, water column, coastal waters, beach sediments, wetlands, marshes, and organisms; and the science of ecosystem recovery.
4. Technology developments for improved response, mitigation, detection, characterization, and remediation associated with oil spills and gas releases.
5. Impact of oil spills on public health.

***Finding 6.2.*** *The current research directions have the potential to greatly enhance our understanding of the GoM ecosystem. If these efforts directly address the quantification of ecosystem services, then the results of the research will better serve to improve management and the long-term health of the GoM system.*

## Research Background and Plans for the Gulf of Mexico

This section reviews what research is currently being planned and, from the perspective of ecosystem services, what additional efforts might enhance our understanding of the GoM ecosystem services and their value to the well-being of the people of the region.

### Response Technologies

At the advent of the *DWH* oil spill, of immediate concern was how to stop the flow of oil, which was pouring out at a rate of up to 60,000 barrels per day, and possibly more (McNutt et al., 2012), from a wellhead located 1,500 m below the sea surface. Additional concerns centered on how to prevent the spread of oil at the surface and into coastal waters and onto coastal landforms, as well as how to deal with the oil that made landfall. In the history of oil spill cleanup, a variety of response technologies and techniques have been developed. Some have proven more effective than others, and all need to be examined and applied in the context of prevailing environmental, habitat, and oiling conditions.

Chapter 4 of this report provides a detailed examination of the response technologies used for the *DWH* oil spill and their relative effectiveness. The questions and controversy surrounding the use of emerging technologies (e.g., subsurface injection of dispersants and in situ burning) during the spill highlight the need for new research, particularly to accurately predict the impact of dispersants on biodegradation rates and the long-term effects of dispersed oil on the food web and on ecosystem services.

The American Petroleum Institute has issued a request for proposals for the study of the use of dispersants at depth, and one of the eight GoMRI consortium programs (C-MEDS) is focusing entirely on the chemistry of dispersants.[9] Specific topics outlined in the C-MEDS research task include the design of porous particles that would stabilize at the oil-water interface and deliver dispersant and biological nutrients, the use of computational methods to establish the self-assembly of surfactants at conditions of high pressure and low temperature and to understand dispersant configuration at the oil-water interface, and the use of natural biologically derived surfactants that are hydrophobin-based proteins or polysaccharides such as cactus mucilage. A recent paper by Paris et al. (2012) concluded that the injection of dispersants into the wellhead did little to increase dispersion of the oil beyond what was already occurring from effects of shear emulsification. However, the literature review presented in Chapter 4 suggests that this is an unresolved issue in need of further investigation.

The fate and transformations of oil from the *DWH* oil spill will continue to be influenced by natural attenuation processes. Continued observations, monitoring, and research of these processes, especially in relation to technological countermeasures, are warranted. Needed are further studies that use realistic initial dispersed oil concentrations and that avoid boundary effects that increase oil-dispersant concentrations. The effectiveness and net benefits of many of the response technologies varied because of environmental conditions (such as the sea surface conditions, currents, and depth of the wellhead) as well as the timing of their application. Additional research and experience will refine and enhance the effectiveness of any of these technologies—with the ultimate goal of reducing their impacts as well as the spill's impacts on ecosystem services.

## Decision Science Tools

Ultimately, the decisions made about the use of response technologies will be tied into a net environmental benefit analysis (NEBA) (see Chapter 4). The extension of these analyses is the construction of "influence diagrams," a graphical network showing probabilistic relationships and how decisions are likely to affect various outcomes (see Figure 6.2).

In retrospect, influence diagrams could have helped in the consideration of resource value and potential risk during the development of decision models for emergency response prior to the spill (Carriger and Barron, 2011). For such diagrams to be reliable, an understanding of the mechanistic linkages among ecosystem processes is required. The technologies used to respond to the *DWH* oil spill and the data from the associated impacts derived from continued research should assist in the creation of these diagrams.

An interactive, layered-mapping Web application called 'SPECIESMAP' has been designed to help fill knowledge gaps regarding potential *DWH* oil spill impacts to fish species such as relocation of spawning grounds, bioaccumulation of hydrocarbons, altered migration routes, expanding hypoxic dead zones, affected life-history stages, reduced populations, and extinctions (Chakrabarty et al., 2012). Additionally, influence diagraming would have benefited from information about the circulation of water masses within the GoM—another identified knowl-

[9] http://dispersant.tulane.edu/.

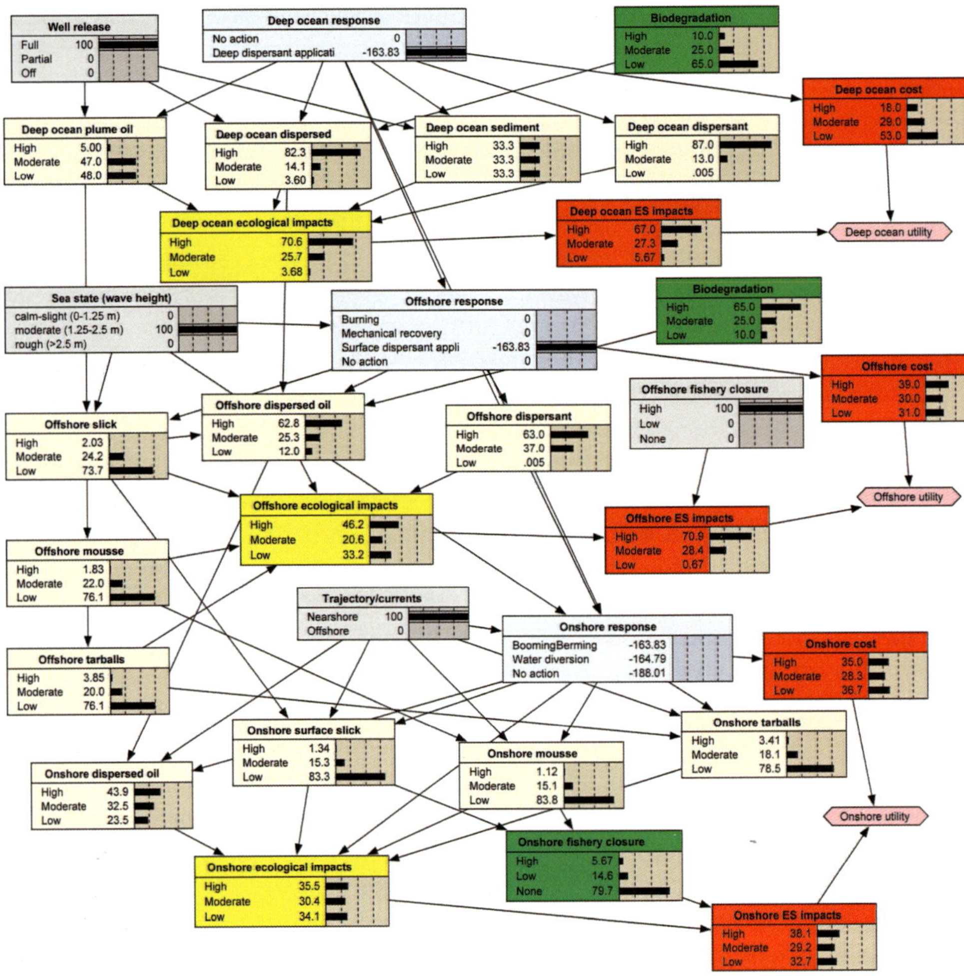


**FIGURE 6.2** Influence diagram for the *DWH* oil spill. SOURCE: Reprinted with permission from Carriger and Barron (2011).

edge gap—because there were fears that some of the oil from the *DWH* oil spill would pass through the Straits of Florida and contaminate the Atlantic. There is, however, no evidence that such oil passage occurred.

The government of Mexico raised concerns that oil residuals would enter Mexican waters. Recent modeling of the spill (Le Henaff et al., 2012) has shown that currents created by southerly winds drove the surface oil toward the northern shores of the GoM. These winds prevailed for the first 15 days after the blowout on April 20, 2010, and returned in June and July, which minimized the amount of oil entrained by the Loop Current and drawn into the central Gulf, so no oil impacted the shores of southern Florida or Mexico. Had information of this sort been appropriately incorporated into the decision-making process at the time of the spill, many unwarranted concerns could have been addressed.

Consequently, along with research on the specific impact of a particular response technology (e.g., the effect of dispersants on biodegradation), research is needed to fill the knowledge gaps associated with our understanding of ecosystem dynamics, which is discussed in the next section.

> ***Finding 6.3.*** *The response technologies applied during the* DWH *oil spill resulted in both beneficial and detrimental effects on ecosystems and the services they provide to humans. The development of additional guidelines to improve the effectiveness of various technologies, and protocols to monitor the impacts on ecosystems and the services they provide, are of paramount importance. The decisions that determine when and where to apply these technologies are guided by decision science tools such as net environmental benefit analyses and influence diagrams, which can only be effective if there is an understanding of the mechanistic linkages among the ecosystem processes.*

## Background Research and Data

At the onset of the *DWH* oil spill, there was great concern about the lack of background research and data, particularly for the blue-water GoM and the deep-water environments of the blowout area. Subsequent efforts to collect data revealed that, for some areas and some variables, a significant amount of pre-spill data were available, but for others, pre-spill data were severely lacking. In some cases, the lack of an organizing structure and aggregated body of knowledge to make sense of the data was more limiting than the lack of data per se.

The former Minerals Management Service (now Bureau of Ocean Energy Management, BOEM) has supported deep-water GoM research since the early 1970s. The Department of the Navy funded earlier zoological expeditions to the deep GoM. Although found primarily in gray-literature reports, this information is available. As a result of the BOEM Environmental Studies Program's historic and continuing efforts, the GoM is one of the more extensively sampled regions of the world's oceans, yet this sampling represents only a tiny fraction of the region's volume. Although many data gaps and inadequately addressed phenomena exist, a basic level of understanding has been established for the continental shelf and the deep basin (see Chapter 5, "The Deep Gulf of Mexico"). Similarly, research on methane seeps, related processes, and the biodiversity of organisms on escarpments, salt domes, and hard banks throughout the deeper GoM is further advanced than for many other deep oceans and is continuing. Much of the research on oil and methane seeps has been funded by NSF as well as BOEM. Genomic data from biodiversity studies of the shelf edge and slope banks are stored at GenBank[10] and BOLD (Barcode of Life)[11] and are fully available to the public.

---

[10] http://www.ncbi.nlm.nih.gov/genbank/GenbankSearch.html.
[11] http://www.barcodinglife.org.

Requirements for data collection, data quality control and assurance, and transfer of data with appropriate metadata to funding agencies have changed dramatically since the early 1970s, when baseline data for the GoM became an area of emphasis in the advent of oil and gas drilling and development. There was no consistent mechanism for reporting of data or metadata for many survey-type data collections such as BOEM-funded work, the fishery-independent data of the Southeast Area Monitoring and Assessment Program (SEAMAP), and many other programs in the GoM. There have been improvements in data quality assurance/quality control, methods to relay data to designated national data repository systems, Federal Geographic Data Committee metadata requirements, Data Management and Control methods to submit oceanographic data to the Integrated Ocean Observing System and the numerous regional associations, such as the Gulf of Mexico Coastal Ocean Observing System Regional Association (GCOOS-RA) and the Southeast Coastal Ocean Observing Regional Association (SECOORA), and specific mechanisms for shared open-source modeling components. It would be difficult to incorporate historic data into the new data compliance systems, but the opportunity to do so may exist in the overall need for historic data to be made available.

Less is known about the open-water ecosystem and the deep GoM, but NOAA and special BOEM-funded studies have collected long-term observational data on marine mammals. Planktonic data from NOAA fisheries surveys are part of the National Ocean Data Center or other federal data repositories. More recent tagging studies of bluefin tuna (Census of Marine Life[12] and listed publications) revealed migration patterns and a breeding area in the GoM, but much remains unknown about their larval and early life history. Background data are missing for many components of the open GoM; for example, mesopelagic fish populations and their ecology are poorly documented.

SEAMAP is a state-federal-university program for collection, management, and dissemination of fishery-independent data and information in the southeastern United States. Major activities in the GoM include surveys of shrimp, groundfish, and ichthyoplankton, collection of environmental and benthic data, mapping of live and hardbottom areas, investigations of low oxygen and other environmental perturbations, and establishment of a regional, user-friendly data management system that can be accessed by all marine management agencies in the southeastern United States. The GoM SEAMAP effort was implemented in 1981 (Eldridge, 1988), and large databases archiving the results of the field and laboratory efforts are maintained by SEAMAP in Pascagoula, Mississippi.[13] These data provide a long-term assessment of numerous organisms in the GoM.

BOEM requires ADCP (acoustic Doppler current profiler) data from any floating platform in the U.S. Outer Continental Shelf leasing areas in water depths of 400 m or greater. These data provide important input for circulation models. Other coastal ocean observing platforms exist in the GoM (see Gulf of Mexico Coastal Ocean Observing System[14] and SECOORA[15] for the location and data products generated from these systems). Furthermore, there are many satellite

[12] http://www.coml.org.

[13] http://www.gsmfc.org/seamap.html.

[14] http://www.gcoos.org.

[15] http://www.secooraedu.

observing systems, ocean modeling efforts, and interpretation activities in the GoM. Over the past several years, SURA (Southeastern Universities Research Association) has funded a "Super-Regional Testbed to Improve Models of Environmental Processes on the U.S. Atlantic and Gulf of Mexico Coasts,"[16] specifically research of and transition to operations models of estuarine hypoxia, shelf hypoxia, and coastal inundation. The shelf hypoxia effort used multiple physical and coupled physical-biogeochemical models, including the U.S. Navy's regional circulation prediction system for the GoM and Caribbean, as a baseline operational capability, along with other physical models for the northern GoM. The findings of this project will be published in the *Journal of Geophysical Research* in 2013.

Some of the best data taken as background to the oil spill, at least in the near-shore area as pre-exposure conditions, were collected before the oil reached the shoreline and served as a basis for the eventual NRDA by the Trustees under the Oil Pollution Act. As noted earlier, these data collections continue, but many of the results are being held in confidence until litigation is complete. The NRDA of *DWH* oil spill extends beyond previous deep habitat descriptions of the GoM (e.g., benthic faunal inventory and biodiversity). The NRDA collection of data in the deep sea has been intensive and spatially dense for the benthos as well as for full water column faunal inventories, resulting in much better information about these previously understudied systems. Figures 6.3 and 6.4 illustrate the numerous types and distributions of samples that will support the NRDA process as well as provide a large reservoir of background data, because not all sites will have been exposed to Macondo oil. As indicated earlier, NRDAs are primarily based upon traditional damage assessment approaches and may not be useful for enhancing the understanding of ecological production functions and ecosystem services. The eventual deposition of these data and the data generated by the GoMRI research programs into national ocean data repositories or their equivalent will provide much-needed "background" data for the northern GoM.

> ***Finding 6.4.*** *A broad spectrum of background pre-spill data has been collected as part of the NRDA process. The NRDA data, once released to the public, should contribute significantly to the body of background data that can help to characterize much of the GoM.*

Recently, Murawski and Hogarth (2013) evaluated the state of current observational programs in the GoM and concluded that some programs provide "adequate" baseline data for the following parameters: annual extent of hypoxia off Louisiana; some water quality and pathogen metrics; population abundances of some fishery and protected species; sea level height; conductivity, temperature, dissolved oxygen, and nutrient levels; and primary productivity, land-use, and wetland landscape change trends from satellite imagery. They also concluded that adequate observational data do not exist for contaminants in water and sediment; polycyclic aromatic hydrocarbons (PAHs) and metabolites in seafood; fundamental health data on fish, mammals, turtles, invertebrates, and deep ocean benthic communities; and the economic, social, and public health of the Gulf Coast. Although they pointed to a lack of fishery-indepen-

[16] http://www.itsc.uah.edu/node/248.

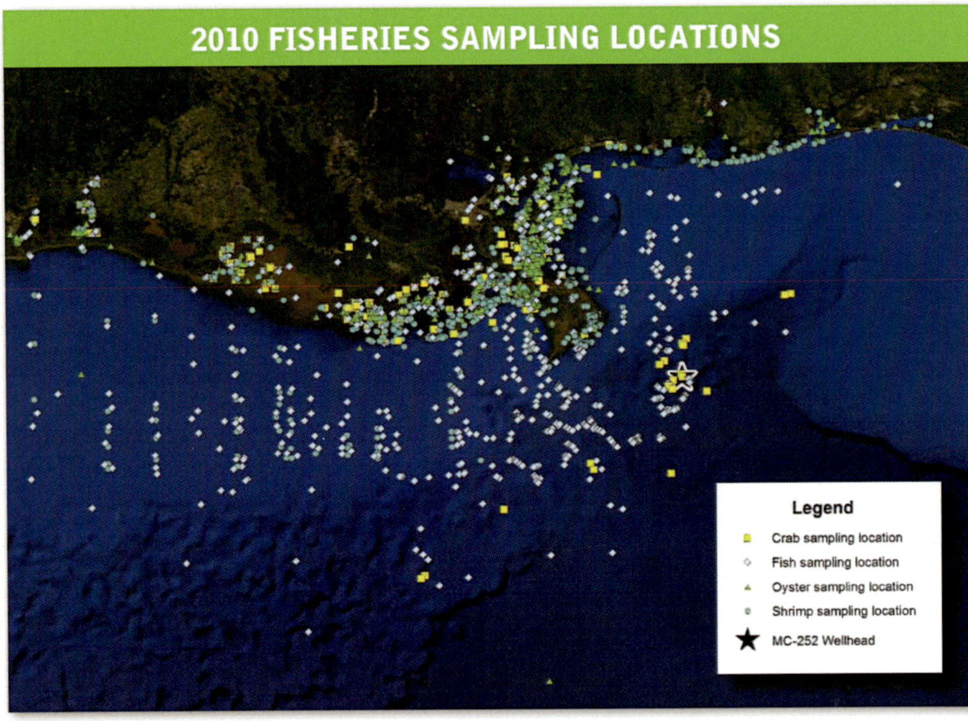


**FIGURE 6.3** Overview of offshore and deep-water stations sampled on recent (2010) *Gyre* and *Ocean Veritas* cruises and prior (2000–2002) MMS (now BOEM)-sponsored cruises (Rowe and Kennicutt 2009, Deep Gulf of Mexico Benthos sites). NOTES: Rings centered around the wellhead are 25 km apart. Sediment samples were collected with multicorers at all sites. SOURCE: Deepwater Benthic Communities Technical Working Group (2011).

dent, population abundance data and distribution data for turtles and marine mammals, these data do exist—that is, from SEAMAP,[17] NOAA,[18] and the NRDA process. Additionally, fundamental oceanographic and meteorological data are provided by moorings of the National Data Buoy Center[19] and the ADCP current meter measurements from offshore oil and gas platforms.

In summary, many groups in the GoM are building on existing data sets, ocean observing capabilities, and continuing basic observational data. For some parameters, observational data are adequate, but for others, more robust measurements are needed. In addition, current and future funding will not be sufficient to collect a full suite of background data to support mitigation of any potential disaster. Given the oil and gas reserves of the GoM, however, certain basic

[17] http://sero.nmfs.noaa.gov/grants/seamap.htm.
[18] https://www.st.nmfs.noaa.gov/marine-mammals-turtles/index.
[19] http://www.ndbc.noaa.gov/.

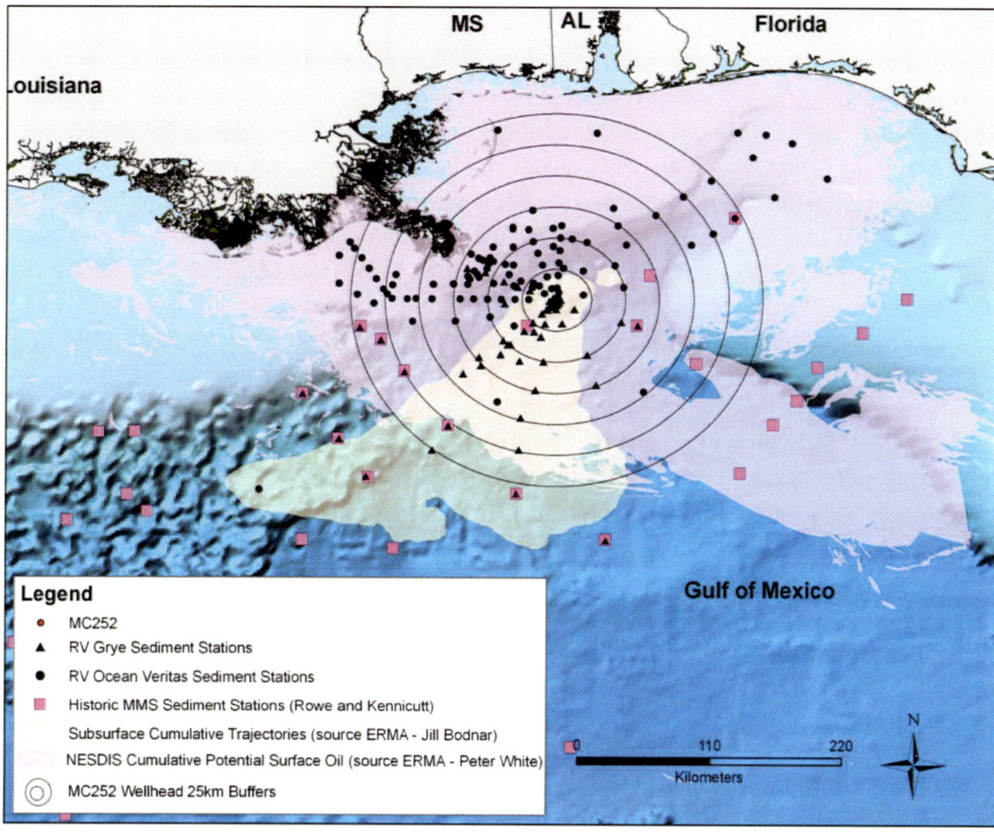


**FIGURE 6.4** Priority stations from fall 2010 response cruises (*Gyre* and *Ocean Veritas*) that were selected for the initial suite of macrofaunal and meiofaunal analyses. NOTE: Rings centered around the wellhead are 25 km apart. SOURCE: Deepwater Benthic Communities Technical Working Group (2011).

data should be collected as a part of a systematic monitoring system. There have been several efforts to compile background information during and after the oil spill, including the NOAA Deepwater Horizon Archive,[20] the Joint Analysis Group for Surface and Sub-Surface Oceanography, Oil and Dispersant Data,[21] and Naval Oceanographic Office Special Support-GOMEX Mississippi Canyon 252 Oil Spill.[22] There is, however, no integrative superstructure or roadmap to available data, the development of which will be even more essential as new data arrive from current and future research activities.

Despite the tremendous amount of funding that will be available to GoM researchers, there will likely never be enough to support all of the potentially important assessment and

[20] http://www.noaa.gov/deepwaterhorizon/.

[21] http://www.ncddc.noaa.gov/activities/healthy-oceans/jag/.

[22] http://ecowatch.ncddc.noaa.gov/JAG/Navy/.

monitoring activities needed to capture the state and function of the GoM's natural and social systems. The necessary types and amounts of baseline data that may be suddenly needed when there is an environmental disaster cannot be fully anticipated. However, as seen with the *DWH* oil spill, although there was an existing cache of long-term and baseline data, it was neither tabulated and synthesized nor accessible from a unified data system. With the advent of the *DWH* oil spill, build-out plans for physical, chemical, and biological monitoring, such as documented by the GCOOS-RA, provide a structure upon which to begin building the necessary framework for identifying baselines, physical forcing, biological communities, and ecosystem functions (Jochens and Watson, 2013). The structure and synthetic nature of such a database calls for (1) a mechanism to create, populate, and maintain relevant data sources in a public format as assessment, restoration, and monitoring activities continue and (2) a long-term mechanism for additional input and maintenance with dedicated funding.

Thus although past, current, and future data collection and monitoring efforts will likely provide much of the needed baseline data from which the impact of future events such as the *DWH* oil spill on ecosystem function and structure could be discerned, the need for an overarching structure to integrate the wealth of data that has been, is being, and will continue to be collected in the GoM is viewed by this committee as a critical research need.

> ***Finding 6.5.*** *Several federal agencies, such as NOAA and the U.S. Navy, already maintain a number of background data repositories. There is, however, no integrative superstructure or roadmap to available data. The development of such a capacity will be even more essential as new data arrive from current and future research activities in the GoM. Needed is a database that can create, populate, and maintain relevant data sources in a public format as assessment, restoration, and monitoring activities continue. Creating such a database can only be successful if there is a long-term mechanism for additional input and maintenance with dedicated funding.*

Over the past several decades, there have been several efforts to develop assessments and research and monitoring plans for the GoM. Much of the information derived from these efforts is relevant to better understanding of the GoM ecosystem, its components, and its processes, but is not entirely applicable to the needs of research concerning the fate and effect of the *DWH* oil spill. As stated throughout this report, however, a fundamental understanding of the GoM ecosystem is critical to application of an ecosystem services approach to damage assessment and restoration. Therefore, these research directions, while not explicit in their identification of ecosystem services, may enlighten the bigger picture.

The most recent process to develop a stakeholder-driven plan that identifies and addresses regional research priorities began with the Gulf of Mexico Alliance, with funding from the National Sea Grant Program, to assist GoM groups that conduct or use marine and coastal research. This regional plan is similar to seven others around the United States that also address the Ocean Research Priorities Plan (NSTC Joint Subcommittee on Ocean Science and Technology, 2007) resulting from the U.S. Ocean Commission and subsequent National Ocean Policy. Initial efforts to bring stakeholders together were set back by Hurricanes Katrina and

Rita in 2005, but eventually five workshops were held across the Gulf region to identify existing research plans and the needs of the Gulf constituency at all levels. The plan (see Box 6.1) was complete at the time of the *DWH* oil spill, and the four Gulf of Mexico Sea Grant Programs reexamined the plan immediately after the spill.[23]

In addition to the GoMRI initiatives, over the next 2 years, the GoM Regional Sea Grant Consortium, the EPA Gulf of Mexico Program, and NOAA are sponsoring $1.3 million in research aimed at improving approaches for valuing the GoM ecosystem services. In particular, the research will focus on the services provided by marshes, mangroves, and oyster reefs. The GCOOS-RA[24] has a detailed plan for "A Sustained, Integrated Ocean Observing System for the Gulf of Mexico (GCOOS): Infrastructure for Decision-making" (the Build-Out Plan; see also Jochens and Watson, 2013). The GCOOS-RA has worked over the past decade to identify stakeholder needs for data, information, and products about the GoM, its resources, and its ecosystem. These results were used by the GCOOS-RA Board to identify the key elements of the GCOOS Build-Out Plan, the first version of which is complete. However, this plan will continue to develop and evolve, especially in view of the oil spill and developing restoration plans.

NOAA and collaborators have developed the Gulf of Mexico Hypoxia Monitoring Implementation Plan for an Operational Observation System (2009, revised 2012). The Gulf of Mexico Alliance and the GCOOS-RA are developing the Harmful Algal Bloom Integrated Observing System. All of these monitoring plans were constructed with the input of multiple stakeholder groups, including representatives from academia, state and federal agencies, industry, recreational and commercial fishers, tourism interests, boaters, health service agencies, resource managers, and members of the public. Thus, the plans, and necessary implementation, serve to improve not only long-term and broad data collection and thus our understanding of the GoM, but also societal needs.

Multiple observatory systems will establish critical time series that can be used as baseline data should another event such as the *DWH* oil spill occur. The Ocean Research Priorities Plan (NSTC Joint Subcommittee on Ocean Science and Technology, 2013) emphasizes the critical need for ocean observations at scales that are temporally and spatially relevant for decisions concerning effective adaptation, management, and mitigation. At the same time, consideration should be given to maintaining vessel and satellite capability. Both science and policy call for improving integrated ocean observing capability across the United States and internationally.

## Updated Research Priorities Report

Six years have passed since the Ocean Research Priorities Plan (2007) was developed. An updated report (NSTC Joint Subcommittee on Ocean Science and Technology, 2013) says the *DWH* oil spill highlighted the "complex interplay of society and human health with chemical, physical, and biological aspects of the ocean and our coasts." The report emphasizes the "wide

[23] The revised version of the Gulf of Mexico Research Plan is available at http://masgc.org/gmrp.

[24] http://www.gcoos.org.

**BOX 6.1**
**The Gulf of Mexico Research Plan: An Opportunity to Improve Our Understanding of Ecosystem Services**

The range and scope of research needs for the Gulf of Mexico are extensive and require prioritization in order to be effective and responsible with the available funding. Fortunately, some 1,500 people from over 250 universities, government agencies, businesses, NGOs, and other organizations worked together to provide broad constituent input in the development of the Gulf of Mexico Research Plan (GMRP). The goal of the plan is to assist the GoM research community, including those who conduct or administer research or use research findings.

The highest-rated GoM research priorities were organized into five theme areas: (a) ecosystem health indicators; (b) freshwater input and hydrology; (c) habitats and living resources; (d) sea-level change, subsidence, and storm surge; and (e) water quality and nutrients. The committee found many of these priorities to be consistent with those we have identified for advancing our understanding of ecosystem services in the GoM and that would be useful for enhancing community and ecosystem resilience in preparation for future environmental challenges, whether they are episodic (i.e., oil spills or hurricanes) or chronic (i.e., sea level rise and wetland loss) events.

Of the many research priorities presented in the GMRP, only a few mention ecosystem services specifically, but most of them provide a starting point for research that would support a better understanding of ecosystem services in the GoM. Following are some key examples of how the scope of ongoing work might be enlarged toward this end. The relevant Ocean Research Priorities Plan (ORPP) priority number is identified for each focus area.

- **Understand and predict the impact of natural and anthropogenic processes on ecosystems** *(ORPP Research Priority 14).*

Understand how the current level of regional development and human-engineered infrastructure influences the flow of freshwater into the GoM and how much of that inflow and sediment discharge is needed to support healthy and functional wetlands in light of challenges such as sea level rise and habitat loss due to erosion.

There is also a suite of factors that relate to land use—pollution and storm water runoff, wastewater, and nutrient loads from agriculture—that contribute to a number of chronic problems in the GoM, such as harmful algal blooms and hypoxia zones. A key priority is to examine their impacts on ecosystem health, sea grasses, biodiversity, and higher trophic organisms.

- **Develop socioeconomic assessments and models to evaluate the impact of multiple human uses on ecosystems** *(ORPP RP15).*

Identify the social and economic drivers that influence how communities use, conserve, and protect their natural resources. It would be useful to identify and estimate the socioeconomic tradeoffs associated with changes in protecting and restoring resources and ecosystems at various scales.

- **Develop appropriate indicators and metrics for sustainable use and effective management of GoM marine resources and ecosystems** *(ORPP RP16).*

Identify appropriate tools and techniques that can evaluate the status and health of ecosystems and resources, specifically those that can be used readily and quickly in the field for the detection of harmful levels of bacteria, contaminants, toxins, and pathogenic organisms.

- **Stewardship of Natural and Cultural Resources: Understand interspecies and habitat/species relationships to support forecasting resource stability and sustainability** *(ORPP RP2).*

Identify and model the connections among and between habitats and living marine resources for the goal of supporting sound ecosystem-based resource management decisions. This would include tracking changes in habitat quality (and quantity) and in ecosystem structure and function in order to be able to identify the effects of these changes on marine organisms and their populations.

- **Understand human use patterns that may influence resource stability and sustainability** *(ORPP RP3).*

Establish the value of GoM ecosystem services, including depleted and renewable resources, to allow for informed decisions on the placement, construction, development, and expansion of large-scale commercial and industrial enterprises.

- **Increase resilience to natural hazards by developing multihazard risk assessments and models, policies, and strategies for hazard mitigation** *(ORPP RP7).*

Determine and predict the physical impacts of climate change on coastal and upland areas in terms of factors such as sea level change, rate of elevation change, shoreline change, loss of barrier islands, and change in regional hydrology and apply this knowledge in habitat restoration efforts. It will be important to examine the public's perception of sea level change and find effective ways to communicate the issue at the individual, community, and government levels.

Predict socioeconomic impacts of climate and sea level change on population dynamics, community infrastructure, and on short- and long-term community demographic shifts.

Characterize and model community and ecological resilience to natural hazards, considering the ecological footprint and level of vulnerability of the built environment, and identify methods to reduce losses. Model the attributes, factors, and strategies that contribute to making a community successfully resilient.

- **Understand response of coastal and marine systems to natural hazards and apply that understanding to assessments of future vulnerability to natural hazards** *(ORPP RP6).*

Many of the same challenges for increasing resilience for human communities exist for the natural ecosystems of the GoM, so there is a need to determine how storm surge, subsidence, and sea level change affects ecosystems.

Identify the optimal use and allocation of sediment and evaluate the rates of shoreline change from anthropogenic and natural impacts, including sediment mobilization, transport, and deposition from major storm events.

Analyze how coastal and near-shore morphology and biota protect inland areas from hurricane impacts by absorbing wind and storm surge energy and determine the economic costs and benefits of this protection.

- **Understand the impact of climate variability and change on the biogeochemistry of the ocean and implications for its ecosystems** *(ORPP PR12).*

Determine changes in freshwater, nutrient, pollution, groundwater, and sediment input due to changes in pattern and quantity of precipitation and predict the subsequent impact of these inputs on geochemical and physical coastal processes and biological communities.

and reverberating impacts that the loss of ocean and coastal resources can have on our country" as well as on local coastal communities, both ecological and sociological.

The report identifies "significant areas in need of improvement, including our ability to monitor and observe the ocean in real time; to rapidly share information; to understand, model, and predict the consequences of extreme events in the context of a changing ocean; and to apply those insights to a rapid and effective response." It also identifies an important overarching societal theme—global change—that requires a more realistic and forward-thinking treatment of restoration responses. The 2013 update also illustrates the direct link between the research priorities and the application of basic research to sound policy and management decisions. Fundamental research continues to be the anchor for issues that have the potential to alter our economy, our security, our environment, and our daily lives. In this vein, the research priorities related to the effects of major oil spills and the recovery of ecological and social systems still require more emphasis on ecosystem services and how they are defined, measured, and equated to ecosystem processes and functions.

There are numerous research plans for the GoM, well-planned scenarios for observation systems, and numerous documents on how restoration should proceed. More plans will be added as various agencies, councils, and states (as outlined in the RESTORE Act) develop guidance for measuring the effectiveness of restoration and monitoring efforts. These plans often do not clearly identify ecosystem services, but in attempting to capture the workings of the GoM ecosystem, they enhance the understanding of GoM ecosystem processes that is essential to the understanding of ecosystem services. The next section discusses how research efforts can be enhanced to specifically address ecosystem services.

## Research Needs Specific to Ecosystem Services and Their Valuation

As discussed in Chapters 2 and 3, the fundamental challenges to application of an ecosystem services approach to damage assessment in the GoM are related to understanding of baseline levels in a shifting or dynamic regime, the lack of comprehensive mechanistic models of GoM ecosystem processes, the uncertainty of and limited data associated with characterizing the human benefits that flow from ecosystem services, and finally, relating ecosystem services to long-term resilience.

Given the substantial ongoing data collection efforts undertaken in response to the *DWH* oil spill, along with the planned research and monitoring efforts, the baseline data needed to determine the impact on ecosystem function and structure of a future event such as the *DWH* oil spill will likely be available. Although an important component of the ecosystem services approach, appropriate baseline data alone are not sufficient for full implementation of an ecosystem services approach to damage assessment, particularly in a regime as dynamic as the GoM. Critical to the success of an ecosystem services approach is a better understanding of the mechanistic linkages between processes in the ecosystem—the ecological production functions described in the Interim Report (NRC, 2011) and Chapter 2. Development of these ecological production functions for the GoM will be challenging, but they are essential to any effort to implement a full ecosystem services approach.

## End-to-End Ecosystem Models

Ideally, the best solution to being able to quantify ecosystem services in the GoM (and thus the impact of an event such as the *DWH* oil spill on those services) is to develop end-to-end models for assessing the interplay of chemical, physical, biological, and socioeconomic conditions in the GoM. Such models would provide mechanisms to more fully understand the potential impacts of the spill on GoM ecosystem services. End-to-end models are complete system models that can relate drivers of change, such as climate change, to effects on ecosystem structure and function, such as the biomass of a particular subcomponent of the ocean biota (Fulton, 2010; Rose et al., 2010; Travers et al., 2007). In traditional compartmentalized thinking, these two ends may be viewed as being unconnected or there is a failure to investigate links that effectively connect them. Although a single end-to-end model that incorporates all of the relevant processes—biophysical and socioeconomic dimensions—and describes the dynamics of the provision of all GoM ecosystem services would be ideal, it is probably unrealistic to expect such a model to be developed for a system as complex at the GoM.

Currently, no single ecosystem model incorporates the full range of physical, chemical, or biological processes, let alone one that also fully incorporates an ecosystem services approach that further integrates socioeconomic dimensions. The models presented in Chapter 2 (Ecopath with Ecosim, Atlantis, Marine InVEST, MIMES, ARIES), however, allow for evaluation of the impact of an event such as the *DWH* oil spill on components of the ecosystem and the extrapolation of those impacts to a subset of ecosystem services. The Marine InVEST (Integrated Valuation of Environmental Services and Tradeoffs)[25] models provide the ability to explore some of these dependencies, trace the spill's impact on the capacity of systems to provide ecosystem services, and valuate the changes in ecosystem services to various groups in society. InVEST models are based on production functions that define how an ecosystem's structure and function affect the flows and values of ecosystem services. With validation, these models can also provide useful estimates of the magnitude and value of services provided. Marine InVEST could also be used to analyze tradeoffs among bundles of services provided under different management alternatives—for example, differences in rebuilding a barrier island for wave surge protection versus one for tourism.

Ecosystem models, such as Atlantis, could be modified to support an ecosystem services approach because they include both shallow- and deep-water components of the ecosystem, chemical and physical processes, and certain components of human dimensions.

***Finding 6.6.*** *Future efforts should be focused on expanding the current generation of ecosystem models, or coupling them with other models, so that they may better approach the "ideal"*

[25] http://naturalcapitalproject.org/InVEST.html.

*goal of an end-to-end model. Inclusion of the human dimension and economics into these types of models is essential to their capability to value changes in ecosystem services following disturbances such as oil spills. As these models are further developed and verified, they will help to improve our understanding of the full value of ecosystem services provided by the GoM ecosystem, as well as inform national efforts to prepare management plans to improve and protect the services for future generations. The development of a single end-to-end ecosystem/socioeconomic system model for the GoM may never be fully realized; however, the development of submodels of key components of the system is a critical research area that needs to be addressed.*

## Submodels

In the absence of completed and validated end-to-end models, submodels of components of the GoM ecosystem can be used to establish the value of ecosystem services within a distinct resource use, location, habitat, and user community. Because they only model part of the system, submodels may miss important connections with the unmodeled portions of the system. If constructed in a compatible manner, however, these submodels could be joined to build a more complete representation of GoM ecosystem interactions.

Fishery ecosystem models (FEMs) are an example of carefully detailed submodels that consider the biological components of the system as well as the socioeconomic aspects of the fishery. The FEM structure reflects the priorities of fishery managers and harvest activities. Thus the model includes the monitoring and management processes related to fisheries. The state of biological stocks can be integrated with the value of economic benefits derived from the fishery. FEMs allow researchers to narrow the set of possible hypotheses to be tested when an ecosystem is disturbed by environmental change or human activities. These models can help in the design of studies to collect the necessary data to more closely estimate the effects of an environmental impact on future fishery and ecosystem productivity. One type of FEM, Ecopath with Ecosim (EwE), is a modeling framework that integrates a wide range of biological and fisheries dynamics for multiple species and functional groups (biomass pools) over long time periods, using a trophic mass-balance approach (Christensen and Walters, 2004). Although originally designed as a FEM, EwE is used to address many aspects of marine ecosystem dynamics, including analyses of impacts from the 1989 *Exxon Valdez* oil spill in the Prince William Sound ecosystem (Okey and Pauly, 1999). EwE could be used to identify the components of the model ecosystem most likely to be affected by the *DWH* oil spill.

## Merging Social Benefits to Ecosystem Models

This report has described the plethora of past and current data collection activities. The vast majority of data collection activities are directed to generating sufficient baseline data about ecosystem function and structure or effects of particular response technologies. Effective implementation of an ecosystem services approach also requires collection of socioeconomic data. "Making the public whole" after a disturbance such as the *DWH* oil spill requires an improved understanding of its impacts on the value of various ecosystem services and on the populations and communities most affected. In many of the research programs mentioned above, with the exception of the GoM Sea Grant Consortium, the impacts on human communities are not rigorously integrated with ecosystem research. Although a stated goal of many of the programs is that their results will benefit society, a formal linkage between their biophysical output and the impact on human well-being is missing.

The public is not homogeneous. Individuals and communities are affected by disturbances that alter the flow of ecosystem services in different ways. In addition, they differ in their preferences for what services are important and what services should be restored (or compensated for in some way). For example, an individual with a seafood packing and distribution company in Houma, Louisiana, and a subsistence fisher living in a subsiding marsh landscape south of Houma in Terrebonne Bay will have different perceptions of how the loss of a seafood supply affects their livelihoods.

Submodels of a fishery, management decisions, and economics of a particular product could be developed that include the different types of social structures that depend on the fishery resources. Some values derived from fisheries, such as the profits of a commercial fishery, are much easier to estimate than others, such as the value of subsistence fishing. A comprehensive analysis should strive to understand important benefits, even those that are difficult to assess. The data needed to assign a value to the loss of the provisioning services of fisheries (see Table 5.6) would include length of period that the fishery was closed, estimated loss of catch, value of catch, alternative revenue streams, public perception of seafood safety and economic losses from buyer concern, unemployment costs, and loss of licenses that would link the fishery catch reduction and economic hardship to the various groups in society that depend on that fishery.

Tracking these variables offers a rather straightforward approach to monitoring the recovery of a community from an event such as the *DWH* oil spill if the goal is simply to return the community to its pre-spill state (the concept of engineering resilience discussed in Chapter 3). If, however, the event causes the system to undergo a regime change, then it will be much more difficult to determine the nature of the recovery (the concept of ecological resilience discussed in Chapter 3). Future research efforts should collect data on the benefits derived from ecosystem services by various groups and communities so that appropriate models can be built that include the full range of socioeconomic impacts including, as discussed below, those that may involve regime changes.

> ***Finding 6.7.*** *A necessary step in modifying submodels that characterize specific components of the GoM ecosystem—such as the fishery ecosystem models (FEMs)—is to integrate a socioeconomic evaluation of a spill impact such as a reduction in fishery productivity and the subsequent economic loss to coastal communities. FEMs should be extended to include additional data that would allow estimation of ecosystem service values for the affected fishery. Other subcomponents of the GoM ecosystem and the services they provide (e.g., wetlands, deep sea,*

*etc.) need to be modeled, which is a more realizable goal in the short term than constructing an end-to-end model.*

## Merging the Concept of Resilience to Ecosystem Service Models

As discussed in Chapter 3, the concept of resilience, the capacity of a system to respond to disturbances, is a potentially important component of ecosystem management to provide for a continuing flow of valuable ecosystem services. That chapter also discussed the importance of viewing resilience from an integrated social-ecological perspective that incorporates the links among human actions, ecosystem structure and function, and human well-being. Achieving social-ecological system resilience is, in reality, quite difficult and requires considerable additional research into the determinants of resilience both in principle and in its application in particular systems.

***Finding 6.8.*** *There is a lack of integrated systems research capable of analyzing the links between socioeconomic systems and ecosystems, as well as a lack of understanding of the complex feedbacks between components of systems necessary for understanding overall system dynamics and resilience. It may be possible to find key indicators that signal when systems are approaching thresholds. Improved methods to detect and avoid crossing thresholds that lead to reductions in valued ecosystem services could provide decision makers with the information they need to better manage the GoM system.*

Whether or not management of social-ecological system resilience is consistent with current resource management laws is not fully clear. If management for resilience represents a significant change from current management and decision-making processes, then congressional authorization in the form of new or amended laws would be required to make it part of the process (see e.g., Ruhl, 2012). To fully implement a resilience approach into practice, an agency would need to (a) craft a firm definition of resilience for the system in question, (b) identify a means of measuring the current state of that system's resilience, (c) build a model capable of predicting to some degree of certainty the resilience effects of specific decisions (or to design management experiments that could lead to the development of predictive capacity), and (d) develop these tools such that it can explain both the process and the results to the public and to the judges likely to review its decisions (Allen et al., 2011). The current state of knowledge and practice falls far short of these standards.

## Valuing Ecosystem Services

Ascribing value to ecosystem services is typically done using economic valuation methods. These methods are capable of generating estimates of the value of services in terms of a common (monetary) metric. One of the challenges to applying an ecosystem services approach to the *DWH* oil spill, particularly with respect to whether it can supplement the existing NRDA process, is that for many ecosystem services, there is either a lack of values associated with functions, a lack of data, or both, preventing accurate quantification of ecosystem services as a function of ecosystem condition. Approaches and tools for assigning values to the ecosystem services impaired by the *DWH* oil spill will need to be developed and applied appropriately, particularly for ecosystem services for which data on human preference and use are lacking.

Some services are more readily quantified than others. For example, several studies have estimated the value of coastal marshes in providing hazard mitigation during hurricane wave surges on coastlines (see Chapter 5). The valuation may consider changes in costs and availability of home insurance coverage or estimates of the likely reduction in damages with intact marsh versus without intact marsh. The monetary values of provisioning services, such as commercial and recreational fishing, or regulating services, such as nutrient and pollution removal, can also be fairly easily estimated. More challenging is the assignment of economic value to other, less tangible impacts on services, such as changes in community vitality, way of life, or aesthetics or spirituality, or biological interactions in the mesopelagic portion of the northern GoM. Although these valuation issues are challenging, they must be addressed to enable comprehensive and consistent measurements of losses or changes of ecosystems resulting from environmental harm.

> ***Finding 6.9.*** *Consideration of the impact on human well-being of the* DWH *oil spill is of great importance, but the ability to analyze these impacts is still partial and incomplete. Consistent frameworks exist for inclusion of these impacts on human well-being (e.g., NOAA's National Centers for Coastal Ocean Science "Human Dimensions Strategic Plan" or the Marine InVEST models), yet our ability to integrate socioeconomic analysis with biophysical analysis lags behind. Models in which the stakeholders are identified, the impacts they experience are described, and the methods for assigning values to the loss or change in ecosystem services they experience should be explored.*

## SUMMARY

As discussed throughout this report and the Interim Report (NRC, 2011), the committee believes that an ecosystem services approach to damage assessment offers several advantages over current practice. However, the research, data, and models needed to more fully implement this approach are lacking.

The research funding and the release of NRDA data that will be available through the settlement process offer an unprecedented opportunity to establish a comprehensive baseline measurement and and greater fundamental understanding of the GoM, both which are critical components of an ecosystem services approach. The committee has identified several key areas of research that are needed to fully implement an ecosystem services approach for the GoM.

1. Although past, current, and future data collection and monitoring efforts, with appropriate modifications, could likely provide much of the baseline data needed to discern the impact of events such as the *DWH* oil spill on ecosystem function and structure, an overarching structure, supported by long-term and stable funding mechanisms, to integrate all of these efforts in the GoM is critically needed.
2. The development of a single end-to-end biophysical-social-economic system model for the GoM would be ideal, but is probably not likely for a system as complex as the GoM. Instead, the short-term focus should be on the development of submodels of key components of the GoM ecosystem and the services it provides (e.g., fisheries, wetlands, deep sea, etc.). These submodels must consider the impacts on human well-being and include approaches for valuing the ecosystem services. Ultimately, these submodels could be combined in work to establish a more complete ecosystem model for the GoM.
3. The committee believes that plans for current and future data collection may provide sufficient baseline data about ecosystem function and structure. The ecosystem services approach, however, is limited by a lack of socioeconomic data, which are needed to better inform understanding of the human dependencies on natural systems and therefore the socioeconomic impacts of disturbances such as the *DWH* oil spill. Future research efforts must include the collection and synthesis of data on relevant human factors to enable development of models that consider the full range of socioeconomic impacts.
4. Understanding of resilience in a socioecological context is critical to this approach, but it requires considerable additional research into the determinants of resilience, both in principle and in application, in particular systems. Proper integration of resilience theory into management decisions—for example, criteria for success in restoration—will require resolution of current legislative barriers.
5. With respect to response technologies, the need to accurately predict the effectiveness of injection of dispersant at depth, the impact of dispersants on biodegradation rates, and the long-term effects of dispersed oil on the food web and other ecosystem services is most critical.

The risk of additional deep-sea reservoir oil spills in the GoM increases as the oil and gas industry moves more and more into deeper waters. Consequently, stakeholders should consider how the nation should prepare for an environmental event of the magnitude of the *DWH* oil spill. For instance, what is the unifying framework for addressing research, monitoring and modeling needs for baseline conditions, mitigation approaches, valuation of environmental impacts and loss of ecosystem services, and restoration of natural and social systems? How would success or effectiveness of these efforts be estimated?

Statements were made after the *Exxon Valdez* grounding and oil spill that lessons learned from that spill would serve the nation in its response to the next large oil spill (Kurtz, 1995; Peterson et al., 2003). Legislation, such as the 1990 Oil Pollution Act, was passed to better ensure cleanup and responsibility for damages to the ecosystem. Although some people familiar with the complexity of the deep ocean may have anticipated the failure to contain a spill there, the nation as a whole was not prepared for the magnitude of new and critical issues presented by the explosion and sinking of the *Deepwater Horizon* and wellhead rupture of the Macondo well. There are countless opportunities, at present, to ensure that the nation is better prepared for a similar incident in the future. These opportunities should not be wasted.

The NRDA guidelines continue to be used to assess damages from the *DWH* oil spill. Just as better mechanisms for assessing and assigning responsibility for damages resulted from the *Exxon Valdez* grounding, the opportunity exists now to incorporate the concepts of ecosystem services, in practice and in law, into oil spill damage assessment and recovery strategies. The evolving understanding of human-ecosystem interactions as articulated by ecosystem-based management, and particularly the concept of ecosystem services, offer an opportunity to address some of the challenges faced during the current NRDA process. By taking a more holistic view of ecosystem interactions, and by following these interactions through all relevant trophic levels and spatial connections to their ultimate impact on human well-being, an ecosystem services approach to damage assessment can capture a more complete picture of potential impacts of a damaging event and offer a broader range of restoration options. By coupling the concept of ecosystem resilience with socioeconomic resilience, we would further ensure that an ecosystem services approach would come closer to restoring social-ecological systems after damage from a human-caused or natural disturbance.

In closing, GoM communities and natural resource managers face many challenges as they contemplate and set research priorities and select restoration plans. It will be important for all stakeholders to think about "how best to manage multiple ecosystem services across a diverse natural and sociological marine ecosystem?" As many in the region have already realized, past decisions to enhance a particular ecosystem service to maximize a particular benefit—for example, energy development, fisheries, tourism—may have resulted in tradeoffs that diminished the ecosystem's capacity to provide other services. The ecosystem services approach could be used to more fully capture the value of assorted services in the GoM, which would assist GoM communities in identifying what is important to their way of life and would inform management decisions regarding the balance among ecosystem service priorities.

# References


Abbriano, R. M., M. M. Carranza, S. L. Hogle, R. A. Levin, A. N. Netburn, K. L. Seto, S. M. Snyder, and P. J. S. Franks. 2011. *Deepwater Horizon* oil spill: A review of the planktonic response. Oceanography 24:294-301.

Adger, W. N. 2000. Social and ecological resilience: Are they related? Progress in Human Geography 24:347-364.

Adger, W. N., T. P. Hughes, C. Folke, S. Carpenter, and J. Rockström. 2005. Social-ecological resilience to coastal disasters. Science 309:1036-1039.

AGCCVB (Alabama Gulf Coast Convention and Visitors Bureau). 2012. Gulf Shores and Orange Beach Tourism. Available at http://www.gulfshores.com/stats/2_a.%20Destination%20Growth%20Indicators.pdf, accessed November 7, 2012.

Aldhous, P., and J. Hecht. 2010. Beware long-term damage when cleaning up the *Deepwater Horizon* spill. New Scientist 206:2765.

Aldhous, P., and P. McKenna. 2010. When the oil stops gushing. New Scientist 206:2764.

Allen, A. A., N. J. Mabile, D. Jaeger, and D. Costanzo. 2011. The use of controlled burning during the Gulf of Mexico *Deepwater Horizon* (MC-252) oil spill response. In International Oil Spill Conference Proceedings, Vol. 2011. Washington, DC: American Petroleum Institute.

Allen, C. R., L. Gunderson, and A. R. Johnson. 2005. The use of discontinuities and functional groups to assess relative resilience in complex systems. Ecosystems 8:958-966.

Allison, M. A., C. R. Demas, B. A. Ebersole, B. A. Kleiss, C. D. Little, E. A. Meselhe, N. J. Powell, T. C. Pratt, and B. M. Vosburg. 2012. A water and sediment budget for the lower Mississippi–Atchafalaya River in flood years 2008–2010: Implications for sediment discharge to the oceans and coastal restoration in Louisiana. Journal of Hydrology 432-433:84-97.

American Society for Testing and Materials. 2010. ASTM E1943-98(2010): Standard Guide for Remediation of Ground Water by Natural Attenuation at Petroleum Release Sites, Vol. 11.05. West Conshohocken, PA: ASTM International.

American Academy of Microbiology. 2011. Microbes & Oil Spills. Washington, DC: American Society for Microbiology. Available at http://academy.asm.org/images/stories/documents/Microbes_and_Oil_Spills.pdf.

Amos, A. F., S. C. Rabalais, and R. S. Scalan. 1983. Oil-filled Callianassa burrows on a Texas barrier-island beach. Journal of Sedimentary Research 53(2):411-416.

Anderies, J. M., B. H. Walker, and A. P. Kinzig. 2006. Fifteen weddings and a funeral: Case studies and resilience-based management. Ecology and Society 11(1): art. 21. Available at http://www.ecologyandsociety.org/vol11/iss1/art21/.

Anokhin, P. K. 1935. The problem of center and periphery in the contemporary physiology of nervous activity. Pp. 9-70 in The Problem of Center and Periphery in the Physiology of Nervous Activity. P. K. Anokhin (Ed.). Moscow: Gosizdat, Gorky.

Ardron, J. A. 2008. The challenges of assessing whether the OSPAR network of marine protected areas is ecologically coherent. Hydrobiologica 606:45-53.

Ariely, D. 2009. Predictably Irrational: The Hidden Forces That Shape Our Decisions. New York: Harper Perrennial.

Armstrong, C. W., N. Foley, R. Tinch, and S. van den Hove. 2011. Ecosystem Goods and Services of the Deep Sea. Southampton, UK: National Oceanography Centre. Available at http://median-web.eu/IMG/pdf/ecosystem_goods_and_services.pdf.

Arranz, P., N. Aguilar Soto, P. T. Madsen, A. Brito, F. Bordes, and M. P. Johnson. 2011. Following a foraging fishfinder: Diel habitat use of Blainville's beaked whales revealed by echolocation. PLOS One 6(12):e28353.

Arreguin-Sanchez, F., and S. Manickchand-Heilman. 1998. The trophic role of lutjanid fish and impacts of their fisheries in two ecosystems in the Gulf of Mexico. Journal of Fish Biology 53:143-153.

Arreguin-Sanchez, S., M. Zetina-Rejon, S. Manickchand-Heileman, M. Ramirez-Rodriguez, and L. Vidal. 2004. Simulated response to harvesting strategies in an exploited ecosystem in the southwestern Gulf of Mexico. Ecological Modelling 172:421-432.

Atlas, R. M. 1981. Microbial degradation of petroleum hydrocarbons: An environmental perspective. Microbiological Reviews 45(1):180-209.

Atlas, R. M. 1984. Petroleum Microbiology. New York: Macmillan.

Atlas, R. M. 1991. Microbial hydrocarbon degradation—bioremediation of oil spills. Journal of Chemical Technology and Biotechnology 52(2):149-156.

Atlas, R. M. 1995. Petroleum biodegradation and oil spill bioremediation. Marine Pollution Bulletin 31:178-182.

Atlas, R. M., and R. Bartha. 1992. Hydrocarbon biodegradation and oil-spill bioremediation. Advances in Microbial Ecology 12:287-338.

Atlas, R. M., and T. C. Hazen. 2011. Oil biodegradation and bioremediation: a tale of the two worst spills in U. S. History. Environmental Science & Technology 45:6709-6715.

Aurell, J., and B. K. Gullett. 2010. Aerostat sampling of PCDD/PCDF emissions from the Gulf oil spill in-situ burns. Environmental Science & Technology 44:9431-9437.

Ayres Associates. 2008. Summary of Work Performed by Ayres Associates in Support of URS Storm Surge Modeling for FEMA Region 4. Available at http://www.fema.gov/library/file?type=publishedFile&file=summary_of_work_perform.ed_in_support_of_storm_surge_modeling.pdf&fileid=7198e050-e9aa-11de-ae85-001cc456982e, accessed June 24, 2013.

Ayres, D. R., D. L. Smith, K. Zaremba, S. Klohr, and D. R. Strong. 2004. Spread of exotic cordgrasses and hybrids (*Spartina* sp.) in the tidal marshes of San Francisco Bay, California, USA. Biological Invasions 6:221-231.

Badola, R., and S. A. Hussain. 2005. Valuing ecosystem functions: An empirical study on the storm protection function of Bhitarkanika mangrove ecosystem, India. Environmental Conservation 32(1):85-92.

Baek, K. H., H. S. Sim, H. M. Oh, B. D. Yoon, J. Kim, and I. S. Lee. 2004. Effects of crude oil, oil components, and bioremediation on plant growth. Journal of Environmental Science and Health A39:2465-2472.

Bakun, A., and S. J. Weeks. 2006. Adverse feedback sequences in exploited marine systems: Are deliberate interruptive actions warranted? Fish and Fisheries 7:316-333.

Barbier, E. B. 1994. Valuing environmental functions: Tropical wetlands. Land Economics 70(2):155-173.

Barbier, E. B. 2000. Valuing the environment as input: Applications to mangrove-fishery linkages. Ecological Economics 35:47-61.

Barbier, E. B. 2003. Habitat-fishery linkages and mangrove loss in Thailand. Contemporary Economic Policy 21:59-77.

Barbier, E. B. 2007. Land conversion, interspecific competition, and bioinvasion in a tropical ecosystem. Journal of Agricultural and Applied Economics 39:133-147.

Barbier, E. B., and I. Strand. 1998. Valuing mangrove-fishery linkages: A case study of Campeche, Mexico. Environmental and Resource Economics 12:151-166.

Barbier, E. B., I. Strand, and S. Sathirathai. 2002. Do open access conditions affect the valuation of an externality? Estimating the welfare effects of mangrove-fishery linkages in Thailand. Environmental and Resource Economics 21:343-367.

Barbier, E. B., E. W. Koch, B. R. Silliman, S. D. Hacker, E. Wolanski, J. Primavera, E. F. Granek, S. Polasky, S. Aswani, L. A. Cramer, D. Stoms, C. J. Kennedy, D. Bael, C. V. Kappel, G. M. E. Perillo, and D. J. Reed. 2008. Coastal ecosystem-based management with non-linear ecological functions and values. Science 319:321-323.

Barbier, E. B., S. D. Hacker, C. Kennedy, E. W. Koch, A. C. Stier, and B. R. Silliman. 2011. The value of estuarine and coastal ecosystem services. Ecological Monographs 81:169-193.

Barkenbus, J. N. 1979. Deep seabed resources: Politics and technology. New York: Free Press.

Barnas, K., and S. L. Katz. 2010. The challenges of tracking habitat restoration at various spatial scales. Fisheries 35(5):232-241.

Barras, J. A., S. Beville, D. Britsch, S. Hartley, S. Hawes, J. Johnston, P. Kemp, P. Kinler, A. Martucci, J. Porthouse, D. Reed, K. Roy, S. Sapkota, and J. Suhayda. 2003. Historical and Projected Coastal Louisiana Land Changes: 1978-2050. Open-File Report 03-334. Reston, VA: U.S. Geological Survey.

Barron, M. G. 2012. Ecological impacts of the *Deepwater Horizon* oil spill. Implications for Immunotoxicity. Toxicologic Pathology 40:315-320.

Barron, M. G., T. Podrabsky, S. Ogle, and R. W. Ricker. 1999. Are aromatic hydrocarbons the primary determinant of petroleum toxicity to aquatic organisms? Aquatic Toxicology 46:253-268.

Bassin, N. J., and T. Ichiye. 1977. Flocculation behavior of suspended sediments and oil emulsions. Journal of Sedimentary Petrology 47:671-677.

Baud, R. D. 2002. Deepwater Gulf of Mexico 2002: America's Expanding Frontier. OCS Study MMS 2002-021. New Orleans, LA: U.S. Department of the Interior, Minerals Management Service, Gulf of Mexico OCS Region.

Baumann, U., M. Benz, E. Pletscher, K. Breuker, and R. Zenobi. 1999. Biodegradation of sugar alcohol ethoxylates. Tenside Surfactants Detergents 36:288-293.

Baustian, J., I. Mendelssohn, Q. Lin, and J. Rapp. 2010. In-situ burning restores the ecological function and structure of an oil-impacted coastal marsh. Environmental Management 46:781-789.

Becker, E. L., E. E. Cordes, S. A. Macko, and C. R. Fisher. 2009. Importance of seep primary production to *Lophelia pertusa* and associated fauna in the Gulf of Mexico. Deep Sea Research Part I: Oceanographic Research Papers 56:786-800.

Beisner, B. E., D. T. Haydon, and K. Cuddington. 2003. Alternative stable states in ecology. Frontiers in Ecology and the Environment 1:376-382.

Bender, L. C., and S. F. DiMarco. 2008. Quality control and analysis of acoustic Doppler current profiler data collected on offshore platforms of the Gulf of Mexico. OCS Study MMS 2009-010. New Orleans, LA: U.S. Department of the Interior, Minerals Management Service, Gulf of Mexico OCS Region. Available at http://tabs.gerg.tamu.edu/~norman/pdf/OCS_Study_MMS_2009_010.pdf, accessed March 22, 2013.

Benson, M. H., and A. S. Garmestani. 2011. Embracing panarchy, building resilience and integrating adaptive management through a rebirth of the National Environmental Policy Act. Journal of Environmental Management 92(5):1420-1427.

Berkes, F., and C. Folke (Eds.). 1998. Linking Social and Ecological Systems: Management Practices and Social Mechanisms for Building Resilience. Cambridge, UK: Cambridge University Press.

Berkes, F., J. Colding, and C. Folke (Eds.). 2003. Navigating Social-Ecological Systems: Building Resilience for Complexity and Change. Cambridge, UK: Cambridge University Press.

Berna, J. L., G. Cassani, C.-D. Hager, N. Rehman, I. Lopez, D. Schowanek, J. Steber, K. Taeger, and T. Wind. 2007. Anaerobic biodegradation of surfactants: Scientific review. Tenside Surfactants Detergents 44:312-347.

Bernier, N. B., K. R. Thompson, J. Ou, and C. H. Ritchie. 2007. Mapping the return periods of extreme sea levels: Allowing for short sea level records, seasonality, and climate change. Global and Planetary Change 57(1-2):139-150.

Beverton, R. 1998. Fish, fact and fantasy: A long view. Reviews in Fish Biology and Fisheries 8:229-249.

Bianchi, T. S., R. L. Cook, E. M. Perdue, P. E. Kolic, N. Green, Y. Zhang, R. W. Smith, A. S. Kolker, A. Ameen, G. King, L. M. Ojwang, C. L. Schneider, A. E. Normand, and R. Hetland. 2011. Impacts of diverted freshwater on dissolved organic matter and microbial communities in Barataria Bay, Louisiana, USA. Marine Environmental Research 72:248-257.

Biggs, D. C., C. Hu, and F. Müller-Karger. 2008. Remotely sensed sea surface chlorophyll and POC flux at Deep Gulf of Mexico Benthos sampling stations. Deep Sea Research Part II: Topical Studies in Oceanography 55:2555-2562.

Biggs, R., S. R. Carpenter, and W. A. Brock. 2009. Turning back from the brink: Detecting an impending regime shift in time to avert it. Proceedings of the National Academy of Sciences of the United States of America 106(3):826-831.

Biggs, R., T. Blenckner, C. Folke, L. Gordon, A. Norström, M. Nyström, and G. D. Peterson. 2012a. Regime shifts. Pp. 609-617 in Encyclopedia of Theoretical Ecology, A. Hastings and L. Gross (Eds.). Ewing, NJ: University of California Press.

Biggs, R., M. Schluter, D. Biggs, E. L. Bohensky, S. BurnSilver, G. Cundill, V. Dakos, T. M. Daw, L. S. Evans, K. Kotschy, A. M. Leitch, C. Meek, A. Quinlan, C. Raudsepp-Hearne, M. D. Robards, M. L. Schoon, L. Schultz, and P. C. West. 2012b. Toward principles for enhancing the resilience of ecosystem services. Annual Review of Environment and Resources 37:421-448.

Boehler, W., V. M. Bordas, and A. Marbs. 2003. Investigating laser scanner accuracy. Pp 696-701 in Proceedings of the XIXth CIPA Symposium, Antalya, Turkey, O. Atlan (Ed.). International Committee for Documentation of Cultural Heritage.

Boehm, P. D. 1987. Transport and transformation processes regarding hydrocarbon and metal pollutants in offshore sedimentary environments. Pp. 233-287 in Transport and Transformation Processes Regarding Hydrocarbon and Metal Pollutants in Offshore Sedimentary Environments, D. F. Boesch and N. N. Rabalais (Eds.). London and New York: Elsevier Applied Science.

Boehm, P. D., and P. D. Carragher. 2012. Location of natural oil seep and chemical fingerprinting suggest alternative explanation for deep sea coral observations. Proceedings of the National Academy of Sciences of the United States of America 109:E2647.

BOEM (Bureau of Ocean Energy Management). 2010. Historic Shipwrecks in the Gulf of Mexico. Available online at http://www.BOEM.gov/Shipwrecks/.

BOEM. 2012. BOEM NTL No. 2012-G01 Notice to Lessees and Operators of Federal Oil, Gas, and Sulfur Leases, Outer Continental Shelf, Gulf of Mexico OCS Region. Available at http://www.BOEM.gov/Regulations/Notices-to-Lessees/2012/BOEM-21012-G01.aspx.

Boesch, D. F., R. E. Turner, and J. W. Day, Jr. 1984. Deterioration of coastal environments in the Mississippi Deltaic Plain: Options for riverine and wetland management. Pp. 447-466 in The Estuary as a Filter, V. Kennedy (Ed.). New York: Academic Press.

Boesch, D. F., M. N. Josselyn, A. J. Mehta, J. T. Morris, W. K. Nuttle, C. A. Simenstad, and J. P. Swift. 1994. Scientific assessment of coastal wetland loss, restoration and management in Louisiana. Journal of Coastal Research, Special Issue 20:1-103.

Boesch, D. F., L. Shabman, L. Antle, J. Day, R. Dean, G. Galloway, C. Groat, S. Laska, R. Luettich, W. Mitsch, N. Rabalais, D. Reed, C. Simenstad, B. Streever, R. Taylor, R. Twilley, C. Watson, J. Wells, and D. Whigham. 2006. A New Framework for Planning the Future of Coastal Louisiana after the Hurricanes of 2005. Working Group for Post-Hurricane Planning for Greater New Orleans and the Louisiana Coast. Cambridge, MD: University of Maryland, Integrated Application Network. Available at http://www.umces.edu/la-restore/.

Boopathy, R., S. Shields, and S. Nunna. 2012. Biodegradation of crude oil from the BP oil spill in the marsh sediments of southeast Louisiana, USA. Applied Biochemistry and Biotechnology 167(6):1560-1568.

Boufadel, M. C., R. Bechtel, and J. Weaver. 2006. The movement of oil under non-breaking waves. Marine Pollution Bulletin 52(9):1056-1065.

Bourgeau-Chavez, L. L., K. Riordan, R. B. Powell, N. Miller, and M. Nowels. 2009. Improving wetland characterization with multi-sensor, multi-temporal SAR and optical/infrared data fusion. Chap. 33 in Advances in Geoscience and Remote Sensing, G. Jedlovec (Ed.). New York: InTech. Available at http://www.intechopen.com/books/advances-in-geoscience-and-remote-sensing, accessed March 25, 2013.

Boxall, P. C., W. L. Adamowicz, M. Olar, G. E. West, and G. Cantin. 2012. Analysis of the economic benefits associated with ecovery of threatened marine mammal species in the Canadian St. Lawrence Estuary. Marine Policy 36:189-197.

Boyer, T., and S. Polasky. 2004. Valuing urban wetlands: A review of non-market valuation studies. Wetlands 24(4):744-755.

Bragg, J. R., and S. H. Yang. 1995. Clay-oil flocculation and its effects on the rate of natural cleansing in Prince William Sound following the *Exxon Valdez* oil spill. Pp. 178-214 in *Exxon Valdez* Oil Spill: Fate and Effects in Alaskan Waters, P. G. Wells, J. N. Butler, and J. S. Hughes (Eds.). Philadelphia: American Society for Testing and Materials.

Bragg, J. R., R. C. Prince, E. J. Harner, and R. M. Atlas. 1994. Effectiveness of bioremediation for the *Exxon Valdez* oil spill. Nature 368:413-418.

Brandvik, P. J., O. Johansen, F. Leirvik, U. Farooq, and P. S. Daling. 2013. Droplet breakup in sub-surface oil releases—Part 1: Experimental study of droplet breakup and effectiveness of dispersant injection. Marine Pollution Bulletin 73:319-326.

Breaux, A., S. Farber, and J. Day. 1995. Using natural coastal wetlands systems for wastewater treatment: An economic benefit analysis. Journal of Environmental Management 44(3):285-291.

Briguglio, L., N. Farrugia, and S. Vella. 2009. Economic vulnerability and resilience: concepts and measurements. Oxford Development Studies 37(3):229-247.

Britsch, L. D., and E. B. Kemp III. 1990. Land Loss Rates: Mississippi River Deltaic Plain. Technical Report GL-90-2. New Orleans, LA: U.S. Army Engineer District, New Orleans.

Brodie, J., and J. Waterhouse. 2012. A critical review of environmental management of the "not so Great" Barrier Reef. Estuarine, Coastal and Shelf Science 104-105:1-22.

Bustamante, M., F. J. Tajadura-Martín, and J. I. Saiz-Salinas. 2010. Temporal and spatial variability on rocky intertidal macrofaunal assemblages affected by an oil spill (Basque coast, northern Spain). Journal of the Marine Biological Association of the United Kingdom 90:1305-1317.

Butterworth, D. S., and E. E. Plagányi. 2004. A brief introduction to some approaches to multispecies/ecosystem modeling in the context of their possible application in the management of South African fisheries. African Journal of Marine Science 26:53-61.

Caddy, J., and A. Bakun. 1995. Marine catchment basins and anthropogenic effects on coastal fishery ecosystems. Pp. 119-133 in Effects of Riverine Inputs on Coastal Ecosystems and Fisheries Resources, FAO Fisheries Department Technical Paper 349. Rome, Italy: Food and Agriculture Organization of the United Nations.

Camilli, R., C. M. Reddy, D. R. Yoerger, B. A. S. Van Mooy, M. V. Jakuba, J. C. Kinsey, C. P. McIntyre, S. P. Sylva. and J. V. Maloney. 2010. Tracking hydrocarbon plume transport and biodegradation at *Deepwater Horizon*. Science 330:201-204.

Campo, P., A. D. Venosa, and M. T. Suidan. 2013. Biodegradability of Corexit 9500 and dispersed south Louisiana crude oil at 5 and 25 °C. Environmental Science & Technology 47(4):1960-1967.

Canfield, D. E., E. Kristensen, and B. Thamdrup. 2005. Microbial ecosystems. Chap. 13 in Aquatic Geomicrobiology, Vol. 48 (Advances in Marine Biology). Sandiego, CA, and London, UK: Elsevier Academic Press.

Cardinale, B. J., K. L. Matulich, D. U. Hooper, J. E. Byrnes, E. Duffy, L. Gamfeldt, and A. Gonzalez. 2011. The functional role of producer diversity in ecosystems. American Journal of Botany 98(3):572-592.

Carlarne, C., and J. Eagle. 2012. Food security, fisheries, and ecosystems. Chap. 23 in The Law of Adaptation to Climate Change: United States and International Aspects, M. B. Gerrard (Ed.). Washington, DC: American Bar Association.

Carney, R. S. 1995. On the adequacy and improvement of marine benthic pre-impacts surveys: Examples from the Gulf of Mexico outer continental shelf. Pp. 295-316 in Detecting Ecological Impacts: Concepts and Applications in Coastal Habitats, R. J. Schmitt and C. W. Osenberg (Eds.). San Diego, CA: Elsevier Academic Press.

Carney, R. S. (Ed.). 1998. Workshop on Environmental Issues Surrounding Deepwater Oil and Gas Development: Final Report. OCS Study MMS 98-0022. New Orleans, LA: U.S. Department of the Interior, Minerals Management Service, Gulf of Mexico OCS Region.

Carney, R. S. 2010. Stable isotope trophic patterns in echinoderm megafauna in close proximity and remote from Gulf of Mexico lower-slope hydrocarbon seeps. Deep Sea Research Part II: Topical Studies in Oceanography 57:1965-1971.

Carpenter, S. R. 2003. Regime Shifts in Lake Ecosystems: Patterns and Variation. Oldendorf, Luhe, Germany: International Ecology Institute.

Carpenter, S., B. Walker, J. M. Anderies, and N. Abel. 2001. From metaphor to measurement: Resilience of what to what? Ecosystems 4:765-781.

Carpenter, S. H., A. Mooney, J. Agard, D. Capistrano, R. S. DeFries, S. Diaz, T. Dietz, A. K. Duraippah, A. Oteng-Yeboah, H. M. Pereira, C. Perrings, W. V. Reid, J. Surukhan, R. J. Scholes, and A. Whyte. 2009. Science for managing ecosystem services: Beyond the Millennium Ecosystem Assessment. Proceedings of the National Academy of Sciences of the United States of America 106:1305-1312.

Carriger, J. F., and M. G. Barron. 2011. Minimizing risks from spilled oil to ecosystem services using influence diagrams: The *Deepwater Horizon* spill response. Environmental Science & Technology 45:7631-7639.

Carruthers, T. R., M. K. McAllister, and N. G. Taylor. 2011. Spatial surplus production modeling of Atlantic tunas and billfish. Ecological Applications 21:2734-2755.

Carscadden, J. E., K. T. Frank, and W. C. Leggett. 2001. Ecosystem changes and the effects on capelin (*Mallotus villosus*), a major forage species. Canadian Journal of Fisheries and Aquatic Sciences 58:73-85.

Carson, R., R. Mitchell, M. Hanemann, R. Kopp, S. Presser, and P. Ruud. 2003. Contingent valuation and lost passive use: Damages from the *Exxon Valdez* oil spill. Environmental and Resource Economics 25:257-286.

CDC (Centers for Disease Control and Prevention). 2010. Oil Spill Dispersant (COREXIT ®EC9500A and EC9527A) Information for Health Professionals. Available at http://www.cdc.gov/nceh/oil_spill/docs/Oil%20Spill%20Dispersant.pdf, accessed June 20, 2013.

Cermak, J. H., G. L. Stephenson, D. Birkholz, Z. Wang, and D. G. Dixon. 2010. Toxicity of petroleum hydrocarbon distillates to soil organisms. Environmental Toxicology and Chemistry 12:2685-2694.

Chaineau, C. H., J. L. Morel, and J. Oudot 1997. Phytotoxicity and plant uptake of fuel oil hydrocarbons. Journal of Environmental Quality 26:1478-1483.

Chakrabarty, P., C. Lam, J. Hardman, J. Aaronson, P. H. House, and D. A. Janies. 2012. SPECIESMAP: A web-based application for visualizing the overlap of distributions and pollution events with a list of fishes put at risk by the 2010 Gulf of Mexico oil spill. Biodiversity and Conservation 21:1865-1876.

Champ, P., K. Boyle, and T. C. Brown. 2003. A Primer of Nonmarket Valuation. Berlin: Springer Verlag.

Chandrasekar, S., G. A. Sorial, and J. W. Weaver. 2005. Dispersant effectiveness on three oils under various simulated environmental conditions. Environmental Engineering Science 22:324-336.

Chandrasekar, S., G. A. Sorial, and J. W. Weaver. 2006. Dispersant effectiveness on oil spills: Impact of salinity. Journal of Marine Science 63:1418-1430.

Chapin, F. S. III, J. Lubchenco, and H. L. Reynolds. 1995. Biodiversity effects on patterns and processes of communities and ecosystems. Pp. 289-301 in Global Biodiversity Assessment, V. H. Heywood (Ed.). Cambridge, UK: Cambridge University Press.

Chapman, H., K. Purnell, R. J. Law, and M. F. Kirby. 2007. The use of chemical dispersants to combat oil spills at sea: A review of practice and research needs in Europe. Marine Pollution Bulletin 54(7):827-838.

Christensen, N. L., A. M. Bartuska, J. H. Brown, S. Carpenter, C. D'Antonio, R. Francis, J. F. Franklin, J. A. MacMahon, R. F. Noss, D. J. Parsons, C. H. Peterson, M. G. Turner, and R. G. Woodmansee. 1996. The report of the Ecological Society of America Committee on the Scientific Basis for Ecosystem Management. Ecological Applications 6:665-691.

Christensen, V., and C. J. Walters. 2004. Ecopath with Ecosim: Methods, capabilities and limitations. Ecological Modelling 172:109-139.

Church, R., D. Warren, R. Cullimore, L. Johnston, W. Schroeder, W. Patterson, T. Shirley. M. Kilgour, N. Morris, and J. Moore. 2007. Archaeological and Biological Analysis of World War II Shipwrecks in the Gulf of Mexico: Artificial Reef Effect in Deep Water. OCS Study MMS 2007-015. New Orleans, LA: U.S. Department of the Interior, Minerals Management Service. Available at http://www.data.boem.gov/PI/PDFImages/ESPIS/4/4239.pdf, accessed June 24, 2013.

Church, R. A., D. J. Warren, and J. B. Irion. 2009. Analysis of deepwater shipwrecks in the Gulf of Mexico: Artificial reef effect of Six World War II shipwrecks. Oceanography 22(2):50-63. Available at http://www.tos.org/oceanography/archive/22-2_church.pdf, accessed June 24, 2013.

Clement, T. P., J. S. Hayworth, V. Mulabagal, G. F. John, and F. Yin. 2012. Research Brief-II: Impact of Hurricane Isaac on Mobilizing *Deepwater Horizon* Oil Spill Residues Along Alabama's Coastline—A Physicochemical Characterization Study. Available at http://www. eng. auburn.edu/files/acad_depts/civil/oil-research-hurricane-isaac.pdf, accessed March 25, 2013.

Clements, F. 1936. Nature and structure of the climax. Journal of Ecology 24:252-284.

Collins, C. D., and J. H. Wlosinski. 1983. Coefficients for Use in the U.S. Army Corps of Engineers Reservoir Model, CE-QUAL-R1. Vicksburg, MS: Environmental Laboratory, U.S. Army Engineer Waterways Experiment Station.

Connell, J. H., and R. O. Slatyer. 1977. Mechanisms of succession in natural communities and their role in community stability and organization. American Naturalist 111:1119-1144.

Conway-Cranos, L. L. 2012. Geographic variation in resilience: An experimental evaluation of four rocky intertidal assemblages. Marine Ecology Progress Series 457:67-83.

Cordes, E. E., D. C. Bergquist, and C. R. Fisher. 2009. Macro-ecology of Gulf of Mexico cold seeps. Annual Review of Marine Science 1:143-168.

Cordes, E. E., E. L. Becker, and C. R. Fisher. 2010. Temporal shift in nutrient input to cold-seep food webs revealed by stable-isotope signatures of associated communities. Limnology and Oceanography 55:2357-2548.

Costanza, R. 2008. Ecosystem services: Multiple classification systems are needed. Biological Conservation 141:350-352.

Costanza, R., R. d'Arge, R. de Groot, S. Farber, M. Grasso, B. Hannon, S. Naeem, K. Limburg, J. Paruelo, R. V. O'Neill, R. Raskin, P. Sutton, and M. van den Belt. 1997. The value of the world's ecosystem services and natural capital. Nature 387:253-260.

Costanza, R., W. J. Mitsch, and J. W. Day, Jr. 2006. A new vision for New Orleans and the Mississippi delta: Applying ecological economics and ecological engineering. Frontiers in Ecology and the Environment 4(9):465-472.

Costanza, R., O. Perez-Maqueo, M. L. Martinez, P. Sutton, S. J. Anderson, and K. Mulder. 2008. The value of coastal wetlands for hurricane protection. Ambio 37(4):241-248.

Couillard, C. M., K. Lee, B. Legare, and T. L. King. 2005. Effect of dispersant on the composition of the water-accommodated fraction of crude oil and its toxicity to larval marine fish. Environmental Toxicology and Chemistry 24:1496-1504.

Cowen, R. K., C. B. Paris, and A. Srinivasan. 2006. Scaling of connectivity in marine populations. Science 311:522-527.

CPRA (Coastal Protection and Restoration Authority). 2012. Louisiana's Comprehensive Master Plan for a Sustainable Coast. Baton Rouge, LA: CPRA. Available at http://coastal.louisiana.gov/index.cfm?md=pagebuilder&tmp=home&nid=24&pnid=0&pid=28&, accessed March 25, 2013.

Cranswick, D., and J. Regg. 1997. Deepwater in the Gulf of Mexico: America's New Frontier. OCS Report MMS 97-0004. New Orleans, LA: U.S. Department of the Interior, Minerals Management Service, Gulf of Mexico OCS Region.

Cubukcu, N., R. L. Pfeffer, and D. E. Dietrich. 2000. Simulation of the effects of bathymetry and land-sea contrasts on hurricane development using a coupled ocean-atmosphere model. Journal of the Atmospheric Sciences 57(4):481-492.

Cumming, G. S., and M. F. Child. 2009. Contrasting spatial patterns of taxonomic and functional richness offer insights into potential loss of ecosystem services. Philosophical Transactions of the Royal Society of London Series B: Biological Sciences 364:1683-1692.

Cutter, S. L., L. Barnes, M. Berry, C. Burton, E. Evans, E. Tate, and J. Webb. 2008. A place-based model for understanding community resilience to natural disasters. Global Environmental Change 18(4):598-606.

Daane, L. L., I. Harjono, G. J. Zylstra, and M. M. Häggblom. 2001. Isolation and characterization of polycyclic aromatic hydrocarbon-degrading bacteria associated with the rhizosphere of salt marsh plants. Applied and Environmental Microbiology 67:2683-2691.

Daily, G. C. (Ed.). 1997. Nature's Services: Societal Dependence on Natural Ecosystems. Washington, DC: Island Press.

Daily, G. C., P. Stephen, G. Joshua, K. M. Peter, M. A. Harold, P. Liba, R. H. Taylor, S. James, and S. Robert. 2009. Ecosystem services in decision making: Time to deliver. Frontiers in Ecology and the Environment 7(1):21-28.

Danielsen, F., M. K. Sørensen, M. F. Olwig, V. Selvam, F. Parish, N. D. Burgess, T. Hiraishi, V. M. Karunagaran, M. S. Rasmussen, L. B. Hansen, A. Quarto, and N. Suryadiputra. 2005. The Asian tsunami: A protective role for coastal vegetation. Science 310(5748):643.

Darby, F. A., and R. E. Turner. 2008. Effects of eutrophication on salt marsh root and rhizome biomass accumulation. Marine Ecology Progress Series 363:63-70.

Das, S., and J. R. Vincent. 2009. Mangroves protected villages and reduced death toll during Indian super cyclone. Proceedings of the National Academy of Sciences of the United States of America 106(40):7357-7360.

Davidson, W. F., K. Lee, and A. Cogswell (Eds.). 2008. Oil Spill Response: A Global Perspective. Dordrecht, Netherlands: Springer.

Day, J. W., Jr., J. Barras, E. Clairain, J. Johnston, D. Justic, G. P. Kemp, J.-Y. Ko, R. Lane, W. J. Mitsch, G. Steyer, P. Templet, and A. Yañez-Arancibia. 2005. Implications of global climatic change and energy cost and availability for the restoration of the Mississippi delta. Ecological Engineering 24(4):253-265.

Day, J. W., Jr., D. F. Boesch, E. J. Clairain, G. P. Kemp, S. B. Laska, W. J. Mitsch, K. Orth, H. Mashriqui, D. J. Reed, L. Shabman, C. A. Simenstad, B. J. Steever, R. R. Twilley, C. C. Watson, J. T. Wells, and D. F. Whigham. 2007. Restoration of the Mississippi Delta: Lessons from Hurricanes Katrina and Rita. Science 315:1679-1684.

Day, J. W., J. E. Cable, J. H. Cowan, Jr., R. DeLaune, K. de Mutsert, B. Fry, H. Mashriqui, D. Justic, P. Kemp, R. R. Lane, J. Rick, S. Rick, L. P. Rozas, G. Snedden, E. Swenson, R. R. Twilley, and B. Wissel. 2009. The impacts of pulsed reintroduction of river water on a Mississippi Delta coastal basin. Journal of Coastal Research 54(Special Issue):225-243.

de Groot, R. S. 1987. Environmental functions as a unifying concept for ecology and economics. The Environmentalist 7(2):105-109.

de Groot, R. S., M. A. Wilson, and R. M. J. Boumans. 2002. A typology for the classification, description and valuation of ecosystem functions, goods and services. Ecological Economics 41:393-408.

de Soysa, T. Y., A. Ulrich, T. Friedrich, D. Pite, S. L. Compton, D. Ok, R. L. Bernardos, G. B. Downes, S. Hsieh, R. Stein, M. C. Lagdameo, K. Halvorsen, L.-R. Kesich, and M. J. F. Barresi. 2012. Macondo crude oil from the *Deepwater Horizon* oil spill disrupts specific developmental processes during zebrafish embryogenesis. BMC Biology 10:1-24.DeAngelis, D. L. 1980. Energy flow, nutrient cycling, and ecosystem resilience. Ecology 61(4):764-771.

Deegan, L. A., D. S. Johnson, R. S. Warren, B. J. Peterson, J. W. Fleeger, S. Fagherazzi, and W. M. Wollheim. 2012. Coastal eutrophication as a driver of salt marsh loss. Nature 490:388-392.

Deepwater Benthic Communities Technical Working Group. 2011. Deepwater Sediment Sampling to Assess Post-Spill Benthic Impacts from the *Deepwater Horizon* Oil Spill. Study Plan for NRDA-Phase II Project. Silver Spring, MD: National Oceanic and Atmospheric Administration. Available at http://www.doi.gov/deepwaterhorizon/adminrecord/upload/DeepBenthicSedimentSampling_5-20-2011-allsigned-redacted-1.pdf, accessed March 21, 2013.

DeLaune, R. D., and A. L. Wright. 2011. Projected impact of the *Deepwater Horizon* oil spill on U.S. Gulf Coast Wetlands. Soil Science Society of America 75(5):1602-1612.

DeLaune, R. D., C. J. Smith, and M. D. Tolley. 1984. The effect of sediment redox potential on nitrogen uptake, anaerobic root respiration and growth of *Spartina alterniflora loisel.* Aquatic Botany 18(3):223-230.

Delille, D., E. Pelletier, A. Rodriguez-Blanco, and J.-F. Ghiglione. 2009. Effects of nutrient and temperature on degradation of petroleum hydrocarbons in sub-Antarctic coastal seawater. Polar Biology 32:1521-1528.

Dell'Apa, A., L. Schiavinato, and R. Rulifson. 2012. The Magnuson-Stevens Act (1976) and its reauthorizations: Failure or success for the implementation of fishery sustainability and management in the US? Marine Policy 36:673-680.

Deming, J. W., L. K. Somers, W. L. Straube, D. G. Swartz, and M. T. MacDonell. 1988. Isolation of an obligately barophilic bacterium and description of a new genus, *Colwellia* gen. nov. Systematic and Applied Microbiology 10:152-160.

Demopoulos, A. J., D. Gualtieri, and K. Kovacs. 2010. Food-web structure of seep sediment macrobenthos from the Gulf of Mexico. Deep Sea Research Part II: Topical Studies in Oceanography 57:1972-1981.

Deriso, R. B., Maunder, M. N., and Pearson, W. H. 2008. Incorporating covariates into fisheries stock assessment models with application to Pacific herring. Ecological Applications 18:1270-1286.

Desai, J. D., and I. M. Banat. 1997. Microbial production of surfactants and their commercial potential. Microbiology and Molecular Biology Reviews 61:47-64.

Die, D. J. 2009. Design and implementation of management plans. Pp. 205-220 in A Fishery Manager's Handbook, K. Cohrane and S. Garcia (Eds.). Chichester, UK: Wiley-Blackwell.

Diercks, A. R., R. C. Highsmith, V. L. Asper, D. Joung, Z. Zhou, L. Guo, S. Joye, A. P. Teske, N. Guinasso, T. Wade, and S. E. Lohrenz. 2010. Characterization of subsurface polycyclic aromatic hydrocarbons at the *Deepwater Horizon* site. Geophysical Research Letters 37:160-164.

Díez, I., A. Secilla, A. Santolaria, and J. M. Gorostiaga. 2009. Ecological monitoring of intertidal phytobenthic communities of the Basque Coast (N. Spain) following the *Prestige* oil spill. Environmental Monitoring and Assessment 159:555-575.

Dingler, J. R., S. A. Hsu, and A. L. Foote. 1995. Wind shear stress measurements in a coastal marsh during Hurricane Andrew. Journal of Coastal Research 21 (Special Issue):295-305.

Ditz, D. W. 1988. Hazardous waste inceneration at sea: EPA decision making on risk. Risk Analysis 8:499-508.

DOE (Department of Energy). 2010. Animal Energy Outlook. 2010. Available at http://www.eia.doc.gov/2010.

Dokka, R. K., G. F. Sella, and T. H. Dixon. 2006. Tectonic control of subsidence and southward displacement of southeast Louisiana with respect to stable North America. Geophysical Research Letters 33(L23308), doi:10.1029/2006GL027250.

Donoghue, J. F. 2011. Sea-level history of the northern Gulf of Mexico and sea-level rise scenarios for the near future. Climatic Change 107(1):17-33.

Doremus, H. 2001. The Endangered Species Act and the institutional challenges of "new age" environmental protection. Washburn Law Journal 41:50-89.

Douglass, S. L., B. M. Webb, C. R. Dixon, and T. Buhring. 2011. Beach profile and island cross-section manipulations in response to an offshore oil spill. Shore and Beach 79:3-10.

Duignan, P. J., C. H. House, D. K. Odell, R. S. Wells, L. J. Hanson, M. T. Walsh, D. J. St. Aubin, B. K. Rima, and J. R. Geraci. 1996. Morbillivirus infection in bottlenose dolphins: Evidence for recurrent epizootics in the Western Atlantic and Gulf of Mexico. Marine Mammal Science 12:499-515.

Dulvy, N. K., R. P. Freckleton, and N. V. Polunin. 2004. Coral reef cascades and the indirect effects of predator removal by exploitation. Ecology Letters 7(5):410-416.

Dunbar, J. B., L. D. Britsch, and E. B. Kemp III. 1992. Land Loss Rates Report 3, Louisiana Coastal Plain. Technical Report GL-90-2. New Orleans, LA: U.S. Army Engineer District.

Dyksterhouse, S. E., J. P. Gray, R. P. Herwig, J. C. Lara, and J. T. Staley. 1995. *Cycloclasticus pugetii* gen. nov., sp. nov., an aromatic hydrocarbon-degrading bacterium from marine sediments. International Journal of Systematic and Evolutionary Microbiology 45:116-123.

Eagle, J. 2006. Regional ocean governance: The perils of multiple-use management and the promise of agency diversity. Duke Environmental Law & Policy Forum 16:143-177.

Ehrenfeld, J. G., and L. A. Toth. 1997. Restoration ecology and the ecosystem perspective. Restoration Ecology 5:307-317.

Ehrlich, P. R., and H. A. Mooney. 1983. Extinction, substitution, and ecosystem services. BioScience 33:248-254.

Eide, A. 2008. An integrated study of economic effects of and vulnerabilities to global warming on the Barents Sea cod fisheries. Climatic Change 87:251-262.

Eide, A. 2009. Economic principles: An economic perspective on fishing. Pp. 75-101 in A Fishery Manager's Handbook, K. Cohrane and S. Garcia (Eds.). Chichester, UK: Wiley-Blackwell.

Eldridge, P. J. 1988. The Southeast Area Monitoring and Assessment Program (SEAMAP): A state-federal-university program for collection, management, and dissemination of fishery-independent data and information in the southeastern United States. Marine Fisheries Review 50:29-39.

Elliot, A. J., N. Hurford, and C. J. Penn. 1986. Shear diffusion and the spreading of oil slicks. Marine Pollution Bulletin 17(7):308-313.

Ellis, G. M., and A. C. Fisher. 1987. Valuing the environment as an input. Journal of Environmental Management 25:149-156.

Elmqvist, T., C. Folke, M. Nyström, G. Peterson, J. Bengtsson, B. Walker, and J. Norberg. 2003. Response diversity, ecosystem change, and resilience. Frontiers in Ecology and the Environment 1(9):488-494.

Emanuel, K. A. 1987. The dependence of hurricane intensity on climate. Nature 326:483-485.

EOSCA (European Oilfield Speciality Chemicals Association). 2000. Bioaccumulation Potential of Surfactants: A Review. Aberdeen, Scotland: EOSCA.

EPA (U.S. Environmental Protection Agency), Science Advisory Board. 2009. Valuing the Protection of Ecological Systems and Services. EPA-SAB-09-012. Washington, DC: EPA.

EPA. 2010. Dispersant Monitoring and Assessment Directive for Subsurface Dispersant Application. Washington, DC: EPA.

Estes, J. A., D. F. Doak, A. M. Springer, and T. M. Williams. 2009. Causes and consequences of marine mammal population declines in southwest Alaska: A food-web perspective. Philosophical Transactions of the Royal Society of London, Series B: Biological Sciences 364:1647-1658.

Fagherazzi, S., M. Marani, and L. Blum. 2004. The Ecogeomorphology of Tidal Marshes, Vol. 59 (Coastal and Esturine Studies). Washington, DC: American Geophysical Union.

Feagin, R. A., W. K. Smith, N. P. Psuty, D. R. Young, M. L. Martínez, G. A. Carter, K. L. Lucas, J. C. Gibeaut, J. N. Gemma, and R. E. Koske. 2010. Barrier islands: Coupling anthropogenic stability with ecological sustainability. Journal of Coastal Research 26(6):987-992.

Federal Interagency Solutions Group. 2010. Oil Budget Calculator: *Deepwater Horizon*, November 2010. Oil Budget Calculator Science and Engineering Team. Available at http://www.restorethegulf.gov/sites/default/files/documents/pdf/OilBudgetCalc_Full_HQ-Print_111110.pdf, accessed July 16, 2011.

FEMA (Federal Emergency Management Agency). 1985. Flood Insurance Study for Pacific County, Washington.

Feral, F. 2009. The fishery management institutions. Pp. 135-163 in A Fishery Manager's Handbook, K. Cohrane and S. Garcia (Eds.). Chichester, UK: Wiley-Blackwell.

Ferenc, S. A., and J. A. Foran. 2000. Multiple stressors in ecological risk and impact assessment: Approaches to risk estimation. Pensacola, FL: SETAC Press.

Field, J. C., and R. C. Francis. 2006. Considering ecosystem-based fisheries management in the California Current. Marine Policy 30:552-569.

Finch, B. E., K. J. Wooten, and P. N. Smith. 2011. Embryotoxicity of weathered crude oil from the Gulf of Mexico in mallard ducks (*Anas platyrhynchos*). Environmental Toxicology and Chemistry 30:1885-1891.

Finch, B. E., K. J. Wooten, D. R. Faust, and P. N. Smith. 2012. Embryotoxicity of mixtures of weathered crude oil collected from the Gulf of Mexico and Corexit 9500 in mallard ducks (*Anas platyrhynchos*). Science of the Total Environment 426:155-159.

Fingas, M. 2011a. An overview of in-situ burning. Pp. 737-903 in Oil Spill Science and Technology: Prevention, Response, and Cleanup, M. Fingas (Ed.). Burlington, MA: Elsevier.

Fingas, M. (Ed.) 2011b. Oil Spill Science and Technology: Prevention, Response, and Cleanup. Burlington, MA: Elsevier.

Finke, N., V. Vandieken, and B. B. Jorgensen. 2007. Acetate, lactate, propionate, and isobutyrate as electron donors for iron and sulfate reduction in Arctic marine sediments, Svalbard. FEMS Microbiology Ecology 59:10-22.

Fire, S. E., and F. M. Van Dolah. 2012. Marine biotoxins: Emergence of harmful algal blooms as health threats to marine wildlife. Pp. 374-389 in New Directions in Conservation Medicine: Applied Cases of Ecological Health, A. A. Aguirre, R. S. Ostfeld, and P. Daszak (Eds.). New York: Oxford University Press.

Fisher, C. R., H. Roberts, E. Cordes, and B. Bernard. 2007. Cold seeps and associated communities of the Gulf of Mexico. Oceanography 20(4):68-79.

Florida Department of Environmental Protection. 2011. *Deepwater Horizon* oil spill: One year later. Available at http://www.dep.state.fl.us/deepwaterhorizon/files2/041211_one_anniversary.pdf, accessed March 26, 2013.

Fodrie, F. J., and K. L. J. Heck. 2011. Response of coastal fishes to the Gulf of Mexico oil disaster. PLoS ONE 6:e21609.

Foght, J. M., and D. W. S. Westlake. 1982. Effect of the dispersant Corexit 9527 on the microbial degradation of Prudhoe Bay oil. Canadian Journal of Microbiology 28:117-122.

Folke, C., S. Carpenter, T. Elmqvist, L. Gunderson, C. S. Holling, B. Walker, J. Bengtsson, F. Berkes, J. Colding, K. Danell, M. Falkenmark, L. Gordon, R. Kasperson, N. Kautsky, A. Kinzig, S. Levin, K.-G. Mäler, F. Moberg, L. Ohlsson, P. Olsson, E. Ostrom, W. Reid, J. Rockström, H. Savenije, and U. Svedin. 2002. Resilience and sustainable development: Building adaptive capacity in a world of transformations. Scientific background paper on Resilience for the Process of the World Summit on Sustainable Development on behalf of the Environmental Advisory Council to the Swedish Government, Johannesburg, South Africa, August 26-September 4, 2002.

Folke, C., S. Carpenter, B. Walker, M. Scheffer, T. Elmqvist, L. Gunderson, and C. S. Holling. 2004. Regime shifts, resilience, and biodiversity in ecosystem management. Annual Review of Ecology, Evolution, and Systematics 35:557-581.

Folke, C., S. R. Carpenter, B. Walker, M. Scheffer, T. Chapin, and J. Rockström. 2010. Resilience thinking: Integrating resilience, adaptability and transformability. Ecology and Society 15(4):20

Food Channel Editor. 2011. What Killed the Oysters? It Wasn't the Oil. The Food Channel (blog). March 9, 2011. Available at http://www.foodchannel.com/articles/article/what-killed-oysters-it-wasnt-oil/, accessed May 22, 2013.

Freeman, A. M. III. 2003. The Measurement of Environmental and Resource Values: Theory and Methods. Washington, DC: Resources for the Future.

Freire, J., L. Fernández, and R. Muiño. 2006. Role of the Spanish Scientific Community in the Initial Assessment and Management of the Environmental Damages Caused by the *Prestige* Oil Spill. Marine Policy 30:308-314.

French, L. S., G. E. Richardson, E. G. Kazanis, T. M. Montgomery, C. M. Bohannon, and M. P. Gravois. Deepwater Gulf of Mexico 2006: America's Expanding Frontier. OCS Study MMS 2006-022. New Orleans, LA: U.S. Department of the Interior, Minerals Management Service, Gulf of Mexico OCS Region.

Friedlander, A. M., and E. E. DeMartini. 2002. Contrasts in density, size, and biomass of reef fishes between the northwestern and the main Hawaiian Islands: The effects of fishing down apex predators. Marine Ecology Progress Series. 230:253-264.

Friedrichs, C., and J. Perry. 2001. Tidal salt marsh morphodynamics. Journal of Coastal Research 27(Special Issue):7-37.

Fromentin, J. M., and J. Powers. 2005. Atlantic bluefin tuna: Population dynamics, ecology, fisheries and management. Fish and Fisheries 6:281-306.

Fuller, C., J. Bonner, C. Page, A. Ernest, T. McDonald, and S. McDonald. 2004. Comparative toxicity of oil, dispersant, and oil plus dispersant to several marine species. Environmental Toxicology and Chemistry 23:2941-2949.

Fulton, E. A. 2010. Approaches to end-to-end ecosystem models. Journal of Marine Systems 81:171-183.

Fulton, E. A., J. S. Parslow, A. D. M. Smith, and C. R. Johnson. 2004a. Biogeochemical marine ecosystem models II: The effect of physiological detail on model performance. Ecological Modelling 173:371-406.

Fulton, E. A., A. D. M. Smith, and C. R. Johnson. 2004b. Biogeochemical marine ecosystem models I: IGBEM—a model of marine bay ecosystems. Ecological Modelling 174:267-307.

Fulton, E. A., A. D. M. Smith, and D. C. Smith. 2007. Alternative Management Strategies for Southeast Australian Commonwealth Fisheries, Stage 2: Quantitative Management Strategy Evaluation. Australian Fisheries Management Authority Report. Hobart, Australia: CSIRO Marine and Atmospheric Research.

Fulton, E. A., J. Link, I. C. Kaplan, P. Johnson, M. Savina-Rolland, C. Ainsworth, P. Horne, R. Gorton, R. J. Gamble, T. Smith, and D. Smith. 2011. Lessons in modelling and management of marine ecosystems: The Atlantis experience. Fish and Fisheries 2:171-188.

Gallaway, B. J. 1988. Northern Gulf of Mexico Continental Slope Study Annual Report, Year 4. Volume I: Executive Summary. OCS Study MMS 88-0052. New Orleans, LA: U.S. Department of the Interior, Minerals Management Service, Gulf of Mexico OCS Region.

Gallaway, B. J., L. R. Martin, and R. L. Howard. 1988. Northern Gulf of Mexico Continental Slope Study Annual Report, Year 3, Volume I: Executive Summary. OCS Study MMS 87-0059. New Orleans, LA: U.S. Department of the Interior, Minerals Management Service, Gulf of Mexico OCS Region.

Gallopin, G. C. 2006. Linkages between vulnerability, resilience, and adaptive capacity. Global Environmental Change 16(3):293-303.

Galuardi, B., F. Royer, W. Golet, J. Logan, and J. Neilson. 2010. Complex migration routes of Atlantic bluefin tuna (*Thunnus thynnus*) question current population structure paradigm. Canadian Journal of Fisheries and Aquatic Sciences 67:966-976.

Gan, J., and C. T. Smith. 2006. A comparative analysis of woody biomass and coal for electricity generation under various $CO_2$ emission reductions and taxes. Biomass and Bioenergy 30:296-303.

Garcia, S., and K. Cochrane. 2009. From past management to future governance: A perspective view. Pp. 447-472 in A Fishery Manager's Handbook, K. Cohrane and S. Garcia (Eds.). Chichester, UK: Wiley-Blackwell.

Garcia, S. M., A. Zerbi, C. Aliaume, T. Do Chi, and G. Lasserre. 2003. The Ecosystem Approach to Fisheries. Issues, Terminology, Principles, Institutional Foundations, Implementation and Outlook. FAO Fisheries Department Technical Paper No. 443. Rome, Italy: Food and Agriculture Organization of the United Nations.

Garcia, M. T., E. Campos, A. Marsal, and I. Ribosa. 2009. Biodegradability and toxicity of sulphonate-based surfactants in aerobic and anaerobic aquatic environments. Water Research 43:295-302.

Garza-Gil, M. D., A. Prada-Blanco, and M. X. Vázquez-Rodríguez. 2006. Estimating the short-term economic damages from the *Prestige* oil spill in the Galician fisheries and tourism. Ecological Economics 58:842-849.

Geiselbrecht, A. D., B. P. Hedlund, M. A. Tichi, and J. T. Staley. 1998. Isolation of marine polycyclic aromatic hydrocarbon (PAH)-degrading *Cycloclasticus* strains form the Gulf of Mexico and comparison of their PAH degradation ability with that of Puget Sound *Cycloclasticus* strains. Applied and Environmental Microbiology 64(12):4703-4710.

Geissler, P. H. 1990. Estimation of confidence intervals for federal waterfowl harvest surveys. Journal of Wildlife Management 54(2):201-205.

George-Ares, A., and J. R. Clark. 2000. Aquatic toxicity of two Corexit dispersants. Chemosphere 40:897-906.

Geraci, J. R. 1990. Physiologic and toxic effects on cetaceans. Pp. 167-197 in Sea Mammals and Oil: Confronting the Risks, J. R. Geraci and D. J. St. Aubin (Eds.). San Diego, CA: Academic Press.

Geraci, J. R., D. J. St. Aubin, and R. J. Reisman. 1983. Bottlenose dolphins, *Tursiops truncatus,* can detect oil. Canadian Journal of Aquatic Science 40:1515-1522.

Gerard, R. 1976. Environmental effects of deep-sea mining. Marine Technology Society Journal 10:7-12.

Gerrodette, T., and J. Forcada. 2005. Non-recovery of two spotted and spinner dolphin populations in the eastern tropical Pacific Ocean. Marine Ecology Progress Series 291:1-21.

Gerrodette, T., P. K. Dayton, S. Macinko, and M. J. Fogarty. 2002. Precautionary management of marine fisheries: Moving beyond burden of proof. Bulletin of Marine Science 70:657-668.

Gilde, K., and J. Pinckney. 2012. Sublethal effects of crude oil on the community structure of estuarine phytoplankton. Estuaries and Coasts 35:853-861.

Gilmore, M. S., E. H. Wilson, D. L. Civco, S. Prisloe, J. D. Hurd, and C. Chadwick. 2008. Integrating multi-temporal spectral and structural information to map dominant tidal wetland vegetation in a lower Connecticut River marsh. Remote Sensing of Environment 112:4048-4060.

Giraud, K., B. Turcin, J. Loomis, and J. Cooper. 2002. Economic benefit of the protection program for Stellar sea lion. Marine Policy 26:451-458.

Gjerde, K. M. 2006. Ecosystems and Biodiversity in Deep Waters and High Seas. UNEP Regional Seas Report and Studies No. 178. Geneva, Switzerland: United Nations Environment Programme. Available at http://www.unep.org/pdf/EcosystemBiodiversity_DeepWaters_20060616.pdf, accessed March 26, 2013.

Graham, B., W. K. Reilly, F. G. Beinecke, D. Boesch, T. D. Garcia, C. A. Murray, and F. Ulmer. 2011a. Deep Water: The Gulf Oil Disaster and the Future of Offshore Drilling. Washington, DC: National Commission on the BP *Deepwater Horizon* Oil Spill and Offshore Drilling. Available at http://www.gpo.gov/fdsys/pkg/GPO-OILCOMMISSION/pdf/GPO-OILCOMMISSION.pdf, accessed May 22, 2013.

Graham, B., W. K. Reilly, F. G. Beinecke, D. Boesch, T. D. Garcia, C. A. Murray, and F. Ulmer. 2011b. The Story of the Louisiana Berms Project. Staff Working Paper No. 8. Washington, DC: National Commission on the BP *Deepwater Horizon* Oil Spill and Offshore Drilling. Available at http://www.oilspillcommission.gov/sites/default/files/documents/Updated%20Berms%20Working%20Paper.pdf, accessed May 22, 2013.

Graham, W. M., R. H. Condon, R. H. Carmichael, I. D'Ambra, H. K. Patterson, L. J. Linn, and F. J. Hernandez. 2010. Oil carbon entered the coastal planktonic food web during the *Deepwater Horizon* oil spill. Environmental Research Letters 5:045301.

Greer, C. D., P. V. Hodson, Z. Li, T. King, and K. Lee. 2012. Toxicity of crude oil chemically dispersed in a wave tank to embryos of Atlantic herring (*Clupea harengus*). Environmental Toxicology and Chemistry 31:1-10.

Guerry, A. D., M. H. Ruckelshaus, K. K. Arkema, J. R. Bernhardt, G. Guannel, C.-K. Kim, M. Marsik, M. Papenfus, J. E. Toft, G. M. Verutes, S. A. Wood, M. Beck, F. Chan, K. M. A. Chan, G. Gelfenbaum, B. Gold, B. Halpern, B. Labiosa, S. Lester, P. Levin, M. McField, M. Pinsky, M. Plummer, S. Polasky, P. Ruggiero, D. Sutherland, H. Tallis, A. Day, and J. Spencer. 2012. Modeling benefits from nature: Using ecosystem services to inform coastal and marine spatial planning. International Journal of Biodiversity Science, Ecosystem Services & Management 8(1-2):107-121.

Gulf Coast Ecosystem Restoration Task Force. 2011. Gulf of Mexico Regional Ecosystem Restoration Strategy. Washington, DC: U.S. Environmental Protection Agency.

Gulf Coast Incident Management Team. 2011. Orphan Anchor Phase II Program. Report to the Federal On-scene Commander. Available at http://www.dhs.gov/xlibrary/assets/orphan-anchor-phase-2-program-report.pdf, accessed May 22, 2013.

Gunderson, L. 2000. Ecological resilience—in theory and application. Annual Review of Ecology and Systematics 31:425-439.

Gunderson, L., and C. Folke. 2011. Resilience 2011: Leading transformational change. Ecology and Society 16(2):30.

Gunderson, L. H., and C. S. Holling (Eds.). 2002. Panarchy: Understanding transformations in human and natural systems. Washington, DC: Island Press.

Gunderson, L. H., and L. Pritchard (Eds.). 2002. Resilience and the behavior of large-scale ecosystems. Washington, DC: Island Press.

Gunderson, L. H., C. S. Holling, and S. Light. 1995. Barriers and bridges to renewal of ecosystems and institutions. New York: Columbia University Press.

Hamdan, L. J., and P. A. Fulmer. 2011. Effects of COREXIT EC9500A on bacteria from a beach oiled by the *Deepwater Horizon* spill. Aquatic Microbial Ecology 63:101-109.

Hannam, M. L., S. D. Bamber, J. A. Moody, T. S. Galloway, and M. B. Jones. 2009. Immune function in the Arctic scallop, *Chlamys islandica,* following dispersed oil exposure. Aquatic Toxicology 92:187-194.

Hannington, M. D., J. Jamieson, T. Monecke, and S. Petersen. 2010. Modern sea-floor massive sulfides and base metal resources: Toward a global estimate of sea-floor massive sulfide potential. Pp. 317-338 in The Challenge of Finding New Mineral Resources: Global Metallogeny, Innovative Exploration and New Discoveries, Volume II, Zinc-Lead, Nickel-Copper-PGE, and Uranium, R. J. Goldfarb, E. E. Marsh, and T. Monecke (Eds.), Special Publication 15. Littleton, CO: Society of Economic Geologists.

Harwell, M., and J. H. Gentile. 2006. Ecological significance of residual exposures and effects from the *Exxon Valdez* oil spill. Integrated Environmental Assessment and Management 2:204-246.

Hayworth, J. S., and T. Prabakhar Clement. 2012. Provenance of Corexit-related chemical constituents found in nearshore and inland Gulf Coast waters. Marine Pollution Bulletin 64:2005-2014.

Hazen, T. C., E. A. Dubinsky, T. Z. DeSantis, G. L. Andersen, Y. M. Piceno, N. Singh, J. K. Jansson, A. Probst, S. E. Borglin, J. L. Fortney, W. T. Stringfellow, M. Bill, M. E. Conrad, L. M. Tom, K. L. Chavarria, R. Alusi, R. Lamendella, D. C. Joyner, C. Spier, J. Baelum, M. Auer, M. L. Zemla, R. Chakraborty, E. L. Sonnenthal, P. D'haeseleer, H. Ying, N. Holman, S. Osman, Z. Lu, J. D. Van Nostrand, Y. Deng, J. Zhou, and O. U. Mason. 2010. Deep-sea oil plume enriches indigenous oil-degrading bacteria. Science 330:204-208.

Hemmer, M. J., M. G. Barron, and R. M. Green. 2010. Comparative Toxicity of Louisiana Sweet Crude Oil (LSC) and Chemically Dispersed LSC to Two Gulf of Mexico Aquatic Test Species. Washington, DC: U.S. Environmental Protection Agency.

Hilborn, R. 2004. Ecosystem-based fisheries management: The carrot or the stick? Marine Ecology Progress Series 274:275-278.

Hilborn, R. 2012. The evolution of quantitative marine fisheries management 1985-2010. Natural Resource Modeling 25:122-144.

Hilborn, R., A. Punt, and J. Orensanz. 2004. Beyond band-aids in fisheries management: Fixing world fisheries. Bulletin of Marine Science 74:493-507.

Hladik, C. M., and M. Alber. 2012. Accuracy assessment and correction of a LIDAR-derived salt marsh digital elevation model. Remote Sensing of Environment 121:224-235.

Hobbs, R. J., and V. A. Cramer. 2008. Restoration ecology: Interventionist approaches for restoring and maintaining ecosystem function in the face of rapid environmental change. Annual Review of Environment and Resources 33:39-61.

Hoff, R. Z. 1995. Responding to Oil Spills in Coastal Marshes: The Fine Line Between Help and Hindrance. HAZMAT Report 96-1. Washington, DC: Hazardous Material Response and Assessments Division, NOAA.

Hoffman, D., and P. Albers. 1984. Evaluation of potential embryotoxicity and teratogenicity of 42 herbicides, insecticides, and petroleum contaminants to mallard eggs. Archives of Environmental Contamination and Toxicology 13:15-27.

Holling, C. S. 1965. The functional response of predators to prey density and its role in mimicry and population regulation. Memoirs of the Entomological Society of Canada 97(S45):5-60.

Holling, C. S. 1973. Resilience and stability of ecological systems. Annual Review of Ecology and Systematics 4:1-23.

Holling, C. S. (Ed.). 1978. Adaptive environmental assessment and management. London: John Wiley and Sons.

Holling, C. S. 1996. Engineering resilience versus ecological resilience. In Engineering Within Ecological Constraints, P. C. Schulze (Ed.). Washington, DC: National Academy Press.

Holling, C. S. 2001. Understanding the complexity of economic, ecological, and social systems. Ecosystems 4(5):390-405.

Holling, C. S., and G. K. Meffe. 1996. Command and control and the pathology of natural resource management. Conservation Biology 10(2):328-337.

Horel, A., B. Mortazavi, and P. A. Sobecky. 2012. Responses of microbial community from northern Gulf of Mexico sandy sediments following exposure to *Deepwater Horizon* crude oil. Environmental Toxicology and Chemistry 31:1004-1011.

Houde, M., G. Pacepavicius, R. S. Wells, R. A. Fair, R. J. Letcher, M. Alaee, G. D. Bossart, A. A. Hohn, J. Sweeney, K. R. Solomon, and D. C. G. Muir. 2006. Polychlorinated biphenyls and hydroxylated polychlorinated biphenlys in plasma of bottlenose dolphins (*Tursiops truncatus*) from the Western Atlantic and Gulf of Mexico. Environmental Science & Technology 40(19):5860-5866.

Hu, C., R. H. Weisberg, Y. Liu, L. Zheng, K. L. Daly, D. C. English, J. Zhao, and G. A. Vargo. 2011. Did the northeastern Gulf of Mexico become greener after the *Deepwater Horizon* oil spill? Geophysical Research Letters 38:L09601.

Hughes, T. P., A. H. Baird, D. R. Bellwood, M. Card, S. R. Connolly, C. Folke, R. Grosberg, O. Hoegh-Guldberg, J. B. C. Jackson, J. Kleypas, J. M. Lough, P. Marshall, M. Nyström, S. R. Palumbi, J. M. Pandolfi, B. Rosen, and J. Roughgarden. 2003. Climate change, human impacts, and the resilience of coral reefs. Science 301:929-933.

Huguenin, M. T., D. H. Haury, J. C. Weiss, D. Helton, C. Manen, and E. Reinharz, and J. Michel. 1996. Injury assessment guidance document for natural resource damage assessment under the Oil Pollution Act of 1990. Silver Spring, MD: National Oceanic and Atmospheric Administration Damage Assessment and Restoration Program.

ICCAT (International Commission for the Conservation of Atlantic Tunas). 2010. Report of the 2009 Atlantic Swordfish Stock Assessment Session. Collective Volume of Scientific Papers 65(1):1-123. Available at http://www.iccat.int/Documents/CVSP/CV065_2010/no_1/CV065010001.pdf.

ICCAT. 2011. Report of the 2010 Atlantic Bluefin Tuna Stock Assessment Session. Collective Volume of Scientific Papers 66(2):505-714. Available at http://www.iccat.int/Documents/CVSP/CV066_2011/no_2/CV066020505.pdf.

ICCAT. 2012. Report of the 2011 Atlantic Yellowfin Tuna Stock Assessment Session. Collective Volume of Scientific Papers 67(1):1-113. Available at http://www.iccat.es/Documents/Meetings/Docs/2011_YFT_ASSESS_REP.pdf.

Impact Assessment, Inc. 2007. Preliminary Assessment of the Impacts of Hurricane Katrina on Gulf of Mexico Coastal Fishing Communities. Final Technical Report. St. Petersburg, FL: National Marine Fisheries Service.

IOM (Institute of Medicine). 2010. Assessing the Effects of the Gulf of Mexico Oil Spill on Human Health: Workshop Summary. Washington, DC: National Academies Press.

Irions, J. B. 2002. Cultural resource management of shipwrecks on the Gulf of Mexico Outer Continental Slope. Presentation at 2nd MIT Conference on Technology, Archaeology, and the Deep Sea, April 26-28, 2002. Available at http://web.mit.edu/deeparch/www/events/2002conference/papers/Irion.pdf, accessed June 24, 2013.

Isbell, F., V. Calcagno, A. Hector, J. Connolly, W. S. Harpole, P. B. Reich, M. Scherer-Lorenzen, B. Schmid, D. Tilman, J. van Ruijven, A. Weigelt, B. J. Wilsey, E. S. Zavaleta, and M. Loreau. 2011. High plant diversity is needed to maintain ecosystem services. Nature 477(7363):199-202.

Ives, A. R. 1995. Predicting the response of populations to environmental change. Ecology 76:926-941.

Ives, A. R., and S. R. Carpenter. 2007. Stability and diversity of ecosystems. Science 317:58-62.

Jackson, J. B. C., J. D. Cubit, B. D. Keller, V. Batista, K. Burns, H. M. Caffey, R. L. Caldwell, S. D. Garrity, C. D. Getter, C. Gonzalez, H. M. Guzman, K. W. Kaufmann, A. H. Knap, S. C. Levings, M. J. Marshall, R. Steger, R. C. Thompson, and E. Weil. 1989. Ecological effects of a major oil spill on Panamanian coastal marine communities. Science 243:37-44.

Jackson, J. B. C., M. X. Kirby, W. H. Berger, K. A. Bjorndal, L. W. Botsford, B. J. Bourque, R. H. Bradbury, R. Cooke, Jr., J. Erlandson, J. A. Estes, T. P. Hughes, S. Kidwell, C. B. Lange, H. S. Lenihan, J. M. Pandolfi, C. H. Peterson, R. S. Steneck, M. J. Tegner, and R. R. Warner. 2001. Historical overfishing and the recent collapse of coastal ecosystems. Science 293:629-638.

Jackson, W. A., and J. H. Pardue. 1997. Seasonal variability of crude oil respiration potential in salt and fresh marshes. Journal of Environmental Quality 26:1140-1146.

Jensen, O. P., T. A. Branch, and R. Hilborn. 2012. Marine fisheries as ecological experiments. Theoretical Ecology 5:3-22.

Jewett, S. C., T. A. Dean, B. R. Woodin, M. K. Hoberg, and J. J. Stegeman. 2002. Exposure to hydrocarbons 10 years after the *Exxon Valdez* oil spill: Evidence from cytochrome P4501A expression and biliary FACs in nearshore demersal fishes. Marine Environmental Research 54:21-48.

Jochens, A. E., and S. M. Watson. 2013. The Gulf of Mexico Coastal Ocean Observing System: An integrated approach to building an operational regional observing system. Marine Technology Society Journal 47:118-133.

Jochens, A. E., L. C. Bender, S. F. DiMarco, J. W. Morse, M. C. Kennicutt II, M. K. Howard, and W. D. Nowlin, Jr. 2005. Understanding the Processes That Maintain the Oxygen Levels in the Deep Gulf of Mexico: Synthesis Report. OCS Study MMS 2005-032. New Orleans, LA: U.S. Department of the Interior, Minerals Management Service, Gulf of Mexico OCS Region.

Johansen, Ø. 2000. DeepBlow: A Lagrangian plume model for deep water blowouts. Spill Science & Technology Bulletin 6:103-111.

Johansen, Ø, P. J. Brandvik, and U. Farooq. 2013. Droplet breakup in subsea oil releases—Part 2: Predictions of droplet size distributions with and without injection of chemical dispersants. Marine Pollution Bulletin 73(1):319-326.

Johnston, R. L., E. Y. Besedin, R. Iovanna, C. J. Miller, R. F. Wardwell, and M. H. Ranson. 2005. Systematic variation in willingness to pay for aquatic resource improvements and implications for benefit transfer: A meta-analysis. Canadian Journal of Agricultural Economics 53:221-248.

Joye, S. B. 2013. Impact of the Gulf oil crisis on the seafloor. Invited talk, American Association for the Advancement of Science, Annual Meeting, Boston, MA, February 16.

Joye, S. B., M. Crespo-Medina, K. Hunter, A. Vossmeyer, M. Bowles, V. Asper, A. Diercks, A. Teske, J. Montoya, C. Arnosti, C. Benitez-Nelson, J. Brandes, W. Moore, U. Passow, A. Subramaniam, T. Wade, K. Zeirvogel, and R. Highsmith, 2011. The Microbial Slime Highway: An efficient mechanism of oil transport to the benthos and consequences on microbial dynamics in deep Gulf of Mexico environments. Presentation at American Society of Limnology and Oceanography, Aquatic Sciences Meeting, San Juan, Puerto Rico, February.

Judson, R. S., M. T. Martin, D. M. Reif, K. A. Houck, T. B. Knudson, D. M. Rotroff, M. Xia, S. Sakamuru, R. Huang, P. Shinn, C. P. Austin, R. J. Kavlock, and D. J. Dix. 2010. Analysis of eight oil spill dispersants using rapid, in vitro tests for endocrine and other biological activity. Environmental Science & Technology 44:5979-5985.

Kahn, J. R., and W. M. Kemp. 1985. Economic losses associated with the degradation of an ecosystem: The case of submerged aquatic vegetation in Chesapeake Bay. Journal of Environmental Economics and Management 12:246-263.

Kahneman, D., and A. Tversky. 1979. Prospect theory: An analysis of decisions under risk. Econometrica 47:313-327.

Kaltenberg, A. M., D. C. Biggs, and S. F. DiMarco. 2007. Deep scattering layers of the northern Gulf of Mexico observed with a shipboard 38-kHz acoustic Doppler current profiler (ADCP). Gulf of Mexico Science 2:97-108.

Kannan, K., S. Corsolini, S. Focardi, S. Tanabe, and R. Tatsukawa. 1996. Accumulation pattern of butyltin compounds in dolphin, tuna and shark collected from Italian coastal waters. Archives of Environmental Contamination and Toxicology 31:19-23.

Kaplan, I. C., and J. Leonard. 2012. From krill to convenience stores: Forecasting the economic and ecological effects of fisheries management on the U.S. West Coast. Marine Policy 36:947-954.

Kaplan, I. C., P. S. Levin, M. Burden, and E. A. Fulton. 2010. Fishing catch shares in the face of global change: A framework for integrating cumulative impacts and single species management. Canadian Journal of Fisheries and Aquatic Sciences 67:1968-1982.

Kareiva, P., H. Tallis, T. H. Ricketts, G. C. Daily, and S. Polasky (Eds.). 2011. Natural Capital: Theory and Practice of Mapping Ecosystem Services. Oxford, UK: Oxford University Press.

Karickhoff, S. W. 1981. Semi-empirical estimation of sorption of hydrophobic pollutants on natural sediments and soils. Chemosphere 10:833-846.

Kazmierczak, R. F., Jr. 2001. Economic Linkages Between Coastal Wetlands and Water Quality: A Review of Value Estimates Reported in the Published Literature. Department of Agricultural Economics. Baton Rouge, LA: Natural Resource and Environment Committee, Louisiana State University Agricultural Economics & Agribusiness.

Kell, L. T., and J. M. Fromentin. 2007. Evaluation of the robustness of maximum sustainable yield based management strategies to variations in carrying capacity or migration pattern of Atlantic bluefin tuna (*Thunnus thynnus*). Canadian Journal of Fisheries and Aquatic Sciences 64:837-847.

Kelly, M., and K. Tuxen. 2009. Remote sensing support for tidal wetland vegetation research and management. Pp. 341-363 in Remote Sensing and Geospatial Technologies for Coastal Ecosystem Assessment and Management, X. Yang (Ed.). Berlin: Springer-Verlag.

Kessler, J. D., D. L. Valentine, M. C. Redmond, M. Du, E. W. Chan, S. D. Mendes, E. W. Quiroz, C. J. Villanueva, S. S. Shusta, L. M. Werra, S. A. Yvon-Lewis, and T. C. Weber. 2011. A persistent oxygen anomaly reveals the fate of spilled methane in the deep Gulf of Mexico. Science 331:312-315.

Kim, C., K. Park, and S. P. Powers. 2002. Establishing restoration strategy of Eastern oysters via a coupled biophysical transport model. Restoration Ecology 21(3):353-362.

Kirwan, M. L., G. R. Guntenspergen, A. D'Alpaos, J. T. Morris, S. M. Mudd, and S. Temmerman. 2010. Limits on the adaptability of coastal marshes to rising sea level. Geophysical Research Letters 37:L23401.

Klemas, V. 2001. Remote sensing of landscape-level coastal environmental indicators. Environmental Management 27:47-57.

Klemas, V. 2011. Remote sensing of wetlands: Case studies comparing practical techniques. Journal of Coastal Research 27:418-427.

Kniemeyer, O., F. Musat, S. M. Sievert, K. Knittel, H. Wilkes, M. Blumenberg, W. Michaelis, A. Classen, C. Bolm, and F. Widdel. 2007. Anaerobic oxidation of short-chain hydrocarbons by marine sulphate-reducing bacteria. Nature 449:898-901.

Koch, E. W., E. B. Barbier, B. R. Silliman, D. J. Reed, G. M. Perillo, S. D. Hacker, E. F. Granek, J. H. Primavera, N. Muthiga, S. Polasky, B. S. Halpern, C. J. Kennedy, C. V. Kappel, and E. Wolanski. 2009. Non-linearity in ecosystem services: Temporal and spatial variability in coastal protection. Frontiers in Ecology and the Environment 7(1):29-37.

Kokaly, R. F., D. Heckman, J. Holloway, S. Piazza, B. Couvillion, G. D. Steyer, C. Mills, and T. M. Hoefen. 2011. Shoreline Surveys of Oil-Impacted Marsh in Southern Louisiana, July to August 2010. USGS Open-File Report No. 2011-1022. Reston, VA: U.S. Geological Survey.

Kolker, A. S., M. A. Allison, and S. Hameed. 2011. An evaluation of subsidence rates and sea-level variability in the northern Gulf of Mexico. Geophysical Research Letters 38(21):L21404, doi: 10.1029/2011GL049458.

Krauss, K. W., T. W. Doyle, T. J. Doyle, C. M. Swarzenski, A. S. From, R. H. Day, and W. H. Conner. 2009. Water level observations in mangrove swamps during two hurricanes in Florida. Wetlands 29(1):142-149.

Krone, R. B. 1985. Simulation of marsh growth under rising sea levels. Pp. 106-115 in Hydraulics and Hydrology in the Small Computer Age, W. R. Waldrop (Ed.). Lake Buena Vista, FL: Hydraulics Division, U.S. Army Corps of Engineers.

Kujawinski, E. B., M. C. Kido Soule, D. L. Valentine, A. K. Boysen, K. Longnecker, and M. C. Redmond. 2011. Fate of dispersants associated with the *Deepwater Horizon* oil spill. Environmental Science & Technology 45:1298-1306.

Kurtz, R. S. 1995. Lessons to Be Learned: The NPS Administrative History and Assessment of the *Exxon Valdez* Oil Spill. Anchorage, AK: National Park Service, Alaska Regional Office.

Kyle, J. R. 2002. A century of fire and brimstone: The rise and fall of the Frasch sulphur industry of the Gulf of Mexico Basin. Pp. 189-198 in Industrial Minerals and Extractive Industry Geology, P. W. Scott and C. M. Bristow (Eds.). Bath, England: Geological Society (London) Publishing House.

Lahvis, G. P., R. S. Wells, D. W. Kuehl, J. L. Stewart, H. L. Rhinehart, and C. S. Via. 1995. Decreased lymphocyte responses in free-ranging bottlenose dolphins (*Tursiops truncatus*) are associated with increased concentrations of PCBs and DDT in peripheral blood. Environmental Health Perspectives 103(Suppl. 4):67-72.

Langmuir, I. 1938. Surface motion of water induced by wind. Science 87:119.

Le Henaff, M., V. H. Kourafalou, C. B. Paris, J. Helgers, Z. M. Aman, P. J. Hogan, and A. Srinivasan. 2012. Surface evolution of the *Deepwater Horizon* oil spill patch: Combined effects of circulation and wind-induced drift. Environmental Science & Technology 46:7267-7273.

Leahy, J. G., and R. R. Colwell. 1990. Microbial degradation of hydrocarbons in the environment. Microbiology and Molecular Biology Reviews 54:305-315.

Lee, K. 1993. Compass and Gyroscope. Integrating Science and Politics for the Environment. Washington, DC: Island Press.

Lee, K. 1999. Appraising adaptive management. Conservation Ecology 3(2):3. Available at http://www.consecol.org/vol3/iss2/art3/, accessed June 20, 2013.

Lee, K., C. S. Wong, W. J. Cretney, F. A. Whitney, T. R. Parsons, C. M. Lalli, and J. Wu. 1985. Microbial response to crude oil and Corexit 9527: SEAFLUXES Enclosure Study. Microbial Ecology 11:337-351.

Lee, K., T. Lunel, P. Wood, R. Swannel, and P. Stoffyn-Egli. 1997. Shoreline cleanup by acceleration of clay-oil flocculation process. Pp. 235-240 in Proceedings of the 1997 International Oil Spill Conference. Washington, DC: American Petroleum Institute.

Lee, K., S. E. Cobanli, J. Gauthier, S. St.-Pierre, G. H. Tremblay, and G. D. Wohlgeschaffen. 1999. Evaluating the addition of fine particles to enhance oil degradation. Pp. 433-438 in Proceedings of the 1999 International Oil Spill Conference. Washington, DC: American Petroleum Institute.

Lee, K., T. Nedwed, and R. C. Prince. 2011. Lab tests on the biodegradation rates of chemically dispersed oil must consider natural dilution. In Proceedings of the 2011 International Oil Spill Conference, Portland, Oregon. Washington, DC: American Petroleum Institute.

Lee, M. L., M. V. Novotny, and K. D. Bartle. 1981. Analytical chemistry of polycyclic aromatic compounds. New York: Academic Press.

Lehr, B., S. Bristol, and A. Possolo. 2010. Oil Budget Calculator *Deepwater Horizon*, Technical Documentation. A Report to the National Incident Command, November 2010. Durham, NH: Coastal Response Research Center, University of New Hampshire. Available at http://www.restorethegulf.gov/sites/default/files/documents/pdf/OilBudgetCalc_Full_HQPrint_111110.pdf, accessed October 11, 2012.

Leibovich, S. 1983. The form and dynamics of Langmuir circulations. Annual Review of Fluid Mechanics 15:391-427.

Leidy, G. R., and R. M. Jenkins. 1977. The Development of Fishery Compartments and Population Rate Coefficients for Use in Reservoir and Ecosystem Modeling. USACE Contract Report Y-77-1. Vicksburg, MS: U.S. Army Corps of Engineers Waterways Experiment Station.

Leidy, G. R., and G. R. Ploskey. 1980. Simulation Modeling of Zooplankton and Benthos in Reservoirs: Documentation and Development of Model Constructs. Technical Report E-80-4. Vicksburg, MS: U.S. Army Corps of Engineers Waterways Experiment Station.

Levin, S. 1999. Fragile Dominion. Cambridge, MA: Perseus Books.

Levin, S. A., S. Barrett, S. Aniyar, W. Baumol, C. Bliss, B. Bolin, P. Dasgupta, P. Ehrlich, C. Folke, I.-M. Gren, C. S. Holling, A. Jansson, B.-O. Jansson, K.-G. Mäler, D. Martin, C. Perrings, and E. Sheshinski. 1998. Resilience in natural and socioeconomic systems. Environment and Development Economics 3:221-235.

Lewontin, R. C. 1969. The meaning of stability. Pp. 13-24 in Diversity and Stability in Ecological Systems, G. W. Woodwell and H. H Smith (Eds.. Upton, NY: Brookhaven National Laboratory.

Li, Y., J. T. Morris, and D. C. Yoch. 1990. Chronic low level hydrocarbon amendments stimulate plant and microbial activity in salt-marsh microcosms. Journal of Applied Ecology 27:159-171.

Li, Z., K. Lee, T. King, M. C. Boufadel, and A. Venosa. 2008. Oil droplet size distribution as a function of energy dissipation rate in an experimental wave tank. Pp. 621-624 in Proceedings of the 2008 International Oil Spill Conference. Washington, DC: American Petroleum Institute.

Li, Z., K. Lee, T. King, M. C. Boufadel, and A. D. Venosa. 2009a. Evaluating chemical dispersant efficacy in an experimental wave tank: 2, Significant factors determining in-situ oil droplet size distribution. Environmental Engineering Science 26:1407-1418.

Li, Z., K. Lee, T. King, P. Kepkay, M. C. Boufadel, and A. D. Venosa. 2009b. Evaluating chemical dispersant efficacy in an experimental wave tank: 1, Dispersant effectiveness as a function of energy dissipation rate. Environmental Engineering Science 26:1139-1148.

Lichtenstein, S., and P. Slovic. 2006. The Construction of Preferences. New York: Cambridge University Press.

Lin, Q., and I. A. Mendelssohn. 1998. The combined effects of phytoremediation and biostimulation in enhancing habitat restoration and oil degradation of petroleum contaminated wetlands. Ecological Engineering 10:263-274.

Lin, Q., and I. A. Mendelssohn. 2003. Dispersant Effects on Fresh Marsh Vegetation: Toxicity Evaluation and Oil Remediation. OSRADP Technical Report Series 169-30-4150. Baton Rouge, LA: Louisiana Applied and Educational Oil Spill Research and Development Program.

Lin, Q., and I. A. Mendelssohn. 2012. Impacts and recovery of the *Deepwater Horizon* oil spill on vegetation structure and function of coastal salt marshes in the northern Gulf of Mexico. Environmental Science & Technology 46:3737-3743.

Lindenmayer, D., and M. Hunter, 2010. Some guiding concepts for conservation biology. Conservation Biology 24(6):1459-1468.

Lindstrom, J. E., and J. F. Braddock. 2002. Biodegradation of petroleum hydrocarbons at low temperature in the presence of the dispersant Corexit 9500. Marine Pollution Bulletin 44:739-747.

Lippencott, R. J. 2005. Evaluating rhizodegradation of petroleum hydrocarbons and polycyclic aromatic hydrocarbons in wetlands sediment using *Spartina patens*. Ph.D. Dissertation, New Jersey Institute of Technology.

Lipton, D. W., and I. Strand. 1997. Economic effects of pollution in fish habitats. Transactions of the American Fisheries Society 126:514-518.

Liu, D. 1983. Fate of oil dispersants in aquatic environment. Science of the Total Environment 32:93-98.

Liu, Z., J. Liu, Q. Zhu, and W. Wu. 2012. The weathering of oil after the *Deepwater Horizon* oil spill: Insights from the chemical composition of the oil from the sea surface, salt marshes and sediments. Environmental Research Letters 7(3):035302, doi:10.1088/1748-9326/7/3/035302.

Loder, N. M., J. L. Irish, M. A. Cialone, and T. V. Wamsley. 2009. Sensitivity of hurricane surge to morphological parameters of coastal wetlands. Estuarine, Coastal and Shelf Science 84:625-636.

Loomis, J. 2006. Estimating recreation and existence values of sea otter expansion in California using benefit transfer. Coastal Management 34:387-404.

Louisiana Coastal Wetlands Conservation and Restoration Task Force and Wetlands Conservation and Restoration Authority. 1998. Coast 2050: Toward a Sustainable Coastal Louisiana. Baton Rouge, LA: Louisiana Department of Natural Resources. Available at http://www.coast2050.gov/report.pdf, accessed March 29, 2013.

Louisiana Office of Coastal Protection and Restoration. 2010. State opens additional freshwater diversion canal at Bayou Lamoque in Plaquemines Parish. Press Release, May 12.

Loureiro, M., J. Loomis, and M. Vazquez. 2009. Economic valuation of environmental damages due to the *Prestige* oil spill in Spain. Environmental and Resource Economics 44:537-553.

Lovley, D. R., J. D. Coates, D. A. Saffarini, and D. J. Lonergan. 1997. Dissimilatory iron reduction. Pp. 187-215 in Transition Metals in Microbial Metabolism, G. Winkelmann and C. J. Carrano (Eds.). Geneva, Switzerland: Harwood Academic.

Lu, Z., Y. Deng, J. D. Van Nostrand, Z. He, J. Voordeckers, A. Zhou, Y.-J. Lee, O. U. Mason, E. A. Dubinsky, K. L. Chavarria, L. M. Tom, J. L. Fortney, R. Lamendella, J. K. Jansson, P. D'haeseleer, T. C. Hazen, and J. Zhou. 2011. Microbial gene functions enriched in the *Deepwater Horizon* deep-sea oil plume. ISME Journal 6(2):451-460.

Lubchenco, J., M. K. McNutt, G. Dreyfus, S. A. Murawski, D. M. Kennedy, P. T. Anastas, S. Chu, and T. Hunter. 2012. Science in support of the *Deepwater Horizon* response. Proceedings of the National Academy of Sciences of the United States of America 109(50):20212-20221.

Lumsden, S. E., T. F. Hourigan, A. W. Bruckner, and G. Dorr (Eds.). 2007. The State of Deep Coral Ecosystems of the United States. NOAA Technical Memorandum CRCP-3. Silver Spring, MD: National Oceanic and Atmospheric Administration.

Lunel, T. 1995. Understanding the mechanism of dispersion through oil droplet size measurements at sea. Pp. 240-285 in The Use of Chemicals in Oil Spill Response, P. Lane (Ed.). Philadelphia: American Society for Testing and Materials.

Lunel, T., K. Lee, R. Swannell, P. Wood, J. Rusin, N. Bailey, C. Halliwell, L. Davies, M. Sommerville, A. Dobie, D. Mitchell, and M. McDonagh. 1996. Shoreline cleanup during the *Sea Empress* incident: The role of surf washing (clay-oil flocculation), dispersants and bioremediation. Pp. 1521-1540 in Proceedings of the 19th Arctic and Marine Oilspill Program (AMOP) Technical Seminar. Ottawa, Ontario: Environment Canada.

Lunel, T., R. Swannell, J. Rusin, P. Wood, N. Bailey, C. Halliwell, L. Davies, M. Sommerville, A. Dobie, D. Mitchell, M. McDonagh, and K. Lee. 1997. Monitoring the effectiveness of response operations during the *Sea Empress* incident: A key component of the successful counter-pollution response. Spill Science and Technology Bulletin 2:99-112.

Lynne, G. D, P. Conroy, and F. J. Prochaska. 1981. Economic valuation of marsh areas for marine production processes. Journal of Environmental Economics and Management 8(2):175-186.

Macavoy, S. E., C. R. Fisher, R. S. Carney, and S. A. Macko. 2005. Nutritional associations among fauna at hydrocarbon seep communities in the Gulf of Mexico. Marine Ecology Progress Series 292:51-60.

MacDonald, I. R., J. F. Reilly, Jr., S. E. Best, R. Venkataramaiah, R. Sassen, N. L. Guinasso, Jr., and J. Amos. 1996. Remote sensing inventory of active oil seeps and chemosynthetic communities in the northern Gulf of Mexico. Pp. 27-37 in Hydrocarbon Migration and Its Near-Surface Expression, D. Schumacher and M. A. Abrams (Eds.). Tulsa, OK: American Association of Petroleum Geologists.

Mace. P. M. 2004. In defence of fisheries scientists, single-species models and other scapegoats: Confronting the real problems. Marine Ecology Progress Series 274:285-291.

Mack, M. C., and C. M. D'Antonio. 1998. Impacts of biological invasions on disturbance regimes. Trends in Ecology & Evolution 13(5):195-198.

Mallman, E. P., and M. D. Zoback. 2007. Subsidence in the Louisiana coastal zone due to hydrocarbon production. Journal of Coastal Research 50(Special Issue):443-449.

Martínez, M. L., R. A. Feagin, K. M. Yeager, J. Day, R. Costanza, J. A. Harris, R. J. Hobbs, J. López-Portillo, I. J. Walker, E. Higgs, P. Moreno-Casasola, J. Sheinbaum, and A. Yáñez-Arancibia. 2012. Artificial modifications of the coast in response to the *Deepwater Horizon* oil spill: Quick solutions or long-term liabilities? Frontiers in Ecology and the Environment 10:44-49.

Mascarelli, A. 2010. *Deepwater Horizon*: After the oil. Nature 467:22-24.

Mason, O. U., J. Han, H. Y. Holman, J. Hultman, R. Lamendella, R. Mackelprang, S. Malfatti, L. M. Tom, S. G. Tringe, T. Woyke, J. Zhou, E. M. Rubin, and J. K. Jansson. 2012. Metagenome, metatranscriptome and single-cell sequencing reveal microbial response to *Deepwater Horizon* oil spill. International Society for Microbial Ecology Journal 6(9):1715-1727.

Massel, S. R., K. Furukawa, and R. M. Brinkman. 1999. Surface wave propagation in mangrove forests. Fluid Dynamics Research 24(4):219-249.

Matkin, C. O., E. L. Saulitis, G. M. Ellis, P. Olesiuk, and S. D. Rice. 2008. Ongoing population-level impacts on killer whales *Orcinus orca* following the "*Exxon Valdez*" oil spill in Prince William Sound, Alaska. Marine Ecology Progress Series 356:269-281.

Maunder, M. M., and A. E. Punt. 2004. Standardizing catch and effort data: A review of recent approaches. Fisheries Research 70:141-159.

May, R. M. 1974. Stability and Complexity in Model Ecosystems. Princeton, NJ: Princeton University Press.

May, R. M., S. A. Levin, and G. Sugihara. 2008. Complex systems: Ecology for bankers. Nature 451:893-895.

Mayer, L., Z. Chen, R. Findlay, J. Fang, S. Sampson, L. Self, P. Jumars, C. Quetél, and O. Donard. 1996. Bioavailability of sedimentary contaminants subject to deposit-feeder digestion. Environmental Science & Technology 30:2641-2645.

Mazda, Y., E. Wolanski, B. King, A. Sase, D. Ohtsuka, and M. Magi. 1997. Drag force due to vegetation in mangrove swamps. Mangroves and Salt Marshes 1:193-99.

McCann, K. S. 2000. The diversity-stability debate. Nature 405:228-233.

McConnell, K. E., and I. E. Strand. 1989. Benefits from commercial fisheries when demand and supply depend on water quality. Journal of Environmental Economics and Management 17:284-292.

McCrea-Strub, A., D. Zeller, U. R. Sumaila, J. Nelson, A. Balmford, and D. Pauly. 2011. Understanding the cost of establishing marine protected areas. Marine Policy 35:1-9.

McFarlin, K. M., M. B. Leigh, and R. Perkins. 2011a. Indigenous microorganisms degrade oil in Arctic seawater. Poster presentation at the 2011 International Oil Spill Conference, Portland, Oregon. Washington, DC: American Petroleum Institute.

McFarlin, K. M., R. A. Perkins, W. W. Gardiner, J. D. Word, and J. Q. Word. 2011b. Toxicity of physically and chemically dispersed oil to selected Arctic species. In Proceedings of the 2011 International Oil Spill Conference, Portland, Oregon. Washington, DC: American Petroleum Institute.

McGill, K. 2011. Survey measures post-oil spill seafood attitudes. The Associated Press, January 31. Available at http://www.businessweek.com/ap/financialnews/D9L3IP0O0.htm, accessed March 29, 2013.

McInerney, M. J., C. G. Struchtemeyer, J. Sieber, H. Mouttaki, A. J. M. Stams, B. Schink, L. Rohlin, and R. P. Gunsalus. 2008. Physiology, ecology, phylogeny, and genomics of microorganisms capable of syntrophic metabolism. Pp. 58-72 in Incredible Anaerobes: From Physiology to Genomics to Fuels, J. Wiegel, R. J. Maier, and M. W. W. Adams (Eds.). Chichester, UK: Wiley-Blackwell.

McIntosh, S., T. King, D. Wu, and P. V. Hodson. 2010. Toxicity of dispersed crude oil to early life stages of Atlantic herring (*Clupea harengus*). Environmental Toxicology and Chemistry 29:1160-1167.

McKee, K. L., and W. L. Patrick. 1988. The relationship of smooth cordgrass (*Spartina alterniflora*) to tidal datums: A review. Estuaries 11:143-151.

McNutt, M. K., R. Camilli, T. J. Crone, G. D. Guthrie, P. A. Hsieh, T. B. Ryerson, O. Savas, and F. Shaffer. 2012. Review of flow rate estimates of the *Deepwater Horizon* oil spill. Proceedings of the National Academy of Sciences 109:20260-20267.

MEA (Millennium Ecosystem Assessment). 2005. Ecosystems and Human Well-Being: Synthesis. Washington, DC: Island Press.

Meade, R. H. 1982. Sources, sinks, and storage of river sediment in the Atlantic drainage of the United States. Journal of Geology 90(3):235-252.

Mendelssohn, I. A., G. L. Andersen, D. M. Baltz, R. H. Caffey, K. R. Carman, J. W. Fleeger, S. B. Joye, Q. X. Lin, E. Maltby, E. B. Overton, and L. P. Rozas. 2012. Oil Impacts on coastal wetlands: Implications for the Mississippi River Delta ecosystem after the *Deepwater Horizon* oil spill. Bioscience 62:562-574.

Menot, L., M. Sibuet, R. S. Carney, L. A. Levin, G. T. Rowe, D. S. M. Billett, G. Poore, H. Kiazato, A. Vanreusel, J. Galleron, H. P. Lavrado, J. Sellanes, B. Ingole, and E. Krylova. 2010. New perceptions of continental margin biogeography. Pp 79-101 in Life in the World's Oceans: Diversity, Distribution, and Abundance, A. D. McIntyre (Ed.). Chichester, UK: Wiley-Blackwell.

Mero, J. L. 1952. Manganese nodules. North Dakota Engineer 27:28-32.

Merrick, R., L. Allan, R. Amgliss, G. Antonelis, T. Eagle, S. Epperly, L. Jones, S. Reilly, B. Schroeder, and S. Swartz. 2004. A Requirements Plan for Improving the Understanding of the Status of U.S. Protected Marine Species. Report of the NOAA Fisheries National Task Force for Improving Marine Mamma and Turtle Stock Assessments. NOAA Technical Memorandum NMFS-F/SPO-63. Silver Spring, MD: National Oceanic and Atmospheric Administration.

Mesnil, B. 2012. The hesitant emergence of maximum sustainable yield (MSY) in fisheries policies in Europe. Marine Policy 36:473-480.

Middlebrook, A. M., D. M. Murphy, R. Ahmadov, E. L. Atlas, R. Bahreini, D. R. Blake, J. Brioude, J. A. de Gouw, F. C. Fehsenfeld, G. J. Frost, J. S. Holloway, D. A. Lack, J. M. Langridge, R. A. Lueb, S. A. McKeen, J. F. Meagher, S. Meinardi, J. A. Neuman, J. B. Nowak, D. D. Parrish, J. Peischl, A. E. Perring, I. B. Pollack, J. M. Roberts, T. B. Ryerson, J. P. Schwarz, J. R. Spackman, C. Warneke, and A. R. Ravishankara. 2011. Air quality implications of the *Deepwater Horizon* oil spill. Proceedings of the National Academy of Sciences of the United States of America 109(50):20280-20285.

Mienert, J., C. Berndt, J. S. Laberg, and T. Vorren. 2003. Slope instability of continental margins. Pp. 179-193 in Ocean Margin Systems, G. Wefer, D. Billet, D. Hebbeln, B. B. Jørgensen, M. Schlüter, and T. van Weering (Eds.). Berlin: Springer Verlag.

Mileti, D. S. 1999. Disasters by Design: A Reassessment of Natural Hazards in the United States. Washington, DC: Joseph Henry Press.

Milinkovitch, T. J. G., M. Théron, and H. Thomas-Guyon. 2011. Toxicity of dispersant application: Biomarkers responses in gills of juvenile golden grey mullet (*Liza aurata*). Environmental Pollution 159:2921-2928.

Minello, T. J., R. J. Zimmerman, and R. Medina. 1994. The importance of edge for natant macrofauna in a created salt marsh. Wetlands 14:184-198.

Miselis, J., J. Flocks, N. Plant. J. Bernier, N. DeWitt, K. Kelso, W. Pfeiffer, B. J. Reynolds, and D. Wiese. 2012. Using shallow-water seafloor mapping to understand sediment movement in the northern Chandeleur Islands, Louisiana. Sound Waves 2012(November/December):4-6. Available at http://soundwaves.usgs.gov/2012/12/ fieldwork3.html, accessed March 29, 2013.

Mississippi Development Authority/Division of Tourism. 2012. Reports and Statistics. Available at http://visitmississippi.org/reports-and-statistics.aspx, accessed November 7, 2012.

MMS (Minerals Management Service). 2002. Gulf of Mexico OCS Oil and Gas Lease Sales. Final Environmental Impact Statement. OCS EIS/EA MMS2002-052. Available at http://www.boem.gov/BOEM-newsroom/library/Publications/2002/2002-052-vol1.aspx.

Montane, J. M., and R. Torres. 2006. Accuracy assessment of LIDAR saltmarsh topographic data using RTK GPS. Photogrammetric Engineering and Remote Sensing 72(8):961-967.

Mora, C., O. Aburto-Oropeza, A. A. Bocos, P. M. Ayotte, S. Banks, A. G. Bauman, M. Beger, S. Bessudo, D. J. Booth, E. Brokovich, A. Brooks, P. Chabanet, J. E. Cinner, J. Cortés, J. J. Cruz-Motta, A. C. Magaña, E. E. DeMartini, G. J. Edgar, D. A. Feary, A. M. Friedlander, K. J. Gaston, C. Gough, N. A. J. Graham, A. Green, H. Guzman, M. Hardt, M. Kulbicki, Y. Letourneur, A. L. Pérez, M. Loreau, Y. Loya, C. Martinez, I. Mascareñas-Osorio, T. Morove, M.-O. Nadon, Y. Nakamura, G. Paredes, N. V. C. Polunin, M. S. Pratchett, H. Reyes Bonilla, F. Rivera, E. Sala, S. A. Sandin, G. Soler, R. Stuart-Smith, E. Tessier, D. P. Tittensor, M. Tupper, P. Usseglio, L. Vigliola, L. Wantiez, I. Williams, S. K. Wilson, and F. A. Zapata. 2011. Global human footprint on the linkage between biodiversity and ecosystem functioning in reef fishes. PLoS Biology 9(4):1-9.

Morris, J. T., P. V. Sundareshwar, C. T. Nietch, B. Kjerfve, and D. R. Cahoon. 2002. Responses of coastal wetlands to rising sea level. Ecology 83:2869-2877.

Morton, R. A., N. A. Buster, and M. D. Krohn. 2002. Subsurface controls on historical subsidence rates and associated wetland loss in southcentral Louisiana. Transactions Gulf Coast Association of Geological Societies 52:767-778.

Muhling, B. A., M. A. Roffer, J. T. Lamkin, G. W. Ingram, M. A. Upton, G. Gawlikowski, F. Muller-Karger, S. Habtes, and W. J. Richards. 2012. Overlap between Atlantic bluefin tuna spawning grounds and observed *Deepwater Horizon* surface oil in the northern Gulf of Mexico. Marine Pollution Bulletin 64(4):679-687.

Mukherjee, B., and B. A. Wrenn. 2009. Influence of dynamic mixing energy on dispersant performance: Role of mixing systems. Environmental Engineering Science 26:1725-1737.

Müller-Karger, F. E., J. J. Walsh, R. H. Evans, and M. B. Meyers. 1991. On the seasonal phytoplankton concentration and sea surface temperature cycles of the Gulf of Mexico as determined by satellites. Journal of Geophysical Research 96:12,645-12,665.

Munro, G. R. 2008. Game theory and the development of resource management policy: The case of international fisheries. In Game Theory and Policymaking in Natural Resources and the Environment, A. Dinar, J. Albiac, and J. Sánchez-Soriano (Eds.). New York: Routledge.

Murawski, S. A., and W. T. Hogarth. 2013. Enhancing the ocean observing system to meet restoration challenges in the Gulf of Mexico. Oceanography 26:10-16.

Myers, R. A., and B. Worm. 2003. Rapid worldwide depletion of predatory fish communities. Nature 423:280-283.

Myers, R. M., J. K. Baum, T. D. Shepherd, S. P. Powers, and C. H. Peterson. 2007. Cascading effects of the loss of apex predatory sharks from a coastal ocean. Science 315(5820):1846-1850.

Naeem, S. 2002. Biodiversity: Biodiversity equals instability? Nature 416:23-24.

Naeem, S., and S. Li. 1997. Biodiversity enhances ecosystem reliability. Nature 390:507-509.

Nance J. 2009. Stock Assessment Report 2008 Shrimp Fishery. Report for the Gulf of Mexico Fishery Management Council.

Narayan, J. P., and S. Kumar. 2006. Effects of soil layering on the characteristics of basin-edge induced surface waves. Acta Geophysica 57(2):294-310.

National Commission on the BP Deepwater Horizon Oil Spill and Offshore Drilling. 2011. Deep Water: The Gulf Oil Disaster and the Future of Offshore Drilling. Report to the President. Available at http://www.oilspillcommission.gov/sites/default/files/documents/DEEPWATER_ReporttothePresident_FINAL.pdf, accessed March 22, 2013.

National Ocean Service, National Oceanic and Atmospheric Administration. 2012. Notice of approval of extension of deep sea hard mineral exploration licenses and amended exploration plan. Federal Register 77(132):40586-40587. Available at http://www.gpo.gov/fdsys/pkg/FR-2012-07-10/html/2012-16794.htm, accessed March 26, 2013

Nestlerode, J. A., V. D. Engle, P. Bourgeois, P. T. Heitmuller, J. M. Macauley, and Y. C. Allen. 2008. An integrated approach to assess broad-scale condition of coastal wetlands—the Gulf of Mexico Coastal Wetlands Pilot Survey. Environmental Monitoring and Assessment 150(1-4):21-29.

Neubert, M. G., and H. Caswell. 1997. Alternatives to resilience for measuring the responses of ecological systems to perturbations. Ecology 78(3):653-665.

NMFS (National Marine Fisheries Service). 2010a. Characterization of the U.S. Gulf of Mexico and Southeastern Atlantic Otter Trawl and Bottom Reef Fish Fisheries Observer Training Manual. Galveston, TX: NMFS Southeast Fisheries Science Center, Galveston Laboratory. Available at http://www.st.nmfs.noaa.gov/Assets/Observer-Program/pdf/Shrimp_Reef_fish_Manual_9_22_10.pdf, accessed June 25, 2013.

NMFS. 2010b. Fisheries Economics of the United States, 2009. U.S. Department Commerce, NOAA Technical Memo. NMFS-F/SPO-118. Available at https://www.st.nmfs.noaa.gov/st5/publication/index.html, accessed March 29, 2013.

NMFS. 2012. Status of Stocks: Report on the Status of U. S. Fisheries for 2011. Silver Spring, MD: National Oceanic and Atmospheric Administration.

NOAA (National Oceanic and Atmospheric Administration). 1996. Preassesment Phase: Guidance Document for Natural Resource Damage Assessment Under the Oil Pollution Act of 1990, the Damage Assessment Remediation and Restoration Program. Available at http://www.darrp.noaa.gov/library/1_d.html, accessed June 21, 2013.

NOAA. 2010. *Deepwater Horizon* (DWH) Oil Spill. Protecting the Public from Oil-Contaminated Seafood: Fishery Area Closure and Surveillance Plan. Available at http://docs.lib.noaa.gov/noaa_documents/DWH_IR/reports/Protecting_Public_Fisheries_Closure_Surveill_Plan.pdf, accessed March 22, 2013.

NOAA. 2011a. In Situ Burning. Silver Spring, MD: Office of Response and Restoration, NOAA. Available at http://response.restoration.noaa.gov/oil-and-chemical-spills/oil-spills/resources/in-situ-burning.html, accessed March 29, 2013.

NOAA. 2011b. SEDAR Stock Assessment Report Gulf of Mexico Goliath Grouper. Silver Spring, MD: SEFSC-NMFS, NOAA.

NOAA. 2011c. SEDAR Stock Assessment Report Gulf of Mexico Menhaden. Silver Spring, MD: SEFSC-NMFS, NOAA.

NOAA. 2011d. SEDAR Stock Assessment Report Gulf of Mexico Tilefish. Silver Spring, MD: SEFSC-NMFS, NOAA.

NOAA. 2011e. SEDAR Stock Assessment Report Gulf of Mexico Yellow Edge Grouper. Silver Spring, MD: SEFSC-NMFS, NOAA.

NOAA. 2011f. The Gulf of Mexico at a Glance: A Second Glance. Silver Spring, MD: NOAA.

NOAA. 2012a. *Deepwater Horizon*: A Preliminary Bibliography of Published Research and Expert Commentary. Silver Spring, MD: NOAA Central Library, National Oceanographic Data Center, NOAA. Available at http://www.lib.noaa.gov/researchtools/subjectguides/dwh.html, accessed March 29, 2013.

NOAA. 2012b. NRDA (Natural Resource Damage Assessment): Status Update for the *Deepwater Horizon* Oil Spill. Silver Spring, MD: NOAA.

NOAA. 2012c. NRDA Workplans and Data: NOAA Gulf Spill Restoration. Silver Spring, MD: NOAA. Available at http://www.gulfspillrestoration.noaa.gov/oil-spill/gulf-spill-data/, accessed March 29, 2013.

NOAA. 2013a. Fisheries Economics of the U.S. 2011. Available at http://www.st.nmfs.noaa.gov/economics/publications/feus/fisheries_economics_2011, accessed March 15, 2013.

NOAA. 2013b. State of the Coast: National Coastal Population Report. Available at http://stateofthecoast.noaa.gov/coastalpopulation-report.pdf, accessed June 24, 2013.

NRC (National Research Council). 1975. Mining in Outer Continental Shelf and in the Deep Ocean. Washington, DC: National Academy Press.

NRC. 1984. Deep Seabed Stable Reference Areas. Washington, DC: National Academy Press.

NRC. 1992. Rethinking the Ozone Problem in Urban and Regional Air Pollution. Washington, DC: National Academy Press.

NRC. 2003. Oil in the Sea III: Inputs, Fates, and Effects. Washington, DC: National Academies Press.

NRC. 2005a. Oil Spill Dispersants: Efficacy and Effects. Washington, DC: National Academies Press.

NRC. 2005b. Valuing Ecosystem Services: Toward Better Environmental Decision Making. Washington, DC: National Academies Press.

NRC. 2009. Final Report from the NRC Committee on the Review of the Louisiana Coastal Protection and Restoration (LACPR) Program. Washngton, DC: National Academies Press.

NRC. 2011. Methods and Metrics of Ecosystem Services Valuation for the Gulf of Mexico After the *Deepwater Horizon* Oil Spill: Interim Report. Washington, DC: National Academies Press.

NSTC (National Science and Technology Council) Joint Subcommittee on Ocean Science and Technology. 2007. Charting the Course for Ocean Science in the United States for the Next Decade. An Ocean Research Priorities Plan and Implementation Strategy. Washington, DC: National Science and Technology Council, Office of Science and Technology Policy. Available at http://www.whitehouse.gov/sites/default/files/microsites/ostp/nstc-orppis.pdf, accessed May 20, 2013.

NSTC Joint Subcommittee on Ocean Science and Technology. 2013. Science for an Ocean Nation: Update of the Ocean Research Priorities Plan. Washington, DC: National Science and Technology Council, Office of Science and Technology Policy. Available at http://www.whitehouse.gov/sites/default/files/microsites/ostp/ocean_research_plan_2013.pdf, accessed May 20, 2013.

Nyström, M., A. V. Norström, T. Blenckner, M. De la Torre-Castro, J. S. Eklöf, C. Folke, H. Österblom, R. S. Steneck, M. Thyresson, and M. Troell. 2012. Confronting feedbacks of degraded marine ecosystems. Ecosystems 15:695-710.

O'Connor, S., R. Campbell, H. Cortez, and T. Knowles. 2009. Whale Watching Worldwide: Tourism Numbers, Expenditures and Expanding Economic Benefits. A Special Report from the International Fund for Animal Welfare. Yarmouth, MA: International Fund for Animal Welfare.

Odokuma, L. O., and G. C. Okpokwasili. 1992. Role of composition in degradability of oil spill dispersants. Waste Management 12:39-43.

Odum, E. P. 1969. The strategy of ecosystem development. Science 164:262-270.

Okey, T. A., and D. Pauly. 1999. A mass-balanced model of trophic flows in Prince William Sound: Decompartmentalizing ecosystem knowledge. Ecosystem Approaches for Fisheries Management 16:621-635.

Okey, T. A., G. A. Vargo, S. Mackinson, M. Vasconcellos, B. Mahmoudi, and C. A. Meyer. 2004. Simulating community effects of sea floor shading by plankton blooms over the West Florida Shelf. Ecological Modelling 172:339-359.

Olascoaga, M. J., F. J. Beron-Vera, L. E. Brand, and H. Kocak. 2007. Tracing the early development of harmful algal blooms with the aid of Lagrangian coherent structure. Journal of Geophysical Research: Oceans 113(C12):C12014.

Olsson, P., C. Folke, and T. Hahn. 2004. Social-ecological transformation for ecosystem management: The development of adaptive co-management of a wetland landscape in southern Sweden. Ecology and Society 9(4):2.

Omasa, K., F. Hosoi, and A. Konishi. 2006. 3D lidar imaging for detecting and understanding plant responses and canopy structure. Journal of Experimental Botany 58:881-898.

Orcutt, B., A. Boetius, M. Elvert, V. Samarkin, and S. B. Joye. 2005. Molecular biogeochemistry of sulfate reduction, methanogenesis and the anerobic oxidation of methane at Gulf of Mexico cold seeps. Geochimica et Cosmochimica Acta 17:4267-4281.

ORPP (Ocean Research Priorities Plan). 2012. Science for an Ocean nationa: Update of the Ocean Research Priorities Plan. Available at http://www.whitehouse.gov/sites/default/files/microsites/OSTP/2013_ocean_nation.pdf.

Orth, R. J. 1975. Destruction of eelgrass, *Zostera marina,* by the cownose ray, *Rhinoptera bonasus,* in the Chesapeake Bay. Chesapeake Science 16(3):205-208.

Ortmann, A. C., J. Anders, N. Shelton, L. Gong, A. G. Moss, and R. H. Condon. 2012. Dispersed oil disrupts microbial pathways in pelagic food webs. PLoS ONE 7:e42548.

OSAT (Operational Science Advisory Team). 2010. Summary Report for Sub-sea and Sub-surface Oil and Dispersant Detection: Sampling and Monitoring. New Orleans LA: OSAT, Unified Area Command.

OSAT. 2011a. Summary Report for Fate and Effects of Remnant Oil in the Beach Environment. Tallahassee, FL: Florida Department of Environmental Protection. Available at http://www.dep.state.fl.us/deepwaterhorizon/files2/osat_2_report__10feb.pdf, accessed December 1, 2011.

OSAT. 2011b. Summary Report for Fate and Effects of Remnant Oil Remaining in the Beach Environment. Annex E: Biodegradation Assessment Tool (BIOMARUN). Tallahassee, FL: Florida Department of Environmental Protection. Available at http://www.restorethegulf.gov/sites/default/files/documents/pdf/Annex%20E%20BIOMARUN.pdf, accessed March 29, 2013.

OSPAR Commission. 2010. Status Report on the OSPAR Network of Marine Protected Areas. Available at http://www.ospar.org/documents/dbase/publications/p00493_status%20report%20mpas.pdf, accessed March 29, 2013.

Owens, E. H. 1999. The interaction of fine particles with stranded oil. Pure and Applied Chemistry 71:83-93.

Owens, E. H., and K. Lee. 2003. Interaction of oil and mineral fines on shorelines: Review and assessment. Marine Pollution Bulletin 47:397-405.

Owens, E., E. Taylor, A. Graham, and R. Castle. 2011. Sand beach treatment studies and field trials conducted during the *Deepwater Horizon*-Macondo response operation. In Proceedings of the 2011 International Oil Spill Conference. Washington, DC: American Petroleum Institute.

Ozesmi, S. L., and M. E. Bauer. 2002. Satellite remote sensing of wetlands. Wetlands Ecology and Management 10:381-402.

Paris, C. B., M. Le Hénaff, Z. M. Aman, A. Subramaniam, J. Helgers, D.-P. Wang, V. H. Kourafalou, and A. Srinivasan. 2012. Evolution of the Macondo well blowout: Simulating the effects of the circulation and synthetic dispersants on the subsea oil transport. Environmental Science & Technology 46(24):13293-13302.

Parks, P., and M. Bonifaz. 1994. Nonsustainable use of renewable resources: Mangrove deforestation and mariculture in Ecuador. Marine Resource Economics 9:1-18.

Passow, U., K. Ziervogel, V. Asper, and A. Diercks. 2012. Marine snow formation in the aftermath of the *Deepwater Horizon* oil spill in the Gulf of Mexico. Environmental Research Letters 7:1-11.

Payne, J. R., J. R. Clayton, Jr., G. D. McNabb, B. E. Kirstein, C. L. Clary, R. T. Redding, J. S. Evans, E. Reimnitz, and E. W. Kempema. 1989. Oil-Ice-Sediment Interactions During Freeze-up and Break-up: Final Report. Outer Continental Shelf Environmental Assessment Program, Final Reports of Principal Investigators. Silver Spring, MD: U.S. Department of Commerce, NOAA, OCSEAP.

Penela-Arenaz, M., J. Bellas, and E. Vazquez. 2009. Effects of the *Prestige* oil spill on the biota of NW Spain: 5 years of learning. Pp. 365-396 in Advances in Marine Biology, Vol. 56, D. W. Sims (Ed.). London: Academic Press.

Pequegnat, W. E. 1983. The Ecological Communities of the Continental Slope and Adjacent Regimes of the Northern Gulf of Mexico. Report to U.S. Minerals Management Service, Metairie, LA. College Station, TX: TerEco Corp.

Pequegnat, W. E., R. M. Darnell, B. M. James, L. H. Pequegnat, and J. T. Turner. 1976. Ecological aspects of the upper continental slope of the Gulf of Mexico. Prepared for the Division of Minerals Environmental Assessment, Bureau of Land Management. College Station, TX: TerEco Corp.

Pequegnat, W. E., B. J. Gallaway, and L. H. Pequegnat. 1990. Aspects of the ecology of the deep-water fauna of the Gulf of Mexico. American Zoologist 30:45-64.

Pereira, H. M., B. Reyers, M. Watanabe, E. Bohensky, S. Foale, C. Palm, M. V. Espaldon, D. Armenteras, M. Tapia, A. Rincón, M. J. Lee, A. Patwardhan, and I. Gomes. 2005. Condition and trends of ecosystem services and biodiversity. Pp. 171-203 in Ecosystems and Human Well-Being: Multi Scale Assessments, Vol 4, Findings of the Sub-global Assessments Working Group of the Millennium Ecosystem Assessment, D. Capistrano, C. Samper, M. J. Lee, and C. Raudsepp-Hearne (Eds.). Washington, DC: Island Press.

Perring, A. E., J. P. Schwarz, J. R. Spackman, R. Bahreini, J. A. de Gouw, R. S. Gao, J. S. Holloway, D. A. Lack, J. M. Langridge, J. Peischl, A. M. Middlebrook, T. B. Ryerson, C. Warneke, L. A. Watts, and D. W. Fahey. 2011. Characteristics of black carbon aerosol from a surface oil burn during the *Deepwater Horizon* oil spill. Geophysical Research Letters 38:L17809.

Petchey, O. L., and K. J. Gaston. 2009. Effects on ecosystem resilience of biodiversity, extinctions, and the structure of regional species pools. Theoretical Ecology 2:177-187.

Peterson, C. H. 2001. The *Exxon Valdez* oil spill in Alaska: Acute, indirect, and chronic effects on the ecosystem. Pp 1-103 in Advances in Marine Biology, Vol. 39, A. J. Southward, P. A. Tyler, C. M. Young, and L. A. Fuiman (Eds.): San Diego, CA: Academic Press

Peterson, C. H., S. D. Rice, J. W. Short, D. Esler, J. L. Bodkin, B. E. Ballachey, and D. B. Irons. 2003. Long-term ecosystem response to the *Exxon Valdez* oil spill. Science 302:2082-2086.

Peterson, C. H., S. S. Anderson, G. N. Cherr, R. F. Ambrose, S. Anghera, S. Bay, M. Blum, R. Condon, T. A. Dean, M. Graham, M. Guzy, S. Hampton, S. Joye, J. Lambrinos, B. Mate, D. Meffert, S. P. Powers, P. Somasundaran, R. B. Spies, C. M. Taylor, R. Tjeerdema, and E. E. Adams. 2012. A tale of two spills: Novel science and policy implications of an emerging new oil spill model. BioScience 62:461-469.

Peterson, G. W., and R. E. Turner. 1994. The value of salt marsh edge vs. interior as a habitat for fish and decapod crustaceans in a Louisiana tidal marsh. Estuaries 17:235-262.

Petraitis, P. S., E. Methratta, E. Rhile, N. Vidargas, and S. R. Dudgeon. 2009. Experimental confirmation of multiple community states in a marine ecosystem. Oecologia 161:139-148.

Petterson, J. S., E. Glazier, L. D. Stanley, C. Mencken, K. Eschbach, P. Moore, and P. Goode. 2008. Benefits and Burdens of OCS Activities on States, Labor Market Areas, Coastal Counties, and Selected Communities. OCS Study MMS 2008-052. New Orleans, LA: U.S. Department of the Interior, Minerals Management Service, Gulf of Mexico OCS Region.

Pezeshki, S. R., M. W. Hester, Q. Lin, and J. A. Nyman. 2000. The effects of oil spill and clean-up on dominant US Gulf coast marsh macrophytes: A review. Environmental Pollution 108(2):129-139.

Phinn, S. R., C. Menges, G. J. E. Hill, and M. Stanford. 2000. Optimizing remotely sensed solutions for monitoring, modeling and managing coastal environments. Remote Sensing of Environment 73:117-132.

Picou, J. S., and C. G. Martin. 2006. Community impacts of Hurricane Ivan: A case study of Orange Beach, Alabama. Boulder, CO: Natural Hazards Center.

Pimm, S. L., and J. H. Lawton. 1980. Are food webs divided into compartments? Journal of Animal Ecology 49:879-898.

Plagányi, E. E. 2007. Models for an Ecosystem Approach to Fisheries. FAO Fisheries Technical Paper 477. Rome, Italy: Food and Agriculture Organization of the United Nations.

Poag, C. W. 1991. Rise and demise of the Bahama-Grand Banks gigaplatform, northern margin of the Jurassic proto-Atlantic seaway. Marine Geology 102:63-130.

Polasky, S., and K. Segerson. 2009. Integrating ecology and economics in the study of ecosystem services: Some lessons learned. Annual Review of Resource Economics 1:409-434.

Polasky, S., A. de Zeeuw, and F. Wagener. 2011. Optimal management with potential regime shifts. Journal of Environmental Economics and Management 62:229-240.

Pope, J. 2009. Input and output controls: The practice of fishing effort and catch management in responsible fisheries. Pp. 220-252 in A Fishery Manager's Handbook, K. Cohrane and S. Garcia (Eds.). Chichester, UK: Wiley-Blackwell.

Post, A. M. 1983. Deepsea Mining and the Law of the Sea. Boston: Martimus Nijhoff.

Prince, R. C. 2010. Bioremediation of marine oil spills. Pp. 2617–2630 in Handbook of Hydrocarbon and Lipid Microbiology, K. N. Timmis, T. J. McGenity, J. R. van der Meer, and V. de Lorenzo (Eds.). Berlin: Springer Verlag.

Putorti, A. D., Jr., D. D. Evans, and E. J. Tennyson. 1994. Ignition of weathered and emulsified oils. pp. 657-667 in Environment Canada Arctic and Marine Oil Spill Program (AMOP) Technical Seminar, 17th Proceedings, Vol. 1. Vancouver, BC: Environment Canada.

Quartel, S., A. Kroon, P. G. E. F. Augustinus, P. Van Santen, and N. H. Tri. 2007. Wave attenuation in coastal mangroves in the Red River Delta, Vietnam. Journal of Asian Earth Sciences 29:576-584.

Quinn, T. J., and R. B. Deriso. 1999. Quantitative Fish Dynamics. Oxford, UK: Oxford University Press.

Rabalais, N. N., B. A. McKee, D. J. Reed, and J. C. Means. 1992. Fate and effects of produced water discharges in coastal Louisiana, Gulf of Mexico, USA. Pp. 355-369 in Produced Water, J. P. Ray and F. R. Engelhardt (Eds.). New York: Plenum Press.

Rabalais, N. N., L. E. Smith, C. B. Henry, Jr., P. O. Roberts, and E. B. Overton. 1998. Long-Term Effects of Contaminants from OCS Produced-Water Discharges at Pelican Island Facility, Louisiana. OCS Study MMS 98-0039. New Orleans, LA: U.S. Department of the Interior, Minerals Management Service, Gulf of Mexico OCS Region.

Rabalais, N. N., R. E. Turner, and W. J. Wiseman, Jr. 2002. Hypoxia in the Gulf of Mexico, a.k.a. "the dead zone." Annual Review of Ecology and Systematics 33:235-263.

Rabalais, N. N., R. E. Turner, B. K. Sen Gupta, D. F. Boesch, P. Chapman, and M. C. Murrell. 2007. Characterization and long-term trends of hypoxia in the northern Gulf of Mexico: Does the science support the Action Plan? Estuaries and Coasts 30(5):753-772.

Ramirez-Llodra, E., P. A. Tyler, M. C. Baker, O. A. Bergstad, M. R. Clark, E. Escobar, L. A. Levin, L. Menot, A. A. Rowden, C. R. Smith, and C. L. van Dover. 2011. Man and the last great wilderness: Human impact on the deep sea. PLoS ONE 6:e22588.

Ramsey, E., III, A. Rangoonwala, Y. Suzuoki, and C. E. Jones 2011. Oil detection in a coastal marsh with polarimetric synthetic aperture radar (SAR). Remote Sensing 3:2630-2662.

Ravier, C., and J. M. Fromentin. 2001. Long-term fluctuations in the eastern Atlantic and Mediterranean bluefin tuna population. ICES Journal of Marine Science 58:1299-1317.

Reddy, C. M., J. S. Arey, J. S. Seewald, S. P. Sylva, K. L. Lemkau, R. K. Nelson, C. A. Carmichael, C. P. McIntyre, J. Fenwick, and G. T. Ventura. 2011. Composition and fate of gas and oil released to the water column during the *Deepwater Horizon* oil spill. Proceedings of the National Academy of Sciences of the United States of America 109(50):20229-20234.

Redmond, M. C., and D. L. Valentine. 2012. Natural gas and temperature structured a microbial community response to the *Deepwater Horizon* oil spill. Proceedings of the National Academy of Sciences of the United States of America 109(50):20292-20297.

Reed, D. J. 1995. The response of coastal marshes to sea-level rise: Survival or submergence? Earth Surface Processes and Landforms 20:39-48.

Reed, D. J. 2002. Sea-level rise and coastal marsh sustainability: Geological and ecological factors in the Mississippi delta plain. Geomorphology 48(1-3):233-243.

Reed, D. J., and L. Wilson. 2004. Coast 2050: A new approach to restoration of Louisiana coastal wetlands. Physical Geography 25:4-21.

Reed, W. J. 1978. The steady state of a stochastic harvesting model. Mathematical Biosciences 41(3-4):273-307.

Reed, W. J. 1979. Optimal escapement levels in stochastic and deterministic harvesting models. Journal of Environmental Economics and Management 6(4):350-363.

Reed, W. J. 1984. The effects of the risk of fire on the optimal rotation of a forest. Journal of Environmental Economics and Management 11:180-190.

Reed, W. J. 1988. Optimal harvesting of a fishery subject to random catastrophic collapse. Mathematical Medicine and Biology 5(3):215-235.

Reich, P. B., D. Tilman, F. Isbell, K. Mueller, S. E. Hobbie, D. F. B. Flynn, and N. Eisenhauer. 2012. Impacts of biodiversity loss escalate through time as redundancy fades. Science 336(6081):589-592.

Reilly, S. B., M. A. Donahue, T. Gerrodette, K. Forney, P. Wade, L. Ballance, J. Forcada, P. Fiedler, A. Dizon, W. Perryman, F. Archer, and E. F. Edwards. 2005. Report of the Scientific Research Program Under the International Dolphin Conservation Program Act. NOAA Technical Memorandum NMFS, NOAA-TM-NMFS-SWFSC-372. Silver Spring, MD: National Oceanic and Atmospheric Administration.

Reinharz, E., and J. Michel. 1996. Preassessment Phase. Guidance Document for Natural Damage Resource Assessment Under the Oil Pollution Act of 1990. Silver Spring, MD: National Oceanic and Atmospheric Administration Damage Assessment and Restoration Program.

Reiss, D., and L. Marino. 2001. Mirror self-recognition in the bottlenose dolphin: A case of cognitive convergence. Proceedings of the National Academy of Sciences of the United States of America 98(10):5937-5942.

Rex, M. A., and R. J. Etter. 2010. Deep-Sea Biodiversity: Pattern and Scale. Boston, MA: Harvard University Press.

Ribic, C. A., S. B. Sheavly, D. J. Rugg, and E. Erdmann. 2012. Trends in marine debris along the U.S. Pacific Coast and Hawaii 1998-2007. Marine Pollution Bulletin 64:994-1004.

Richardson, G. E., L. S. French, R. D. Baud, R. H. Peterson, C. D. Roark, T. M. Montgomery, E. G. Kazanis, G. M. Conner, and M. P. Gravois. 2004. Deepwater Gulf of Mexico 2004: America's Expanding Frontier. OCS Report MMS 2004-021. New Orleans, LA: U.S. Department of the Interior, Minerals Management Service, Gulf of Mexico OCS Region.

Richardson, G. E., L. D. Nixon, C. M. Bohannon, E. G. Kazanis, T. M. Montgomery, and M. P. Gravois. 2008. Deepwater Gulf of Mexico 2008: America's Offshore Energy Future. OCS Report MMS 2008-013. New Orleans, LA: U.S. Department of the Interior, Minerals Management Service, Gulf of Mexico OCS Region.

Richardson, L., and J. Loomis. 2009. The total economic value of threatened, endangered and rare species: An updated meta-analysis. Ecological Economics 68:1535-1548.

Rivas, D., A. Badan, and J. Ochoa. 2005. The ventilation of the deep Gulf of Mexico. Journal of Physical Oceanography 35:1763-1781.

Rivas, D., A. Badan, J. Sheinbaum, J. Ochoa, and J. Candela. 2008. Vertical velocity and vertical heat flux observed within loop current eddies in the central Gulf of Mexico. Journal of Physical Oceanography 38(11):2461-2481.

Roberts, H. H., and R. S. Carney. 1997. Evidence of episodic fluid, gas, and sediment venting on the northern Gulf of Mexico continental slope. Economic Geology 9:863-879.

Rodriguez, L. C., U. Pascual, and H. M. Niemeyer. 2006. Local identification and valuation of ecosystem goods and services from Opuntia scrublands of Ayacucho, Peru. Ecological Economics 57:30-44. Available at http://dx.doi.org/10.1016/j.ecolecon.2005.03.022, accessed June 21, 2013.

Röling, W. F. M., M. G. Milner, D. M. Jones, K. Lee, F. Daniel, R. J. P. Swannell, and I. M. Head. 2002. Robust hydrocarbon degradation and dynamics of bacterial communities during nutrient-enhanced oil spill bioremediation. Applied and Environmental Microbiology 68:5537-5548.

Rooney, N., K. McCann, G. Gelner, and J. C. Moore. 2006. Structural asymmetry and the stability of diverse food webs. Nature 442:265-259.

Rorick, R., T. Nedwed, G. DeMarco, and C. Cooper. 2012. Comments on "A tale of two spills: Novel science and policy implications of an emerging new oil spill model." Bioscience 62:1009-1010.

Rosado-Solórzano, R., and S. A. Guzmán Del Próo. 1998. Preliminary trophic structure model for Tampamachoco Lagoon, Veracruz, Mexico. Ecological Modelling 109:141-145.

Rose, A. 2007. Economic resilience to natural and man-made disasters: Multidisciplinary origins and contextual dimensions. Environmental Hazards 7:383-398.

Rose, K. A., J. I. Allen, Y. Artioli, M. Barange, J. Blackford, F. Carlotti, R. Cropp, U. Daewel, K. Edwards, K. Flynn, S. L. Hill, R. H. R. Lambers, G. Huse, S. Mackinson, B. Megrey, A. Moll, R. Rivkin, B. Salihoglu, C. Schrum, L. Shannon, Y.-J. Shin, S. L. Smith, C. Smith, C. Solidoro, M. St. John, and M. Zhou. 2010. End-to-end models for the analysis of marine ecosystems: challenges, issues, and next steps. Marine and Coastal Fisheries: Dynamics, Management, and Ecosystem Science 2:115-130.

Ross, S. W., and M. S. Nizinski. 2007. State of the U.S. deep coral ecosystems in the southeastern United States region: Cape Hatteras to the Florida Straits. Pp. 233-270 in The State of Deep Coral Ecosystems of the United States, S. E. Lumsden, T. F. Hourigan, A. W. Bruckner, and G. Dorr (Eds.). NOAA Technical Memorandum CRCP-3. Silver Spring MD: National Oceanic and Atmospheric Administration.

Rosso, P. H., S. L. Ustin, and A. Hastings. 2006. Use of lidar to study changes associated with *Spartina* invasion in San Francisco Bay marshes. Remote Sensing of the Environment 100:295-306.

Roth, A. M., F. Baltz, and M. Donald. 2009. Short-term effects of an oil spill on marsh-edge fishes and decapod crustaceans. Estuaries and Coasts 32:565-572.

Roughgarden, J., and F. Smith. 1996. Fisheries collapse and what to do about it. Proceedings of the National Academies of Sciences of the United States of America 93:5078-5083.

Rowe, G. T., and J. W. Deming. 2011. An alternative view of the role of heterotrophic microbes in the cycling of organic matter in deep-sea sediments. Marine Biology Research 7(7):629-636, doi:10.1080/17451000.2011.560269.

Rowe, G. T., and M. C. Kennicutt II (Eds.). 2008. The Deep Gulf of Mexico Benthos Program. Deep-Sea Research Part II: Topical Studies in Oceanography 55(24-26).

Rowe, G. T., and M. C. Kennicutt II (Eds.). 2009. Northern Gulf of Mexico Continental Slope Habitats and Benthic Ecology Study: Final Report. OCS Study MMS 2009-039. New Orleans, LA: U.S. Dept. of the Interior, Minerals Management Service, Gulf of Mexico OCS Region.

Rowe, G. T., C. Wei, C. Nunnally, R. Haedrich, P. Montagna, J. G. Baguley, J. M. Bernhard, M. Wicksten, A. Ammons, E. Escobar Briones, Y. Soliman, and J. W. Deming. 2008. Comparative biomass structure and estimated carbon flow in food webs in the deep Gulf of Mexico. Deep Sea Research Part II: Topical Studies in Oceanography 55(24-26):2699-2711.

Rudolf, J. C. 2010. Berms built to stop oil are seen as ineffective. New York Times, December 16, p. A18.

Ruhl, J. B. 2012. Panarchy and the law. Ecology and Society 17(3):31.

Ruttenberg, B. I., S. L. Hamilton, S. M. Walsh, M. K. Donovan, A. Friedlander, E. DeMartini, E. Sala, and S. A. Sandin. 2011. Predator-induced demographic shifts in coral reef fish assemblages. PLoS ONE 6(6):e21062, doi:10.1371/journal.pone.0021062.

Sala, E., C. F. Boudouresque, and M. Harmelin-Vivien. 1998. Fishing, trophic cascades, and the structure of algal assemblages: Evaluation of an old but untested paradigm. Oikos 82:425-439.

Salomon, A. K., S. K. Gaichas, N. T. Shears, J. E. Smith, E. M. P. Madin, and S. D. Gaines. 2010. Key features and context-dependence of fishery-induced trophic cascades. Conservation Biology 24:382-394.

Sanchirico, J. N., and M. Springborn. 2010. How to get there from here: ecological and economic dynamics of ecosystem service provision. Environmental and Resource Economics 48(2):243-267.

Santner, R., M. Cocklan-Vendl, J. Michel, E. Owens, B. Stong, and E. Taylor. 2011. Shoreline treatment during the *Deepwater Horizon*-Macondo response. In Proceedings of the 2011 International Oil Spill Conference. Washington, DC: American Petroleum Institute.

Santschi, P. H., and G. T. Rowe. 2008. Radiocarbon-derived sedimentation rates in the Gulf of Mexico. Deep Sea Research Part II: Topical Studies in Oceanography 5(24-26):2572-2576.

Sathirathai, S., and E. B. Barbier. 2001. Valuing mangrove conservation in Southern Thailand. Contemporary Economic Policy 19:109-122.

Saul, S. E. 2012. An individual-based model to evaluate the effect of fisher behavior on reef fish catch per unit effort. Open Access Dissertations. Available at http://scholarlyrepository.miami.edu/oa_dissertations/872.

SCAT (Shoreline Cleanup Assessment Technique) Team. 2010. MC 252 Stage III SCAT: Shoreline Treatment Implementation Framework for Louisiana. Available at http://bpoilspill.us/wp-content/uploads/2010/10/Louisiana-Stage-III-Shoreline-Treatment-Plan-Final-22Sept2010-DRAFT.pdf, accessed May 22, 2013.

Schaum, J., M. Cohen, S. Perry, R. Artz, R. Draxler, J. B. Frithsen, D. Heist, M. Lorber, and L. Phillips. 2010. Screening level assessment of risks due to dioxin emissions from burning oil from the BP *Deepwater Horizon* Gulf of Mexico spill. Environmental Science & Technology 44:9383-9389.

Scheffer, M., and S. R. Carpenter. 2003. Catastrophic regime shifts in ecosystems: Linking theory to observation. Trends in Ecology and Evolution 18:648-656.

Scheffer, M., S. R. Carpenter, J. A. Foley, C. Folke, and B. Walker. 2001. Catastrophic shifts in ecosystems. Nature 413:591-596.

Scheffer, M., J. Bascompte, W. A. Brock, V. Brovkin, S. R. Carpenter, V. Dakos, H. Held, E. H. van Nes, M. Rietkerk, and G. Sugihara. 2009. Early-warning signals for critical transitions. Nature 461(7260):53-59.

Scheffer, M., S. R. Carpenter, T. M. Lenton, J. Bascompte, W. Brock, V. Dakos, J. van de Koppel, I. A. van de Leemput, S. A. Levin, E. H. van Nes, M. Pascual, and J. Vandermeer. 2012. Anticipating critical transitions. Science 338(6105):344-348.

Schmitt, R. J., and C. W. Osenberg (Eds.). 1996. Detecting Ecological Impacts: Concepts and Applications in Coastal Habitats. San Diego, CA: Academic Press.

Schroder, A., L. Persson, and A. De Roos. 2005. Direct experimental evidence for alternative stable states: A review. Oikos 110:3-19.

Scott-Denton, E., P. F. Cryer, J. P. Gocke, M. R. Harrelson, D. L. Kinsella, J. R. Pulver, R. C. Smith, and J. A. Williams. 2011. Descriptions of the U.S. Gulf of Mexico reef fish bottom longline and vertical line fisheries based on observer data. Marine Fisheries Review 73:1-26.

SEDAR (Southeast Data, Assessment, and Review). 2010. SEDAR 12. Stock assessment Report 1 Gulf of Mexico Red Grouper. Available at http://www.sefsc.nma.gov/sedar/download/S12SAR1%20Gulf%20Red%20Grouper%20completevs.pdf.

Service, R. F. 2010. Bits of good news from the Gulf. ScienceInsider. Available at http://news.sciencemag.org/science insider/2010/08/bitsof-good-news-from-the-gulf.html, accessed March 26, 2012.

Shabman, L. A., and S. Batie. 1978. Economic value of natural coastal wetlands: A critique. Coastal Zone Management Journal 4:231-247.

Shafir, S., J. Van Rijn, and B. Rinkevich. 2007. Short and long term toxicity of crude oil and oil dispersants to two representative coral species. Environmental Science & Technology 41:5571-5574.

Shen, W., I. Ginis, and R. E. Tuleya. 2002. A numerical investigation of land surface water on landfalling hurricanes. Journal of the Atmospheric Sciences 59:789-802.

Shepard, C. C., C. M. Crain, and M. W. Beck. 2011. The protective role of coastal marshes: A systematic review and meta-analysis. PLoS ONE 6:e27374.

Sherman, K., and S. Adams (Eds.). 2010. Sustainable Development of the World's Large Marine Ecosystems during Climate Change: A Commemorative Volume to Advance Sustainable Development on the Occasion of the Presentation of the 2010 Göteborg Award. Gland, Switzerland: IUCN.

Shiller, A. M. 1999. An overview of the marine chemistry of the Gulf of Mexico. Chap. 8 in The Gulf of Mexico Large Marine Ecosystem: Assessment, Sustainability and Management, K. Sherman, H. Kumpf, K. Steidinger (Eds.). Malden, MA: Blackwell Science.

Shin, W. S., and J. H. Pardue. 2001a. Oxygen dynamics in crude oil contaminated salt marshes. I. Aerobic respiration model. Environmental Technology 22:845-854.

Shin, W. S., and J. H. Pardue. 2001b. Oxygen dynamics in crude oil contaminated salt marshes. II. Carbonaceous sediment oxygen demand model. Environmental Technology 22:855-867.

Shin, W. S., J. H. Pardue, and W. A. Jackson. 2000. Oxygen demand and sulfate reduction in petroleum hydrocarbon contaminated salt marsh soils. Water Research 34:1345-1353.

Shin, W. S., J. H. Pardue, W. A. Jackson, and S. J. Choi. 2001. Nutrient enhanced biodegradation of crude oil in tropical salt marshes. Water, Air, and Soil Pollution 131:135-152.

Silliman, B. R., J. van de Koppel, M. W. McCoy, J. Diller, G. N. Kasozi, K. Earl, P. N. Adams, and A. R. Zimmerman. 2012. Degradation and resilience in Louisiana salt marshes after the BP-*Deepwater Horizon* oil spill. Proceedings of the National Academy of Sciences od the United States of America 109(28):11234-11239.

Simmie, J., and R. Martin. 2010. The economic resilience of regions: Towards an evolutionary approach. Cambridge Journal of Regions, Economy and Society 3:27-43.

Sissenwine, M., and S. Murawski. 2004. Moving beyond "intelligent tinkering": Advancing an ecosystem approach to fisheries. Marine Progress Series 274:291-295.

Smith, E. E., J. A. Carr, M. Wages, J. F. Wang, S. Murali, and R. Kendall. 2012. Response of larval frogs to Corexit 9500. Toxicological and Environmental Chemistry 94:1199-1210.

Smith, J. W., and D. S. Vaughan. 2011. Harvest, Effort, and Catch-at-Age for Gulf Menhaden. SEDAR 27-DW05. Silver Spring, MD: SEFSC-NMFS, NOAA.

Smith, T. G., J. R. Geraci, and D. J. St. Aubin. 1983. The reaction of bottlenose dolphins, *Tursiops truncatus*, to a controlled oil spill. Canadian Journal of Fisheries and Aquatic Sciences 40:1522-1527.

Smith, V. K., and W. H. Desvousges. 1986. Measuring Water Quality Benefits. Boston, MA: Kluwer-Nijhoff.

Sommer, U. 2008. Trophic cascades in marine and freshwater plankton. International Review of Hydrobiology 93:506-516.

Sousa, W. P., and J. H. Connell. 1983. On the evidence needed to judge ecological stability or persistence. The American Naturalist 121:789-824.

Sousa, W. P., and J. H. Connell. 1985. Further comments on the evidence for multiple stable points in natural communities. American Naturalist 125:612-615.

Southam, G., M. Whitney, and C. Knickerbocker. 2001. Structural characterization of the hydrocarbon degrading bacteria-oil interface: Implications for bioremediation. International Biodeterioration & Biodegradation 47:197-201.

Speck, O. 2003. Field measurements of wind speed and reconfiguration in *Arundo donax* (Poaceae) with estimates of drag forces. American Journal of Botany 90:1253-1256.

Spies, R. B., J. J. Stegeman, D. E. Hinton, B. Woodin, M. Okihiro, R. Smolowitz, and D. Shea. 1996. Biomarkers of hydrocarbon exposure and sublethal effects in embiotocid fishes from a natural petroleum seep in the Santa Barbara Channel. Aquatic Toxicology 34:195-219.

St. Aubin, D. J., J. R. Geraci, T. G. Smith, and T. G. Friesen. 1985. How do bottlenose dolphins, *Tursiops truncatus*, react to oil films under different light conditions? Canadian Journal of Fisheries Aquatic Sciences 42:430-436.

Stern, R. J., E. Y. Anthoney, R. Minghua, B. E. Lock, I. Norton, J.-I. Kimura, T. Miyazaki, T. Hanyu, Q. Chang, and Y. Hirahara. 2011. Southern Louisiana salt dome xenoliths: First glimpse of Jurassic (ca. 160 Ma) Gulf of Mexico crust. Geology 39:315-318.

Steyer, G. D. 2008. Landscape analysis of vegetation change in coastal Louisiana following hurricanes Katrina and Rita. Ph.D. Dissertation, Louisiana State University.

Steyer, G. D., C. E. Sasser, J. M. Visser, E. M. Swenson, J. A. Nyman, and R. C. Raynie. 2003. A proposed coast-wide reference monitoring system for evaluating wetland restoration trajectories in Louisiana. Environmental Monitoring and Assessment 81:107-117.

Steyer, G. D., K. F. Cretini, S. Piazza, L. A. Sharp, G. A. Snedden, and S. Sapkota. 2010. Hurricane Influences on Vegetation Community Change in Coastal Louisiana. Open-File Report 2010-1105. Reston, VA: U.S. Geological Survey.

Stoffyn-Egli, P., and K. Lee. 2002. Formation and characterization of oil-mineral aggregates. Spill Science and Technology Bulletin 8:31-44.

Sturges, W. 2005. Deep-water exchange between the Atlantic, Caribbean, and Gulf of Mexico. Pp, 263-278 in Circulation of the Gulf of Mexico: Observation and Models, W. Sturges and A. Lugo-Fernandez (Eds.). Geophysical Monograph Series 161. Washington, DC: American Geophysical Union.

Su, T., and E. D. Zhang. 2007. Ecosystem valuation and the conservation of wild lands in vigorous economic regions: A case study in Jiuduansha Wetland, Shanghai. Chinese Science Bulletin 52(19):2664-2674.

Suding, K. N., K. L. Gross, and G. R. Houseman. 2004. Alternative states and positive feedbacks in restoration ecology. Trends in Ecology and Evolution 19:46-53.

Sulak, K. J., M. T. Randall, K. E. Luke, A. D. Norem, and J. M. Miller (Eds.). 2008. Characterization of Northern Gulf of Mexico Deepwater Hard Bottom Communities with Emphasis on *Lophelia* Coral—*Lophelia* Reef Megafaunal Community Structure, Biotopes, Genetics, Microbial Ecology, and Geology. USGS Open-File Report 2008-1148; OCS Study MMS 2008-015. New Orleans, LA: Minerals Management Service, Gulf of Mexico OCS Region.

Swallow, S. K. 1994. Renewable and nonrenewable resource theory applied to coastal agriculture, forest, wetland, and fishery linkages. Marine Resource Economics 9:291-310.

Swannell, R. P., K. Lee, and M. McDonagh. 1996. Field evaluations of marine oil spill bioremediation. Microbiological Reviews 60:342-365.

Swannell, R. P. J., and F. Daniel. 1999. Effect of dispersants on oil biodegradation under simulated marine conditions. Pp. 166-176 in Proceedings of the 1999 International Oil Spill Conference, Seattle, Washington. Washington, DC: American Petroleum Institute.

Swannell, R. P. J., D. Mitchell, G. Lethbridge, D. Jones, D. Heath, M. Hagley, M. Jones, S. Petch, R. Milne, R. Croxford, and K. Lee. 1999. A field demonstration of the efficacy of bioremediation to treat oiled shorelines following the *Sea Empress* incident. Environmental Technology 20:863-873.

Tallis, H., and S. Polasky. 2009. Mapping and valuing ecosystem services as an approach for conservation and natural-resource management. Annals of the New York Academy of Sciences 1162:265-283.

Tang Y. J., S. D. Carpenter, J. W. Deming, and B. Krieger-Brockett. 2006. Depth-related influences on biodegradation rates of phenanthrene in polluted marine sediments of Puget Sound, WA. Marine Pollution Bulletin 52:1431-1440.

TEEB (The Economics of Ecosystems and Biodiversity). 2010. The Economics of Ecosystems and Biodiversity: Mainstreaming the Economics of Nature: A Synthesis of the Approach, Conclusions and Recommendations of TEEB. Nairobi, Kenya: United Nations Environment Programme. Available at http://www.teebweb.org/wp-content/uploads/Study%20and%20Reports/Reports/Synthesis%20report/TEEB%20Synthesis%20Report%202010.pdf, accessed March 29, 2013.

ten Brink, U. S., D. Twichell. P Lynett, E Geist, J. Chaytor, H. Lee, B. Buczkowski, and C Flores. 2009. Regional Assessment of Tsunami Potential in the Gulf of Mexico: Report to the National Tsunami Hazard Mitigation Program. Available at http://woodshole.er.usgs.gov/staffpages/utenbrink/my%20publications/NTHMP_GulfOfMexicoReport.pdf.

Tilman, D. 1996. Biodiversity: Population versus ecosystem stability. Ecology 77(2):350-363.

Tilman, D., and J. A. Downing. 1994. Biodiversity and stability in grasslands. Nature 367:363-365.

Tilman, D., and J. A. Downing. 1996. Biodiversity and stability in grasslands. Pp. 3-7 in Ecosystem Management, F. B. Samson and F. L. Knopf (Eds.). New York: Springer Verlag.

Tilman, D., D. Wedin, and J. Knops. 1996. Productivity and sustainability influenced by biodiversity in grassland ecosystems. Nature 379:718-720.

Tilman, D., S. Polasky, and C. Lehman. 2005. Diversity, productivity and temporal stability in the economies of humans and nature. Journal of Environmental Economics and Management 49(3):405-426.

Tilman, D., P. B. Reich, and J. M. H. Knops. 2006. Biodiversity and stability in a decade-long grassland experiment. Nature 441:629-632.

Tinch, R., S. van den Hove, C. Armstrong, N. Foley, and T. van Rensburg. 2012. Integrating Public Values in Deep Sea Governance. Final Report to DG Research under the HERMIONE project (Deliverable D6.10). Southampton, UK: Hotspot Ecosystem Research and Man's Impact on European Seas.

Tol, R. S. J. 2009. The economic effects of climate change. Journal of Economic Perspectives 23:29-51.

Tolbert, C.M., II (Ed.). 2006. Sustainable Community in Oil and Gas Country: Final Report. OCS Study MMS 2006-011. New Orleans, LA: U.S. Department of the Interior, Minerals Management Service, Gulf of Mexico OCS Region.

Tolls, J., P. J. Kloepper-Sams, and D. T. H. M. Sijm. 1994. Surfactant bioconcentration—a critical review. Chemosphere 29:693-717.

Travers, M., Y. J. Shin, S. Jennings, and P. Cury. 2007. Towards end-to-end models for investigating the effects of climate and fishing in marine ecosystems. Progress in Oceanography 75:751-770.

Traxler, R. W., and L. S. Bhattacharya. 1978. Effect of a chemical dispersant on microbial utilization of petroleum hydrocarbons. Pp. 181-187 in Chemical Dispersants for the Control of Oil Spills, L. T. McCarthy, G. P. Lindblom, and H. F. Walter (Eds.). Philadelphia: American Society for Testing and Materials.

Tulane Institute on Water Resources Law and Policy. 2013. Promise, Purpose, and Challenge: Putting the RESTORE Act into Context for the Communities and Ecosystems of the Gulf of Mexico. Available at http://www.law.tulane.edu/uploadedFiles/Institutes_and_Centers/Water_Resources_Law_and_Policy/Content/TU%20Water%20Institute%20RESTORE%20Act%20White%20Paper%204-8-13(2).pdf.

Turner, R. E. 1997. Wetland loss in the northern Gulf of Mexico: Multiple working hypotheses. Estuaries 20:1-13.

Turner, R. E. 2011. Beneath the salt marsh canopy: Loss of soil strength with increasing nutrient loads. Estuaries and Coasts 34:1084-1093.

Turner, R. E., R. Costanza, and W. Scaife. 1982. Canals and wetland erosion rates in coastal Louisiana. Pp. 73-84 in Proceedings of the Conference on Coastal Erosion and Wetland Modification in Louisiana: Causes, Consequences, and Options, D. Boesch (Ed.). WS/OBS-82/59. Slidell, LA: Office of Biological Services, U.S. Fish and Wildlife Service.

Turner, R. E., E. M. Swenson, and C. S. Milan. 2001. Organic and inorganic contributions to vertical accretion in salt marsh sediments. Pp. 583-595 in Concepts and Controversies in Tidal Marsh Ecology, M. P. Weinstein and D. A. Kreeger (Eds.). Berlin: Springer Verlag.

Turner, R. E., B. L. Howes, J. M. Teal, C. S. Milan, E. M. Swenson, and D. D. Goehringer-Tonerb. 2009. Salt marshes and eutrophication: An unsustainable outcome. Limnology and Oceanography 54:1634-1642.

Una, G. V., and M. J. N. Garcia. 1983. Biodegradation of non-ionic dispersants in sea-water. Journal of Applied Microbiology and Biotechnology 18:315-319.

Upton, H. F. 2011. The *Deepwater Horizon* Oil Spill and the Gulf of Mexico Fishing Industry. Report No. R41640. Washington, DC: Congressional Research Service. Available at http://www.fas.org/sgp/crs/misc/R41640.pdf, accessed April 1, 2013.

USACE (U.S. Army Corps of Engineers). 2009. Draft Louisiana coastal protection and restoration (LACPR) Technical Report, Engineering Appendix. New Orleans, LA: USACE New Orleans District.

USCG (U.S. Coast Guard). 2011. On Scene Coordinator Report, *Deepwater Horizon* Oil Spill. Submitted to the National Response Team, September, 2011. Washington, DC: USCG. Available at http://www.uscg.mil/foia/docs/dwh/fosc_dwh_report.pdf, accessed May 22, 2013.

Utech, D. 2000. Valuing Hawaii's humpback whales: The economic impact of humpbacks on Hawaii's ocean tour boat industry. In The Economic Contribution of Whale Watching to Regional Economies: Perspectives from Two National Marine Sanctuaries. Marine Sanctuaries Conservation Series MSD-00-2. Silver Spring, MD: National Oceanic and Atmospheric Administration.

Valentine, D. L., J. D. Kessler, M. C. Redmond, S. D. Mendes, M. B. Heintz, C. Farwell, L. Hu, F. S. Kinnaman, S. Yvon-Lewis, M. Du, E. W. Chan, F. G. Tigreros, and C. J. Villanueva. 2010. Propane respiration jump-starts microbial response to a deep oil spill. Science 330:208-211.

Valentine, D. L., I. Mezić, S. Maćešić, N. Črnjarić-Žic, S. Ivić, P. J. Hogan, V. A. Fonoberov, and S. Loire. 2012. Dynamic autoinoculation and the microbial ecology of a deep water hydrocarbon irruption. Proceedings of the National Academy of Sciences of the United States of America 109(50):20286-20291.

Valentine, M. M., and M. Benfield. 2013. Characterization of epibenthic and demersal megafauna at Mississippi Canyon 252 shortly after the *Deepwater Horizon* oil spill. Presentation at ASLO 2013 Aquatic Sciences Meeting, February 17-22, New Orleans, LA.

Valentukevičiené, M., and E. Brannvall. 2008. Marine Pollution: An overview. Geologija 50(1):17-23.

Van de Koppel, J., M. Rietkerk, and F. J. Weissing. 1997. Catastrophic vegetation shifts and soil degradation in terrestrial grazing systems. Trends in Ecology and Evolution 12:352-356.

van Nes, E. H., and M. Scheffer. 2007. Slow recovery from perturbations as a generic indicator of a nearby catastrophic shift. American Naturalist 169(6):738-747.

Varadaraj, R., M. L. Robbins, J. Bock, S. Pace, and D. MacDonald. 1995. Dispersion and biodegradation of oil spills on water. Pp. 101-106 in Proceedings of the 1995 International Oil Spill Conference. Washington, DC: American Petroleum Institute.

Vega-Cendejas, M. E., and F. Arreguin-Sanchez. 2001. Energy fluxes in a mangrove ecosystem from a coastal lagoon in Yucatan Peninsula, Mexico. Ecological Modelling 137:119-133.

Venosa, A. D., and E. L. Holder. 2007. Biodegradability of dispersed crude oil at two different temperatures. Marine Pollution Bulletin 54:545-553.

Venosa, A. D., and X. Zhu. 2003. Biodegradation of crude oil contaminating marine shorelines and freshwater wetlands. Spill Science and Technology Bulletin 8:163-178.

Vosse, M. 2008. Wave attenuation over marshlands: Determination of marshland influences on New Orleans' flood protection. M.S. Thesis, University of Twente. Available at http://essay.utwente.nl/58496/1/scriptie_M_Vosse.pdf, accessed March 22, 2013.

Walker, B., and D. Salt. 2006. Resilience Thinking: Sustaining Ecosystems and People in a Changing World. Washington, DC: Island Press.

Walker, B., and D. Salt. 2012. Resilience Practice: Engaging the Sources of Our Sustainability. Washington, DC: Island Press.

Walker, B., C. S. Holling, S. R. Carpenter, and A. Kinzig. 2004. Resilience, adaptability and transformability in social-ecological systems. Ecology and Society 9(2):5.

Walker, B. H., L. H. Gunderson, A. P. Kinzig, C. Folke, S. R. Carpenter, and L. Schultz. 2006. A handful of heuristics and some propositions for understanding resilience in social-ecological systems. Ecology and Society 11(1):13

Walker, H. J., J. M. Coleman, H. H. Roberts, and R. S. Tye. 1987. Wetland loss in Louisiana. Geografiska Annaler 69A:189-200.

Walsh, B. 2011. After the great spill: How the Gulf cleaned itself. Time, January 10. Available at http://www.time.com/time/health/article/0,8599,2041345,00.html, accessed March 26, 2012.

Walter, J., and C. Porch. 2008. Three Different Strategies for Modeling the Terminal-Year Fishing Mortality Rates in Virtual Population Analyses of Western Bluefin Tuna: Retrospective Patterns and Consequences for Projections. SCRS/2008/089. NOAA Fisheries.

Walters, C. J. 1986. Adaptive Management of Renewable Resources. New York: Macmillan.

Walters, C., S. J. D. Martell, V. Christensen, and B. Mahmoudi. 2008. An Ecosim model for exploring Gulf of Mexico ecosystem management options: Implications of including multistanza life-history models for policy predictions. Bulletin of Marine Science 83:251-271.

Walton, W. D., and N. H. Jason (Eds.). 1999. In Situ Burning of Oil Spills Workshop Proceedings. New Orleans, Louisiana, November 2-4, 1998. NIST Special Publication 935. Gaithersburg, MD: Building and Fire Research Laboratory, National Institute of Standards and Technology.

Webb, B. M., S. L. Douglass, C. R. Dixon, and T. Buhring. 2011. Application of coastal engineering principles in response to the *Deepwater Horizon* disaster: Lessons learned in coastal Alabama. Pp. 359-372 in Proceedings of the Conference on Coastal Engineering Practice. Reston, VA: American Society of Civil Engineers.

Webb, J. W., S. K. Alexander, and J. K. Winters. 1985. Effects of autumn application of oil on *Spartina alterniflora* in Texas salt marsh. Environmental Pollution 38:321-337.

Weber, T. C., A. De Robertis, S. F. Greenway, S. Smith, L. Mayer, and G. Rice. 2011. Estimating oil concentration and flow rate with calibrated vessel-mounted acoustic echo sounders. Proceedings of the National Academy of Sciences of the United States of America 108(50) 20240-20245. Available at http://www.pnas.org/content/early/2011/12/07/1108771108.full.pdf+html, accessed April 1, 2013.

Wei, C-L., G. T. Rowe, G. F. Hubbard, E. Escobar-Briones, A. Boetius, T. Soltwedel, M. J. Caley, Y. Soliman, F. Huettmann, F. Qu, Z. Yu, C. R. Pitcher, R. L. Haedrich, M. K. Wicksten, M. A. Rex, J. G. Baguley, J. Sharma., R. Danovaro, I. R. MacDonald, C. C. Nunnally, J. W. Deming, P. Montagna, M. Lévesque, J. M. Weslawski, M. Wlodarska-Kowalczuk, B. S. Ingole, B. J. Bett, D. S. M. Billett, A. Yool, B. A. Bluhm, K. Iken, and B. E. Narayanaswamy. 2010a. Global patterns and predictions of seafloor biomass using random forests. PLoS ONE 5(12):e15323.

Wei, C.-L., G. T. Rowe, G. F. Hubbard, A. H. Scheltema, G. D. F. Wilson, I. Petrescu, J. M. Foster, M. K. Wicksten, M. Chen, R. Davenport, Y. Soliman, and Y. Wang. 2010b. Bathymetric zonation of deep-sea macrofauna in relation to export of surface phytoplankton production. Marine Ecology Progress Series 399:1-14.

Wei, C. L., G. T. Rowe, E. E. Briones, C. Nunnally, Y. Soliman, and N. Ellis. 2012a. Standing stocks and body size of deep-sea macrofauna: Predicting the baseline for 2010 BP oil spill in the northern Gulf of Mexico. Deep Sea Research Part I: Oceanographic Research Papers 69:82-99.

Wei, C. L., G. Rowe, C. Nunnally, and M. K. Wicksten. 2012b. Anthropogenic "litter" and macrophyte detritus in the deep northern Gulf of Mexico. Marine Pollution Bulletin 64:966-973.

Wells, R. J. D., J. H. Cowan, Jr., and W. F. Patterson III. 2008. Habitat use and the effect of shrimp trawling on fish and invertebrate communities over the northern Gulf of Mexico continental shelf. ICES Journal of Marine Science 65:1610-1619.

Wells, R. S. 2003. Dolphin social complexity: Lessons from long-term study and life history. Pp. 32-56 in Animal Social Complexity: Intelligence, Culture, and Individualized Societies, F. B. M. deWaal and P. L. Tyack (Eds.). Cambridge, MA: Harvard University Press.

Wells, R. S. 2009. Learning from nature: Bottlenose dolphin care and husbandry. Zoo Biology 28:1-17.

Welsh, S. E., and M. Inoue. 2002. Lagrangian Study of Circulation, Transport, and Vertical Exchange in the Gulf of Mexico. OCS Study MMS 2002-064. New Orleans, LA: U.S. Department of the Interior, Minerals Management Service, Gulf of Mexico OCS Region.

Wendland, K. J., M. Honzak, R. Portela, B. Vitale, S. Rubinoff, and J. Randrianarisoa. 2010. Targeting and implementing payments for ecosystem services: Opportunities for bundling biodiversity conservation with carbon and water services in Madagascar. Ecological Economics 69:2093-2107.

Westman, W. E. 1977. How much are nature's services worth? Science 197:960-964.

White, H. K., P.-Y. Hsing, W. Cho, T. M. Shank, E. E. Cordes, A. M. Quattrini, R. K. Nelson, R. Camilli, A. W. J. Demopoulos, C. R. German, J. M. Brooks, H. H. Roberts, W. Shedd, C. M. Reddy, and C. R. Fisher. 2012. Impact of the *Deepwater Horizon* oil spill on a deep-water coral community in the Gulf of Mexico. Proceedings of the National Academy of Sciences of the United States of America 109(50):20303-20308.

White, P. A., S. Robitaille, and J. B. Rasmussen. 1999. Heritable reproductive effects of benzo[*a*]pyrene on the fathead minnow (*Pimephales promelas*). Environmental Toxicology and Chemistry 18(8):1843-1847.

Whitehead, A., B. Dubansky, C. Bodinier, T. I. Garcia, S. Miles, C. Pilley, V. Raghunathan, J. L. Roach, N. Walker, R. B. Walter, C. D. Rice, and F. Galvez. 2011. Genomic and physiological footprint of the *Deepwater Horizon* oil spill on resident marsh fishes. Proceedings of the National Academy of Sciences of the United States of America 109(50):20298-20302.

WHOI (Woods Hole Oceanographic Institution). 2012. WHOI Scientists contribute to study on impact to coral communities from *Deepwater Horizon* spill. Available at http://www.whoi.edu/main/news-releases?tid=3622&cid=132889, accessed March 27 2012.

Wiens, J. A. 1995. Recovery of seabirds following the *Exxon Valdez* oil spill: An overview. Pp. 854-893 in *Exxon Valdez* Oil Spill: Fate and Effects in Alaskan Waters, P. G. Wells, J. N. Butler, and J. S. Hughes (Eds.). ASTM Special Technical Publication 1219. Philadelphia: American Society for Testing and Materials.

Wieski, K., H. Guo, C. B. Craft, and S. C. Pennings. 2010. Ecosystem functions of tidal fresh, brackish, and salt marshes on the Georgia Coast. Estuaries and Coasts 33:169.

Wilber, P. 1993. Managing Dredged Material Via Thin-Layer Disposal in Coastal Marshes. EEDP-01-32, 1-14. Vicksburg, MS: U.S. Army Engineer Waterways Experiment Station.

Williams, R., S. Gero, L. Bejder, J. Calambokidid, S. D. Kraus, S. Luuseau, A. J. Read, and J. Robbins. 2011. Underestimating the damage: Interpreting cetacean carcass recoveries in the context of the *Deepwater Horizon*/BP incident. Conservation Letters 4(3):228-233.

Williams, R. N. (Ed.). 2006. Return to the River: Restoring Salmon to the Columbia River. Amsterdam, The Netherlands: Elsevier Academic Press.

Wilson, M. A., and S. R. Carpenter. 1999. Economic valuation of freshwater ecosystems services in the United States: 1971-1997. Ecological Applications 9:772-783.

Wilson, S. G., and T. R. Fischetti, 2010. Coastline Population Trends in the United States: 1960 to 2008. Washington, DC: U.S. Department of Commerce, Economics and Statistics Administration, U.S. Census Bureau.

Woodside, J. M., L. David, A. Frantzis, and S. K. Hooker. 2006. Gouge marks on deep-sea mud volcanoes in the eastern Mediterranean: Caused by Cuvier's beaked whales? Deep Sea Research Part I: Oceanographic Research Papers 53:1762-1771.

Woodward, R. T., and Y. S. Wui. 2001. The economic value of wetland services: A meta-analysis. Ecological Economics 37:257-270.

Wursig, B. 1990. Cetaceans and Oil: Ecological Perspectives. Pp. 167-192 in Sea Mammals and Oil: Confronting the Risks, J. R. Geraci and D. J. St. Aubin (Eds.). San Diego: Academic Press.

Yanez-Arencibia, A., and J. W. Day. 2004. The Gulf of Mexico: Towards an integration of coastal management with large marine ecosystem management. Ocean & Coastal Management 47:537-563.

Ying, G.-G. 2006. Fate, behavior and effects of surfactants and their degradation products in the environment. Environmental International 32:417-431.

Yoskowitz, D., C. Santos, B. Allee, C. Carollo, J. Henderson, S. Jordan, and J. Ritchie. 2010. Proceedings of the Gulf of Mexico Ecosystem Services Workshop, Bay St. Louis, Mississippi, June 16-18, 2010. Corpus Christi: Texas A&M University and Harte Research Institute for Gulf of Mexico Studies.

Zahed, M. A., H. A. Aziz, M. H. Isa, L. Mohajeri, S. Mohajeri, and S. R. M. Kutty. 2011. Kinetic modeling and half life study on bioremediation of crude oil dispersed by Corexit 9500. Journal of Hazardous Materials 185:1027-1031.

Zedler, J. B. 2000. Progress in wetland restoration ecology. Trends in Ecology and Evolution 15:402-407.

Zhang, J., S. L. Ustin, E. Rejmankova, and E. W. Sanderson 1997. Monitoring Pacific Coast salt marshes using remote sensing. Ecological Applications 7:1039-1053.

Zimmerman, R. J., T. J. Minello, and L. P. Rozas. 2002. Salt marsh linkages to productivity of penaeid shrimps and blue crabs in the Northern Gulf of Mexico. Pp. 293-314 in Concepts and Controversies in Tidal Marsh Ecology, M. P. Weinstein and D. A. Kreeger (Eds.). Dordrecht, The Netherlands: Kluwer Academic.

Ziółkowska, A., and M. Wyszkowski. 2010. Toxicity of petroleum substances to microorganisms and plants. Ecological Chemistry and Engineering 17:73-82.

ZoBell, C. E. 1973. Microbial degradation of oil: Present status, problems, and perspectives. Pp 3-16 in The Microbial Degradation of Oil Pollutants, D. G. Ahearn and S. P. Meyers (Eds.). Baton Rouge, LA: Center for Wetland Resources, Louisiana State University.

APPENDIX A

# *Biographical Sketches of Committee Members and Staff*

## COMMITTEE MEMBERS

**Larry A. Mayer** (*Chair*) is the Director of the Center for Coastal and Ocean Mapping, Co-Director of the Joint Hydrographic Center, and Professor of Earth Science and Ocean Engineering at the University of New Hampshire. His research interests include sonar imaging, remote characterization of the seafloor, and advanced applications of 3-D visualization to ocean mapping challenges. Dr. Mayer received his Ph.D. from the Scripps Institution of Oceanography in marine geophysics in 1979, and he graduated magna cum laude with an honors degree in geology from the University of Rhode Island in 1973. At Scripps his future path was determined when he worked with the Marine Physical Laboratory's Deep-Tow Geophysical package but applied this sophisticated acoustic sensor to study the history of climate. Dr. Mayer has participated in more than 50 cruises and has been chief or co-chief scientist of numerous expeditions, including two legs of the Ocean Drilling Program. Recently he has been involved (both at sea and in the lab) with the visualization of environmental data from the *Deepwater Horizon* incident and the application of acoustic techniques to monitor wellhead integrity and the subsurface environment in the region. He brings a strong set of spatial analysis skills and tools to this committee that will be valuable in mapping the affected areas of the Gulf of Mexico (GoM) and understanding impacts. Dr. Mayer served on the President's Panel for Ocean Exploration and chaired the 2004 National Research Council's Committee on National Needs for Coastal Mapping and Charting.

**Michel C. Boufadel** is the Director of the Center for Natural Resources Development and Protection and a Professor in the Department of Civil and Environmental Engineering at the New Jersey Institute of Technology. Previously, he was a Professor and Chair of the Department of Civil and Environmental Engineering at Temple University. Dr. Boufadel's expertise includes investigating the offshore transport and fate of oil, original and dispersed, since 2001. He is very familiar with the Regional Ocean Modeling System (ROMS) and the various windcast and wave models, such as the Joint North Sea Wave Project (JONSWAP). Furthermore, Dr. Boufadel has developed a strong understanding of the physics of waterflow, oil transport, and oil transformation (with and without dispersants), and he has a strong understanding of the role of oil viscosity, surface tension, emulsion, evaporation, droplet formation (i.e., dispersion), and breakup under various energy levels. Dr. Boufadel's skills will be essential when the committee addresses the question of where the oil went and what it will likely do under a broad range of marine and coastal conditions. Dr. Boufadel earned a Ph.D. and an M.S. in environmental engineering from the University of Cincinnati in 1998 and 1992, respectively, and a B.S. in civil engineering and hydraulics from the Jesuit University at Beirut, Lebanon, in 1988.

**Jorge Brenner** is currently the Associate Director of Marine Science at The Nature Conservancy. Dr. Brenner is interested in ecosystem services health assessment, valuation models, and spatial dynamics of biodiversity. He is also working on marine conservation and sustainability sciences. Dr. Brenner has experience working on related issues in Mexico, the Mediterranean, and the Gulf of Mexico regions. He brings an international perspective to the committee in addition to his strength in identifying the relevant ecosystem services that the committee will need to quantify for valuation. He earned a Ph.D. in marine sciences from the Catalonia Polytechnic University in 2007, an M.S. in environmental engineering, and a B.S. in biochemical engineering and aquatic resources from the Monterrey Technology Institute University in 1997 and 1995, respectively.

**Robert S. Carney** is a Professor in Louisiana State University's (LSU's) Department of Oceanography and Coastal Sciences. Dr. Carney's primary research expertise is in deep-ocean biological oceanography, but he is also familiar with shallow systems, having directed the Coastal Ecology Institute of LSU for 9 years. He has been awarded numerous grants for his research since 1978, including multiple awards from the Minerals Management Service and National Oceanic and Atmospheric Administration to support the new sampling as well as reanalysis of archival deep Gulf of Mexico data. He is a principal investigator in the Alfred P. Sloan Foundation Census of Marine Life and co-directs international research on continental margin ecosystems. He is a founding member of INDEEP (International Network for Scientific Investigation of the Deep Sea), which will begin funding by Foundation TOTAL in 2011. In addition to basic science, he has published on the design of oil-related impact studies and information needs of deep ocean management. Dr. Carney will provide critical insights regarding the effects of the spill on benthic biota in the GoM, particularly in the deeper waters near the blowout. In 1977 Dr. Carney earned a Ph.D. in oceanography from Oregon State University; he also earned an M.S. in oceanography from Texas A&M University in 1971 and a B.S. in zoology from Duke University in 1967.

**Cortis K. Cooper** currently serves as Fellow with Chevron Energy Technology Company, a position he has held since 2002. Prior to beginning his service as Fellow, Dr. Cooper was employed as Scientist/Engineer at Chevron Exploration Technology for 12 years. In this position, he was primarily tasked with quantifying winds, waves, and currents for operation and design of offshore facilities worldwide including measuring and modeling oil spill fates; modeling hurricane alleys in the Gulf of Mexico; modeling sea level in the Caspian Sea; forecasting the Loop Current and associated eddies in the Gulf of Mexico; supervising major ocean current models in the Gulf of Mexico, West Africa, Northeast Atlantic, and Northwest Australia; leading a $1.6 million, 32-company joint industry project (JIP) to improve ocean towing; and leading a $2 million, 24-company JIP to investigate the fate of oil and gas from deepwater blowouts. Dr. Cooper was a member of the 2003 National Research Council's *Committee on Oil in the Sea: Inputs, Fates, and Effects*, which initiated and led a field experiment in 2000 that simulated a deepwater blowout off Norway. He has studied the physical oceanography of the Gulf of Mexico for 25 years. Dr. Cooper brings a wealth of relevant skills to the committee, but his grasp of industry standard operating procedure and his understanding of oil dispersion under various oceanographic conditions will be most useful. He earned a Ph.D. in environmental engineering from the University of Maine in 1987, and an M.Sc. and a B.S. in civil engineering from the Massachusetts Institute of Technology in 1977 and 1975, respectively.

**Jody W. Deming** holds the Walters Endowed Professorship in the University of Washington's School of Oceanography. She has also served as Director of the University of Washington's Marine Bioremediation Program. Dr. Deming has made major contributions to the understanding of life in deep sea and polar environments. As a marine microbiologist, Dr. Deming has focused her research efforts on the behavior of bacteria under conditions of extreme temperatures, pressures, and salt concentrations. She has used a combination of observational, experimental, and modeling approaches to explore the role of bacteria in the flow of carbon through deep-sea ecosystems, including in the Gulf of Mexico. Dr. Deming's expertise on marine microbial communities and their role in ecosystem functioning will be essential as the committee assesses the impact of the oil on the lower trophic levels of the Gulf food web. Dr. Deming earned a Ph.D. in microbiology from the University of Maryland, College Park in 1981, and a B.A. in biological sciences from Smith College in 1974. She was elected to the National Academy of Sciences in 2003 and is a current member of the Ocean Studies Board.

**David J. Die** is an Associate Professor at the University of Miami's Rosenstiel School of Marine and Atmospheric Science and the Associate Director of the Cooperative Institute for Marine and Atmospheric Studies. Dr. Die's research focus is on the quantitative evaluation of fishery management strategies, and his current portfolio includes collaborative development of a fishery ecosystem model for the Gulf of Mexico. He also has strong links to the Gulf of Mexico Fishery Management Council and was the founding director of the Center of Independent Experts, a central part of the peer-review process for the National Marine Fisheries Service. Dr. Die is the current chair of the Big-Eye Tuna Working Group of the International Commission for the Conservation of Atlantic Tuna and has recently been asked to serve on the international panel synthesizing tuna and billfish science for the International Union for the Conservation of Nature. He has extensive knowledge of both the ecology and population dynamics of upper trophic levels in the Gulf of Mexico ecosystem and the fishing pressures, and management regimes, to which they are subject. He will bring an international perspective and a strong understanding of the impacts of the spill on the fishing industries in the Gulf. Dr. Die received a Ph.D. in biology and living resources from the University of Miami in 1989, and a B.Sc. in zoology and marine biology from the Universidad de La Laguna (Spain) in 1982.

**Josh Eagle** is an Associate Professor of Law at the University of South Carolina (USC) School of Law and is affiliated with USC's Marine Sciences Program and its School of Earth, Ocean and Environment. His expertise is in ocean and coastal law, natural resources law, environmental law, and property law. He has testified before Congress and the Interagency Ocean Policy Task Force on legal issues related to ocean zoning and the siting of offshore energy facilities. Professor Eagle's expertise in pertinent laws and regulations, including the Natural Resource Damage Assessment (NRDA) process, will help the committee to understand the legal implications of various definitions and assessments of ecosystem services. From 1990 to 1995, Professor Eagle served as a trial attorney for the U.S. Department of Justice in Washington, DC. From 1997 to 1998, he was wildlife counsel in the policy office of the National Audubon Society in Washington, DC. Mr. Eagle received a J.D. from Georgetown University Law Center in 1990, an M.S. in forest sciences from Colorado State University in 1996, and a B.A. from Johns Hopkins University in 1985.

**Joseph R. Geraci** is Professor in the Department of Pathology and in the Program of Comparative Medicine at the University of Maryland's School of Medicine. His research over the past 40 years has focused on understanding how biological and environmental factors underpinning marine mammal health break down to affect the viability of the individual or population. Dr. Geraci has led research teams from the Arctic to the tropics, on studies of factors governing the health of marine mammals and their environment. He has published extensively on the effects of oil on marine mammals. In addition, Dr. Geraci has served as marine mammal health and program advisor to the U.S. Department of Agriculture; Department of Justice; Marine Mammal Commission; National Marine Fisheries Service; U.S. Navy; Canada Department of Fisheries and Oceans; the governments of Argentina, Brazil, Ecuador, Australia, New Zealand, Japan, and Spain and the Caribbean and Pacific Rim nations; the International Whaling Commission; United Nations Environment Program; International Atomic Energy Agency; and nongovernmental organizations, aquariums, and research laboratories internationally. Working with the National Marine Fisheries Service, he has played a founding role in developing U.S. regional marine mammal stranding-recovery programs into a nationwide network with international reach. Dr. Geraci will provide an important set of skills in assessing the pathologies and mortalities of megafauna (such as marine mammals and sea turtles) that are attributed to the oil spill. Dr. Geraci has a V.M.D. from the University of Pennsylvania, a Ph.D. from McGill University in Montreal, Canada, and a B.Sc. from Suffolk University in Boston.

**Barbara A. Knuth** is Vice Provost, Dean of the Graduate School, Professor of Natural Resource Policy and Management, and Associate Director of the Human Dimensions Research Unit in the Department of Natural Resources at Cornell University. Dr. Knuth's research focus, through theory development and empirical studies, advances understanding and practice related to the human dimensions of natural resource management, particularly related to fisheries and wildlife resources, and aims to foster integration of social science and natural/physical science information within natural resource management and policy decision-making processes. Dr. Knuth's expertise will help the committee to accurately identify and quantify the relevant ecosystem services that have significant societal importance. She earned a Ph.D. in fisheries and wildlife sciences from the Virginia Polytechnic Institute and State University in 1986, an M.Eng., a B.Phil., and a B.A. from Miami University in 1982 and 1980, respectively.

**Kenneth Lee** is the Director of the Wealth from Oceans National Research Flagship, part of Australia's Commonwealth Scientific and Industrial Research Organisation. He leads a national multidisciplinary team to promote research and development and the application of emerging technologies in ocean sciences to increase Australia's global competitiveness. Previously, he was Executive Director of the Centre for Offshore Oil, Gas and Energy Research (COOGER), part of Fisheries and Oceans Canada. Dr. Lee's research and project management activities include studies to link organic and inorganic contaminants, marine noise, and alterations in hydrodynamic processes to effects on biota, including commercial fisheries species; chemical/microbiological studies on the biotransformation and biodegradation of contaminants; development of novel approaches to assess the impact of organic pollutants by the development and validation of toxicity assays based on advances in genomics, microbial ecology, and biochemical analysis; and coordination of multidisciplinary studies including the application of numerical models to predict the risk of industrial activities and contaminants on ecosystem health. Dr. Lee is one of the world's leading experts on the effects of dispersants and other spill response technologies: This expertise will be most useful as the committee assesses the impact of the 1.8 million gallons of dispersants used in the *DWH* oil spill. He received a Ph.D. and M.Sc. in botany/environmental studies from the University of Toronto in 1982 and 1977, respectively, and a B.Sc. in biology from Dalhousie University in 1975.

**James T. Morris** is the Director of the Belle Baruch Institute for Marine and Coastal Sciences, Professor of Biological Sciences, Distinguished Professor of Marine Studies at the University of South Carolina, and an American Association for the Advancement of Science Fellow. Dr. Morris has authored more than 80 peer-reviewed publications, largely focused on coastal wetlands. He has served on numerous committees and panels for various agencies, including the U.S. National Science Foundation (NSF), the Irish National Science Foundation, the National Research Council, and the IndoFlux committee of India. Dr. Morris has a long history of funding from NSF for research at North Inlet, South Carolina, on the effects of sea level change on coastal wetlands. Dr. Morris will help the committee assess the impacts of the spill (and spill responses) on the Gulf wetlands—arguably the most critical and complex habitat responsible for many of the ecosystem services under review with this study. He earned a Ph.D. in forestry and environmental studies and am M.S. in biology from Yale University in 1979 and 1975, respectively, and a B.A. in environmental sciences from the University of Virginia in 1973.

**Stephen Polasky** is the Fesler-Lampert Professor of Ecological/Environmental Economics in the Department of Applied Economics at the University of Minnesota. His research interests include ecosystem services, natural capital, biodiversity conservation, endangered species policy, integrating ecological and economic analysis, renewable energy, environmental regulation, and common property resources. Papers authored by Dr. Polasky have been published in *Biological Conservation, Ecological Applications, Journal of Economics Perspectives, Nature,* and *Science,* among others. He has served as co-editor and associate editor for the *Journal of Environmental Economics and Management*. He previously held faculty positions in the Department of Agriculture and Resource Economics at Oregon State University (1993–1999) and the Department of Economics at Boston College (1986–1993). Dr. Polasky was the senior staff economist for environment and resources for the President's Council of Economic Advisors from 1998 to 1999. He was elected to the National Academy of Sciences in 2010. Also, he was elected as a Fellow of the American Academy of Arts and Sciences in 2009 and a Fellow of the American Association for the Advancement of Science in 2007. Dr. Polasky is a leader in the rapidly growing field of ecosystem services valuation, which is one of the core tasks for this study. Dr. Polasky received his Ph.D. in economics from the University of Michigan in 1986.

**Nancy N. Rabalais** is Executive Director and Professor at the Louisiana Universities Marine Consortium. Dr. Rabalais' research includes the dynamics of hypoxic environments, interactions of large rivers with the coastal ocean, estuarine and coastal eutrophication, and environmental effects of habitat alterations and contaminants. Dr. Rabalais is an American Association for the Advancement of Science Fellow, an Aldo Leopold Leadership Program Fellow, a National Associate of the National Academies of Science, a past president of the Estuarine Research Federation, a past vice-chair of the Scientific Steering Committee of Land-Ocean Interactions in the Coastal Zone/International Geosphere-Biosphere Program, and a past chair of the NRC Ocean Studies Board. She is a current member of the University-National Oceanographic Laboratory System (UNOLS) Council, the National Sea Grant Advisory Board, a Trustee for the Consortium for Ocean Leadership, a member of the Governing Board for the Gulf of Mexico Coastal Ocean Observing System, and an NRC committee member for Applying the Clean Water Act across the Mississippi River Basin. She received the 2002 Ketchum Award for coastal research from the Woods Hole Oceanographic Institution and shares the Blasker award with R. E. Turner. She was awarded the American Society of Limnology and Oceanography Ruth Patrick Award and the National Water Research Institute Clarke Prize in the summer of 2008. Her technical familiarity with the GoM and the interface between the deep benthic habitats and habitats along the coastal and continental shelf will be useful in determining the impacts of the oil at various depths. Dr. Rabalais received her Ph.D. in zoology from the University of Texas at Austin in 1983.

**Ralph G. Stahl, Jr.** received his B.S. in marine biology from Texas A&M University (cum laude), his M.S. in Biology from Texas A&M University, and his Ph.D. in Environmental Science and Toxicology from the University of Texas School of Public Health. After receiving his Ph.D., he was a National Institute of Environmental Health Sciences Senior Postdoctoral Fellow in the Department of Pathology at the University of Washington in Seattle where he investigated the impact of genetic toxins on biological systems. Dr. Stahl joined the DuPont Company in 1984 and in the intervening years has held both technical and management positions in the research and internal consulting arenas. His research over the past 25 years has focused primarily on evaluating the effects of chemical stressors on aquatic and terrestrial ecosystems. Since 1993 Dr. Stahl has been responsible for leading DuPont's corporate efforts in ecological risk assessment and natural resource damage assessments for site remediation. He has been involved with oceanographic studies in the Atlantic, Pacific, Gulf of Mexico, and Caribbean Sea; biological and ecological assessments at contaminated sites in the United States, Europe, and Latin America; and numerous toxicological studies with mammals, birds, and aquatic organisms. He has been selected by the U.S. Environmental Protection Agency (EPA), Army Corps of Engineers, Strategic Environmental Research and Development Program, National Institute of Environmental Health Sciences, National Academy of Sciences, the Water Environment Research Foundation, National Oceanic and Atmpspheric Administration, state of Washington, state of Texas, and others to national or state peer-review panels on ecological risk assessment, endocrine disruption in wildlife, or natural resource injury determination. Dr. Stahl has served on EPA's Science Advisory Board (Advisory Council on Clean Air Compliance Analysis, Ecological Effects Subcommittee) and the Department of the Interior's FACA Panel on Natural Resource Damages, and he currently is active in the Society of Environmental Toxicology and Chemistry (SETAC), Ecological Risk Assessment Advisory Group. He is board certified in general toxicology and is a Diplomate of the American Board of Toxicology. He has authored more than 45 peer-reviewed publications on topics in environmental toxicology, ecological risk assessment, and risk management. He recently edited three books stemming from SETAC Education Foundation–sponsored workshops, and he currently serves on the editorial board of the journal *Integrated Environmental Assessment and Management*.

**David Yoskowitz** is the HRI Endowed Chair for Socio-Economics at the Harte Research Institute for Gulf of Mexico Studies, and Professor in the College of Business at Texas A&M University, Corpus Christi. Dr. Yoskowitz's interests include market and nonmarket valuation; ecosystem services; micro and small enterprise development; environmental and water markets; border economics; development microeconomics in Latin America; and socioeconomic environment of the GoM region. Dr. Yoskowitz will bring a strong understanding of the GoM ecosystem services and valuation, as well as a local appreciation of the smaller businesses and enterprises impacted by the spill. He received a Ph.D. in economics and an M.A. in economics from Texas Tech University in 1997 and 1994, respectively, and a B.A. in economics and finance from Bentley College in 1990.

## STAFF

**Kim Waddell** is a senior program officer with the Ocean Studies Board. He received his Ph.D. in the biological sciences from the University of South Carolina and his B.A. in environmental studies from the University of California, Santa Cruz. Dr. Waddell recently rejoined the NRC after a 6-year hiatus during which he was a research associate professor at the University of the Virgin Islands and Texas A&M University working to build marine and environmental research capacity in the Caribbean region. During his previous tenure with the NRC, Dr. Waddell directed a number of studies for the Board on Agriculture and Natural Resources including California Agricultural Research Priorities: Pierce's Disease (2004); Biological Confinement of Genetically Engineered Organisms (2004); Animal Biotechnology; Science-Based Concerns (2002); The Environmental Effects of Transgenic Plants (2002); Exploring Horizons for Domestic Animal Genomics (2002); and The Future Role of Pesticides in U.S. Agriculture (2000).

**Sherrie Forrest** is an associate program officer with the Ocean Studies Board and the Board on Science Education at the National Research Council. She currently supports the work of several projects, including the Roundtable on Climate Change Education and the Effects of the *Deepwater Horizon* Mississippi Canyon-252 Oil Spill on Ecosystem Services in the Gulf of Mexico, and she previously worked on the Conceptual Framework for New Science Education Standards. She is also the study director on the Workshop on Climate Change Education in Formal Settings, K-14. She has a B.A. in English literature from Pepperdine University and an M.S. in biological oceanography from the Institute of Marine and Coastal Sciences at Rutgers University.

**Lauren Harding** was a program assistant with the Ocean Studies Board from August 2011 to October 2012. In 2011, she graduated from High Point University majoring in biology and minoring in chemistry. As an undergraduate, she conducted an independent research project on cave ecosystems. Prior to her position at OSB, Lauren was a marketing and accounting assistant with Webco General Partnership, a company of the U.S. military resale market.

**Heather Chiarello** joined the National Academy of Sciences in July 2008. She graduated magna cum laude from Central Michigan University in 2007 with a B.S. in political science with a concentration in public administration. Ms. Chiarello is currently a senior program assistant with the Ocean Studies Board in the Division on Earth and Life Sciences of the National Academies. She is pursuing a master's degree in sociology and public policy analysis at The Catholic University of America in Washington, D.C.

**Jessica Dutton** received her B.A. from Mount Holyoke College and her Ph.D. in marine biology from the University of California, Santa Barbara. As an ecological physiologist, her doctoral research focused on understanding the relationship between species tolerances and coastal environmental conditions, and how such patterns relate to range distributions and climate change. She was a Fellow in 2009 with the National Sea Grant Knauss Marine Policy Fellowship Program and in 2012 with the Christine Mirzayan Science and Technology Policy Fellowship Program at the National Academy of Sciences. In the latter position, and subsequently as a Research Associate, she has worked with the Ocean Studies Board on several NRC studies including the Effects of the *Deepwater Horizon* Mississippi Canyon-252 Oil Spill on Ecosystem Services in the Gulf of Mexico, Review of the National Ocean Acidification Research Plan, and Evaluating the Effectiveness of Stock Rebuilding Plans of the 2006 Fishery Conservation and Management Reauthorization Act.

**Constance (Stacee) Karras** joined the National Academy of Sciences in September 2012. She received her B.A. in marine affairs and policy with concentrations in biology and political science from the University of Miami in 2007. The following year she received an M.A. in marine affairs and policy from the University of Miami's Rosenstiel School of Marine and Atmospheric Science. Most recently, she earned her J.D. from the University of Virginia School of Law. Ms. Karras is now serving as a post graduate intern with the Ocean Studies Board in the Division of Earth and Life Sciences of the National Academies.